THE FLORA OF LINCOLNSHIRE

LINCOLNSHIRE
NATURAL HISTORY BROCHURE
No. 6

THE FLORA OF LINCOLNSHIRE

by

E. JOAN GIBBONS, F.L.S.

President, Lincolnshire Naturalists' Union
1939 *and* 1974-5

Published by
LINCOLNSHIRE NATURALISTS' UNION
LINCOLN
1975

©Lincolnshire Naturalists' Union

Reprinted 1991

ISBN : O 9500353 4 3

Editor
D. N. ROBINSON, M.SC.

Editor's note
The Lincolnshire Naturalists' Union gratefully acknowledges the gift of £200 from the former Lindsey and Holland Rural Community Council towards the publication of this book. This money was raised from the sale of herbs during World War II and the income from its investment has been the financial basis for this publication.

Cover photograph
Sea Buckthorn (*Hippophae rhamnoides*) by G. S. Phillips, F.C.A.

Printed in Great Britain by
Peter Spiegl & Co., 6 St. George's Street, Stamford, Lincolnshire.

FIGURES

1. Topography and Natural History Divisions of Lincolnshire — *facing page* 1
2. The Geology of Lincolnshire — *pages* 6–7
3. Block diagram of Central and North Lincolnshire — 8
4. Rainfall — 20
5. Temperature, Sunshine, Snow and Wind Direction — 23
6. Lincolnshire Domesday Woodland — 28
7. Present distribution of woodland — 29
8. Small-leaved Lime from Brass in St. Cornelius Church, Linwood — 35
9. Lincolnshire *c.* 1750 — 42
10. Natural Regions and the distribution of Nature Reserves in Lincolnshire — 49
11. Number of species recorded by 1960 and by 1973 — 288

PLATES
Between pages 52 and 53

WOODLANDS
1. Oak woodland, Kirkby Moor
2. Lime woodland, Ivy Wood, Bardney

GRASSLANDS
3. Risby Warren
4. Martin Moor
5. Limestone grassland, Ancaster Valley
6. Chalk grassland, Red Hill, Goulceby

FENLANDS
7. Crowle Waste
8. Baston Fen

COASTLANDS
9. Saltfleetby-Theddlethorpe National Nature Reserve
10. Saltmarshes of the Welland estuary
11. Gibraltar Point Local Nature Reserve

NOTABLE LINCOLNSHIRE BOTANISTS
12. Rev. E. A. Woodruffe-Peacock
13. Canon W. Fowler
14. Rev. W. Wright Mason
15. Miss S. C. Stow

CONTENTS

	Page
Foreword	ix
Preface	x

Chapters

1. INTRODUCTION — 1
2. GEOLOGY AND SCENERY — 9
3. WEATHER AND CLIMATE — 17
4. THE FLORA OF THE MAJOR HABITATS
   - Woodlands — 27
   - Grasslands — 36
   - Fenlands — 43
   - Isle of Axholme — 45
   - Coastlands — 45
   - Notes on Ferns and Sedges — 50
5. BOTANICAL HISTORY AND BOTANISTS — 53
6. LINCOLNSHIRE RECORDERS — 63
   - Location of Lincolnshire Herbaria — 74
7. THE COUNTY DIVISIONS — 75
8. THE COUNTY FLORA
   - Explanation and key — 79
   - Pteridophyta — 80
   - Spermatophyta
     - Gymnospermae — 86
     - Angiospermae
       - Dicotyledones — 88
       - Monocotyledones — 244
   - Appendices:
     - I   First records — 289
     - II Notable additions to Lees' List — 289
     - III Plants extinct or thought to be extinct — 290
     - IV Plants which have been added to the County List since the publication of the Atlas of the British Flora — 291
     - V   Limits of distribution — 292
     - VI Aliens — 293
     - VII Fossil and Peat records — 302
9. THE LINCOLNSHIRE NATURALISTS' UNION — 303
   - Location of Field Meetings 1893-1973 — 304

Bibliography — 310

Index — 323

FOREWORD

In more senses than one the publication of Joan Gibbons' *Flora of Lincolnshire* is a unique occasion. Its appearance marks the end of a remarkable series of books begun over 300 years ago with the publication of John Ray's *Catalogue Plantarum circa Cantabrigiam nascentium*, the first Flora of an English county, and now completed by this volume, the last Flora of an English county. How fitting and yet how strange that these two counties, Lincolnshire and Cambridgeshire, have common boundaries; truly the wheel has turned full circle.

As one author of the last Flora of Cambridgeshire, the joint effort of four males, I salute Joan Gibbons for another unique feature of this Flora, It is the first full Flora for any county in England to have been written by a woman, and she did it almost alone. That Lincolnshire is the second largest English county, after Yorkshire, is some measure of the task she has completed, and her success is all the more remarkable when it is remembered that Yorkshire is covered by three separate Floras.

When one considers the traditional affinity between ladies and botany it is even more surprising that the day which has now arrived should have been so long in coming. Many have aspired to write Floras but few have completed them; with echoes from a thousand school reports they promised more than they achieved.

So it is with enormous satisfaction and wonder that I discover that Joan Gibbons has achieved far more than she ever promised. I knew the account of each species would be thorough and accurate because I acquired the only extant copy of the Flora during ten years of the Botanical Society of the British Isles Distribution Maps Scheme. She sent it to me, species by species, each on a separate sheet of identical paper, with full biographical notes on each recorder, especially if he were a clergyman, which in Lincolnshire many of them were. The biographies are here, of course, gems of condensed observation, but there is so much else beside; the vegetation, extinct plants, aliens and even a complete list of L.N.U. meetings since 1893.

The richness of the account the author has prepared is worthy of the botanical riches of the county she describes and loves, and has lived in virtually all her life. I feel privileged to be able to acknowledge in public the debt that I and all botanists owe to Joan Gibbons for bringing Lincolnshire out of the shade into the forefront of British Botany, where it surely belongs, by the publication of this Flora.

FRANKLYN PERRING.

Oundle Lodge,
Oundle.
August, 1974.

PREFACE

The writing of this Flora has been a hard task for a botanist who feels more at home in the field than in writing. The many hours of research into old records in an attempt to make this, the first Flora for Lincolnshire, as complete as possible have not been as rewarding as the many years spent in field recording. The Rev. W. Fowler and Dr. F. A. Lees made great efforts in 1877 to survey as much as possible of Watson's two vice-counties, and many new records were made. After 1893 the Lincolnshire Naturalists' Union under Rev. E. A. Woodruffe-Peacock started recording in the eighteen natural history Divisions of the County which he had devised. This resulted in the publication of tle *Check List of Lincolnshire Plants* in 1909. Over the years since then fresh records have been added to that list, most of which have been published in the Union's annual *Transactions*. Efforts have been made to mark records with a dagger (†) where specimens are known to exist, but in the case of Sir Joseph Banks's and other large herbaria, opportunities for research have not been sufficient.

In April, 1974 the three historic divisions of Lincolnshire (Lindsey, Kesteven and Holland) disappeared. The new administrative county of Lincolnshire consists of the former Kesteven and Holland, Lincoln City and those parts of Lindsey except the former Rural Districts of Grimsby, Glanford Brigg and Isle of Axholme which together form the southern part of the new Humberside County. However, this book is concerned with the historic county of Lincolnshire which will continue to be the basis for recording by the Lincolnshire Naturalists' Union in the foreseeable future. As I explain later it is because of the wealth of information recorded on the basis of the natural history Divisions of the County that it has been necessary to retain them. To have transferred all records already assembled onto the National Grid would have been an almost impossible task. However, with the aid of related maps provided, it should be possible for the reader or researcher adequately to trace specific plants. Maps are also provided to show the increase in recorded species in each 10 Km. square since the publication of the *Atlas of the British Flora*.

Many botanists have been waiting a long time for this book to appear. In 1948 and 1955 the Union was able to publish the first of its natural history brochures — on Geology and Birds. In his last report as General Secretary/Treasurer, F. T. Baker said in 1960 that "soon it is hoped to follow with The Flora of Lincolnshire", and it was he who stimulated the first moves towards publication. It is now fifteen years since a start was made by Mark R. D. Seaward, then Hon. Secretary of the L.N.U., to devise a scheme for the publication of the records and to produce the first typescripts. I am grateful for this fundamental base without which further progress could not have been made. The work on the typescripts was continued very capably by Mrs. B. Watkinson. Since then difficulties in editing, the writing of special chapters and the desire to incorporate new information as it came to light have caused inevitable delays.

The help given by Mrs. I. Weston has been invaluable in the later stages and I am deeply indebted to her, particularly in connection with the chapter on Major Habitats, the appendix on Aliens and the Bibliography. My particular thanks are also due to D. N. Robinson and F. A. Barnes for contributing special chapters on Geology and Scenery and Weather and Climate, to Miss G. M. Waterhouse for early advice in editing, to A. J. Gray for his meticulous proof reading of the lists, to J. H. Chandler for checking part of the main list and for the Roses list, to E. J. Redshaw for preparing some of the maps, and to Dr. M. R. D. Seaward for preparing the index.

A full list of all who have recorded in Lincolnshire is included, but I should also like to make special mention of the following who have helped me, particularly specialists who have checked critical plants: A. H. G. Alston, F. T. Baker, P. W. Ball (*Salicornias*), Dr. R. W. Butcher, J. E. Dandy (*Potamogetons* and Aliens), Dr. J. G. Dony, E. S. Edees (Brambles), Dr. J. Hope Simpson, R. C. L. and Mrs. B. M. Howitt, J. E. Lousley, Dr. R. Melville (Elms), E. Nelmes (*Carex*), Dr. F. H. Perring, G. S. and Mrs. G. Phillips, Miss M. N. Read, P. D. Sell (*Hieracia*), Dr. S. M. Walters (*Alchemillas*), P. F. Yeo (*Euphrasia*) and Dr. D. P. Young (*Epipactis*).

E. JOAN GIBBONS.

Glentworth 1974.

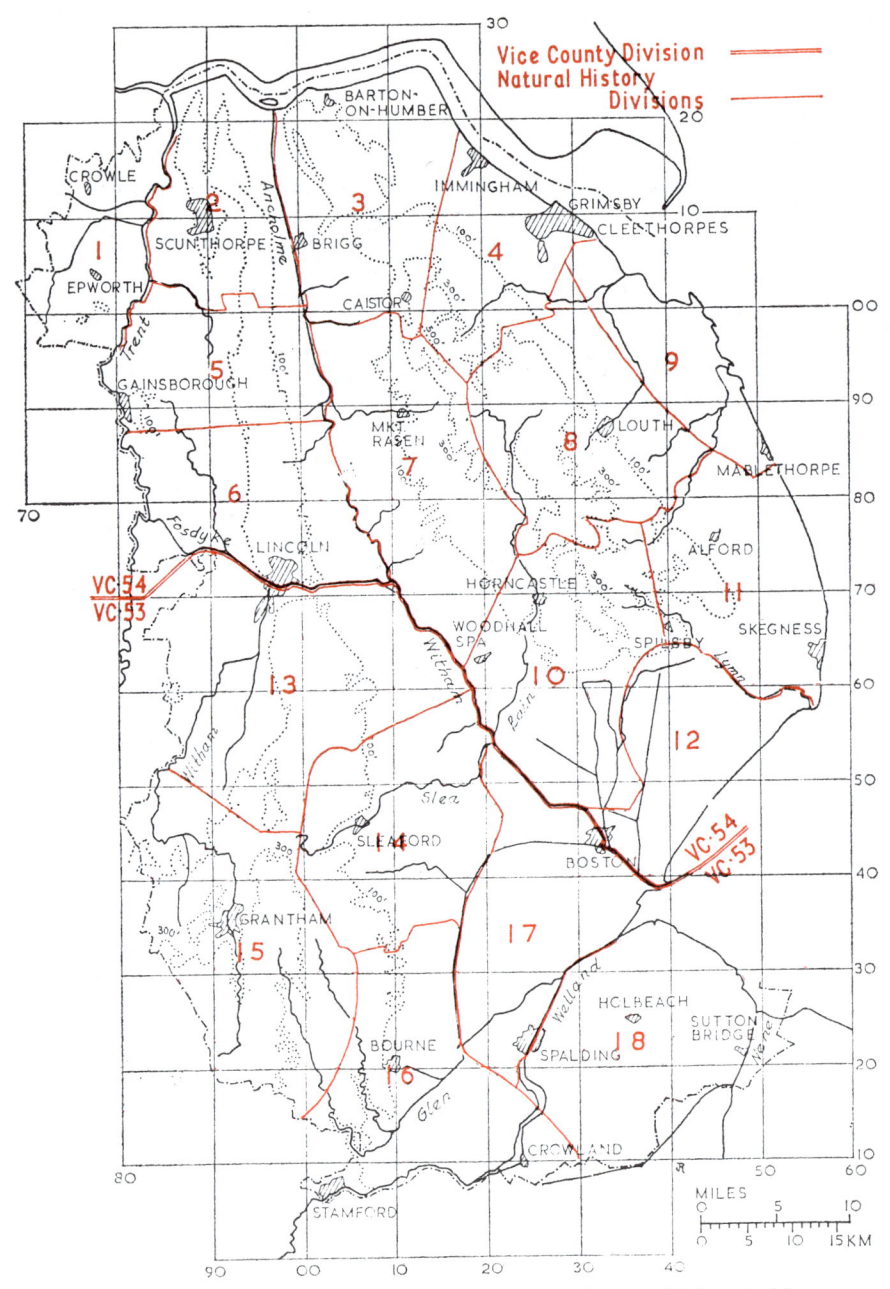

Fig. 1. Topography and Natural History Divisions of Lincolnshire
(*Based upon the ordnance Survey map with the sanction of the Controller of Her Majesty's Stationery Office. Crown Copyright Reserved*)

CHAPTER 1

INTRODUCTION

To understand the present pattern of the Flora of Lincolnshire it is necessary to go back to a time when the land was largely uncultivated. For example the flora of the vast plains of calcareous Heath along which the Romans made Ermine Street was mainly grasses like Tor Grass (*Brachypodium pinnatum*) and Upright Brome (*Bromus erectus*), both unpalatable to cows and sheep, with Field Scabious (*Knautia arvensis*), and Knapweeds (*Centauria scabiosa and C. nemorosa*). There were also pockets of calcareous bog like the former Waddingham Common. Each parish had a strip of higher heathland as well as low fields. At first the cultivated land was close to the villages which were on the spring line of both slopes but more land was taken in as the inhabitants became more efficient farmers.

The Warrens and the Bluestone Heath of the Wolds were similar but not so regular in pattern. The valleys in some cases were wooded where the clays were exposed by strong streams, and produced a very attractive woodland flora including Ramsons (*Allium ursinum*), Wood Forget-me-not (*Myosotis sylvatica*), Red Campion (*Silene dioica*), Bugle (*Ajuga reptans*), Water Avens (*Geum rivale*) and Wood Anemone (*Anemone nemorosa*). More unusual are the two Golden Saxifrages (*Chrysosplenium oppositifolium and C. alternifolium*), Giant Campanula (*C. latifolia*), Small Teasel (*Dipsacus pilosus*), Moschatel (*Adoxa moschatellina*) and Yellow Archangel (*Lamiastrum galeobdolon*) with Great Horsetail (*Equisetum telemateia*) and Marsh Marigold (*Caltha palustris*) in the wet soggy areas. Orchids are scarce and Bluebells and Primroses are only seen occasionally.

The woodland areas of the County have seen great changes and much of the ancient deciduous woods have disappeared. Newer woods have been planted in the last 200 years for fox hunting or as game reserves, including the extensive area round Brocklesby but on a smaller scale elsewhere. These have not yet acquired an interesting flora but have largely ferns, brambles and nettles. Clear felling by the Forestry Commission and the planting of conifers is destroying the ancient flora of the oak woods by producing a dense cover of pine and fir needles, changing the soil and smothering any plants growing below.

Of the old fenland John Cordeaux wrote:

"There is no other formal area in Lincolnshire where the old glories have so entirely vanished as in the fenland, formerly a vast level of peat, moor, morass and bog with league beyond league of shallow mere interspersed with a vast growth of reed and bulrush and various water-loving plants and on the drier portion deep sludge and doubtless some rich pasturage with thickets of sallow willow, birch and sweet gale which before the dawn of history had usurped the place of Oak, Scotch fir and yew." Two hundred years ago much was undrained, the East Fen being the last part to come under cultivation early in the 19th century.

The fens of Wainfleet, Friskney and, perhaps before 1700, Deeping Fen produced Cranberries (*Vaccinium oxycoccus*) in marketable quantities, and so must have had heather though this is not recorded. Crowberry used to grow on these moors too, but has all vanished.

The fenland now all drained and cultivated, is not all ancient, as many acres in Fosdyke Wash, Holbeach, Gedney and Cross Keys Wash were under the sea less than 200 years ago. The flora here is only maritime and weeds of cultivation. The courses of old estuaries can be traced at Grainthorpe, Friskney, Bicker Haven and from Moulton to Tydd by the maritime plants growing along the dykes further inland.

The acid moors of Blown Sand in the Scunthorpe district, Scotton Common and the low lying land west of the Wolds from Wrawby Moor to Linwood Warren have been rabbit warrens for many centuries and grew much heather and birch. Bog land in these areas was rich in plants including Purple Moorgrass (*Molinia caerulea*), Bog Asphodel (*Narthecium ossifragum*), Marsh Gentian (*Gentiana pneumonanthe*) the three Sundews (*Drosera anglica, D. intermedia* and *D. rotundifolia*) and Bog Myrtle (*Myrica gale*), but little now survives. The sand and gravel of Stapleford, Doddington and Boultham and the Woodhall Spa — Tattershall district has much the same flora but it is not so acid.

Whereas the initial composition of the flora depended on natural factors — rock, soil, slope and drainage — the effects of man on the land whether through agriculture, industry, urban expansion or recreation has altered it considerably. Now more than ever pressures are being brought to bear on the countryside, and old habitats are changing and decreasing.

From Roman times widespread settlements and cultivations decreased the deciduous forest areas and by the 17th century much of the woodland had gone. Some of the marshes were grazed. Even though in the 18th and early 19th centuries the Wolds and Heath were ploughed and enclosed there were large areas of varied uncultivated habitats. By the mid-19th century Lincolnshire was highly cultivated but the pressures on the remaining wild lands came during and after the Second World War when the poorer, hitherto unploughed land was cultivated and great areas of heathland were drained and afforested. Some of the classic

INTRODUCTION

county habitats disappeared such as wet peaty flushes on heathland and boggy places on limestone. The overall flora diminished.

When considering the contemporary flora of Lincolnshire it is important to realise that it is the impact of the present day agricultural industry which has so drastically changed the Lincolnshire scene, both semi-natural and man-made. Lincolnshire is essentially a rural county and agriculture its main industry. It is one of the leading agricultural areas in Britain and its soil, particularly in the fen-land districts, has been described as 'agricultural gold'. The peats and silts are very rich and the newly reclaimed areas are valuable for market garden crops and bulbs. The new intensive cultivation and mechanisation have given rise to vast grain prairies on the Wolds and Heath, and even the traditional Marsh grazing is being replaced by arable cropping. Little permanent grassland remains, except where mechanical difficulties arise. There is considerable decrease in numbers of grassland communities. Plants of neutral grassland — Cowslip (*Primula veris*), Green-winged Orchid (*Orchis morio*) and Meadow Saxifrage (*Saxifraga granulata*) have become rare. The encouraged use of fertilisers and herbicides has also reduced the range and quantity of cornfield weeds. Particularly important were those of Eastern distribution — Corn Cockle (*Agrostemma githago*), Cornflower (*Centaurea cyanus*), Venus Looking-glass (*Legousia hybrida*), the two Fluellens (*Kickxia spuria and K. elatine*), Corn Crowfoot (*Ranunculus arvensis*) and Shepherd's Needle (*Scandix pecten-veneris*). Some are now seen sparingly in the country and occasionally occur where there has been no spraying.

The forestry plantations show interesting habitat fluctuations. In a newly felled and replanted area the number of species rises and then falls as the canopy of conifers evolves. Rand and Wickenby Woods in the mid 1960s had a multitude of woodland plants typical of the area including Greater Butterfly Orchid (*Platanthera chlorantha*), Early Purple Orchid (*Orchis mascula*), Herb Paris (*Paris quadrifolia*), Primrose (*Primula vulgaris*), Wood Sorrel (*Oxalis acetosella*) and Wood Anemone (*Anemone nemorosa*), but these are now sparse.

Drastic habitat losses occurred in the late 19th and early 20th centuries due to the exploitation of the iron ore deposits. Old, well documented habitats near Scunthorpe, like Crosby Warren and "the sandy commons about Frodingham", have gone. The topography of the district is altogether changed and chemical fumes have affected vegetation in the surrounding areas. Vast scars and dumps colonised by Coltsfoot (*Tussilago farfara*) and the slag heaps by Oxford Ragwort (*Senecio squalidus*) are a feature of the landscape. Round Colsterworth in the south, ironstone quarrying has also changed the Lincolnshire scene. The industrial development at Immingham and the growth of South Humberside have more recently altered the north county landscape and diminished its flora. Areas like Freshney bog have altered, and Grass of Parnassus (*Parnassia palustris*) has gone. The stone quarries at Hibaldstow on the limestone Heath and the clay quarries on the escarpment now being worked are bare of vegetation, but the older quarries have colonised and contain a wealth of calcicole species. Ploughman's Spikenard (*Inula conyza*), Basil Thyme (*Acinos arvensis*), Marjoram (*Origanum vulgare*), Pyramidal Orchid (*Anacamptis pyramidalis*), Knapweed (*Centaurea scabiosa*), with Tall Broomrape (*Orobanche elatior*) and Bee Orchid (*Ophrys apifera*) have appeared in quantity. Disused stone quarries at Ancaster also provide one of the best habitats in the county for a transition from calcicole grassland to scrub and woodland of this type. Spurge Laurel (*Daphne laureola*), Autumn Gentian (*Gentianella amarella*), Pyramid and Bee Orchids, Woolly Thistle (*Cirsium eriophorum*), Ploughman's Spikenard (*Inula conyza*), Rock-rose (*Helianthemum chamaecistus*) and calcareous grasses (*Brachypodium pinnatum, Bromus erectus, Briza media, Koeleria cristata*) all occur. Similarly the chalk quarries of the Wolds are important; Yellow-wort (*Blackstonia perfoliata*), Centaury (*Centaurium erythraea*) and Kidney-vetch (*Anthyllis vulneraria*) are typical here.

The old clay brickpits near Lincoln, Hatton and Market Rasen, the sand and gravel workings in the Lincoln area — Skellingthorpe, Doddington, Hykeham and Burton, the new sand workings at Messingham and the extensive sand and gravel workings at Woodhall and Tattershall have all destroyed old habitats but are steadily creating new ones both sandy and aquatic. Several rarer plants are already established in some of these — Great Spearwort (*Ranunculus lingua*), Arrowhead (*Sagittaria sagittifolia*), Water Soldier (*Stratiotes aloides*), Flowering Rush (*Butomus umbellatus*) and Galingale (*Cyperus longus*). The gravel in South Rauceby pit yields Glabrous Rupturewort (*Herniaria glabra*). Railway ballast pits, river borrow pits and dykes all provide important man-made habitats. These are threatened by tipping and drainage. Saxilby to Skellingthorpe was well worked by Dr. Lees and Rev. Wm. Fowler in 1877 and was still full of rarities — Great Spearwort (*Ranunculus lingua*) and Lesser Reedmace (*Typha augustifolia*) — in 1968. Killingholme Pit, over 60 years old, has regenerated with a wealth of interesting plants including Blue Broomrape (*Orobanche purpurea*) also *O. minor*, Yellow-wort (*Blackstonia perfoliata*), Orchids and Adder's Tongue (*Ophioglossum vulgatum*). Baston Fen has a similar flora to Saxilby with some additions of fenland type. Bladderwort (*Utricularia vulgaris*) is still to be found in a few of the pits. Spraying of the river banks has reduced their flora, but in the Isle of Axholme the drains are rich in pondweeds, particularly at Craiselound and Dirtness. Large Bitter-cress (*Cardamine amara*), so long unrecorded from Lincolnshire has been found in four separate localities in the Isle, one of them the normal white-flowered form but the others are a most unusual deep lilac pink var *erubescans*.

Other disappearing habitats are hedgerows and road verges, the latter by spraying and road widening. Some of our best limestone and chalk plants are present on roadside verges — Man Orchid (*Aceras anthropophorum*), Horse-shoe Vetch (*Hippocrepis comosa*), Rock-rose

(*Helianthemum chamaecistus*), Clustered Bellflower (*Campanula glomerata*), Pyramidal Orchid (*Anacamptis pyramidalis*), Yellow-wort (*Blackstonia perfoliata*), Felwort (*Gentianella amarella*), Perennial Flax (*Linum anglicum*) and Tall Broomrape (*Orobanche elatior*). Clay plants such as Cowslip and Lady's Smock and acid plants like Butterwort also have roadside habitats in the county, some of which are now scheduled sites.

In his notes and writings Dr. Arnold Lees describes the geographical distribution of Lincolnshire species in the 19th century in relation to habitat and soil type. He attributes the large number of Lincolnshire species to the number of plants we share with Yorkshire, Norfolk and counties west of the Trent and also to "elements of a montane flora still surviving". These limit plants can be readily traced in the *Atlas of the British Flora*. He also quotes several species surprisingly 'absent' from the county at the time of his *Outline Flora*. Notable are *Vaccinium myrtillus*, *Cardamine amara*, *Crepis paludosa*, *Carex pendula* and *Rosa spinosissima*. These have been recorded since, the last only fleetingly.

Old descriptions of Lincolnshire plants like "*Lycopodium alpinum* mingled with *Gentiana pneumonanthe*, *Lastrea thelypteris*, all three Sundews, *Epipactis palustris* and *Anagallis tenella* on the heathy warrens," or "*Selaginella selaginoides* with *Parnassia*," or, from the Fens, "*Sonchus palustris*, *Senecio paludosus* and *S. palustris*, *Cicuta virosa*, *Lathyrus palustris* and *Oxycoccus quadripetalus*" no longer apply, but one can still trace the pattern of the vegetation. One can also detect the wealth of previous species by the small pockets of uncultivated land which are left. These are invaluable as a key to the descriptions of the flora in the old documents. Many of these areas are now nature reserves or sites of special scientific interest. They include small remnants of permanent grassland on the chalk and limestone or in the marshes, the fragments of oak/ash, oak/birch or birch woodlands, hazel coppice, alder carr, wet and dry heathland and turbary. These provide important reservoirs for species often common in other counties but which have become rare in Lincolnshire, for example Sundews, Butterwort (*Pinguicula vulgaris*), Bog Asphodel (*Narthecium ossifragum*), Fen Sedge (*Cladium mariscus*) and Bog Rosemary (*Andromeda polifolia*). The two coastal reserves at Gibraltar Point and Saltfleetby-Theddlethorpe show this particularly well on a holiday coastline where habitats are very much altered. (See Fig. 10, p. 49 for distribution of nature reserves).

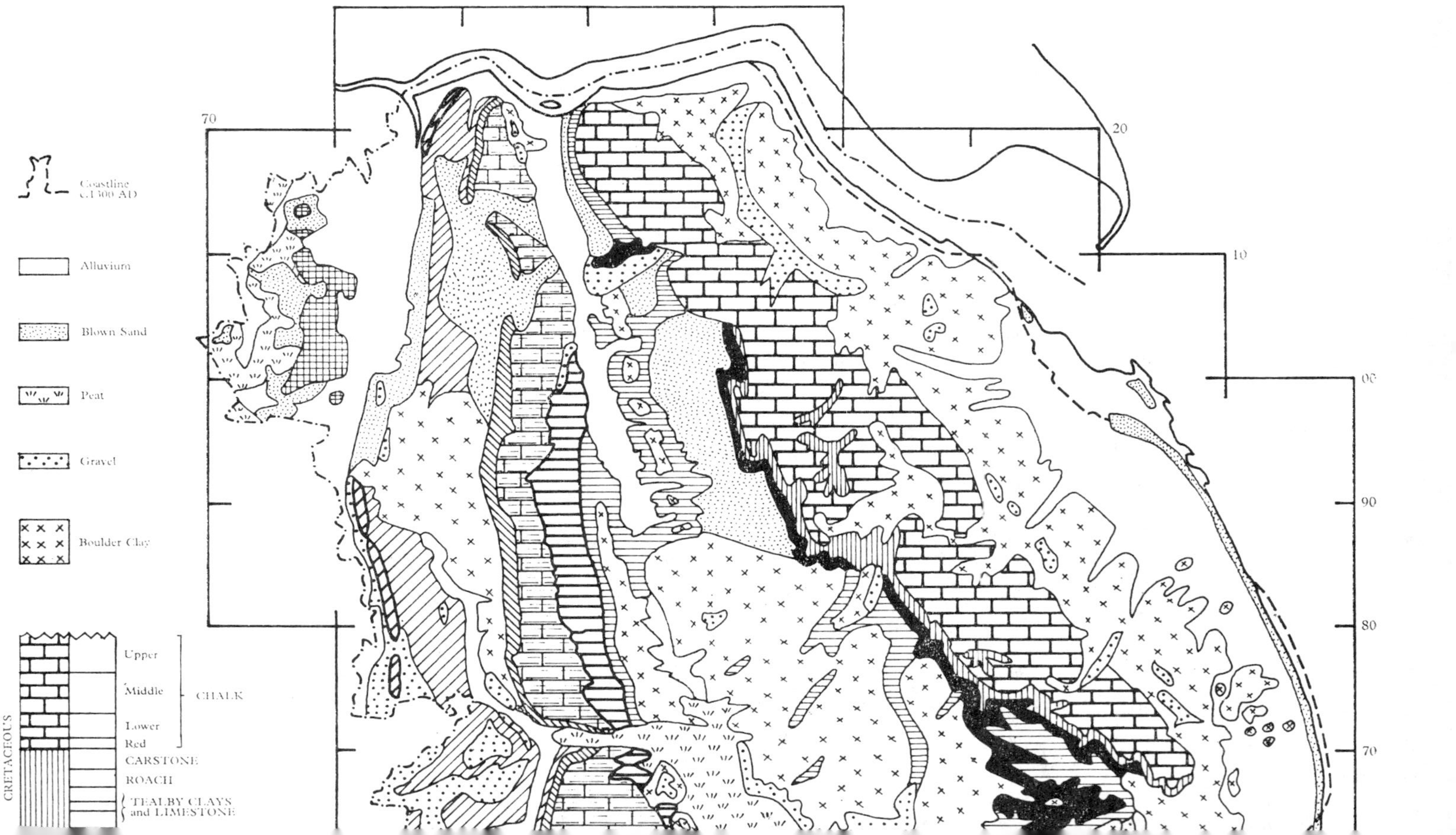
70
20
10
00
90
80
70
Coastline c.1300 AD
Alluvium
Blown Sand
Peat
Gravel
Boulder Clay
Upper
Middle
Lower
Red
CHALK
CARSTONE
ROACH
TEALBY CLAYS and LIMESTONE
CRETACEOUS

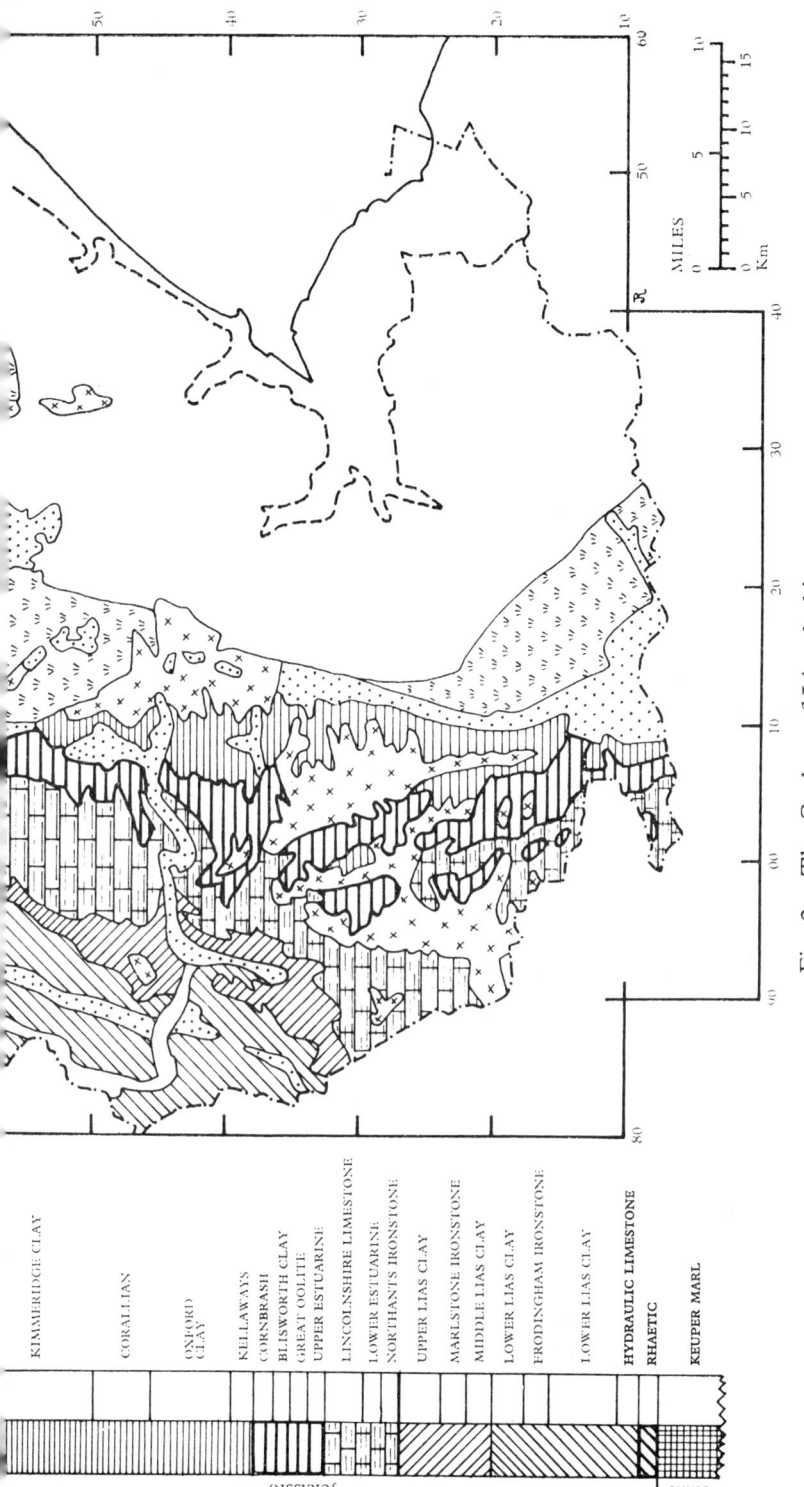

Fig. 2. The Geology of Lincolnshire.
(compiled by D. N. Robinson).

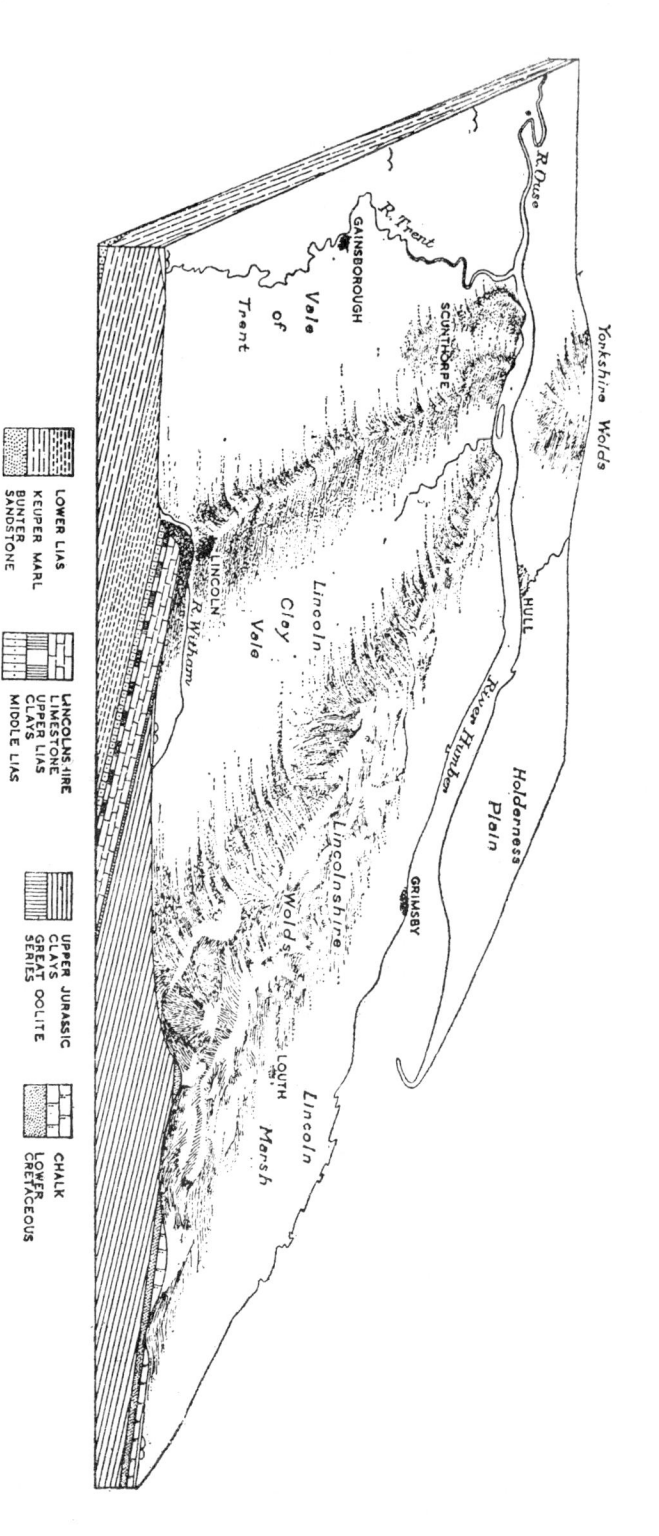

Fig. 3. Block diagram of Central and North Lincolnshire.
(Crown Copyright Geological Survey diagram. Reproduced by permission of the Controller, Her Majesty's Stationery Office).

CHAPTER 2

GEOLOGY AND SCENERY

D. N. Robinson, M.Sc.

Lincolnshire is the second largest, but perhaps one of the least known of English counties. Having a total area of 2,787 sq. miles (nearly 1.8m acres), it is almost as far from Barton on Humber to Stamford as it is from the latter to London. With its long coastline it forms a somewhat blunted peninsular. This insular character was early emphasised by the Fen marshes together with those at the head of the Humber and the lower Trent which restricted entry to the Kesteven Plateau in the south-west. Three-quarters of the county is under 100 feet in altitude and much of this is only a little above sea level, although the central Wolds contain the highest ground (550 ft.) in eastern England between Kent and Yorkshire.

Geologically Lincolnshire forms part of the eastern lowlands of England, where the alternate zones of hard and soft rock trend north-south. There are two lines of hills — the mainly chalk Wolds and the limestone Heath, between which is a clay vale broadening southwards to the peat and silt filled depression of the Fens. To the west lies the Vale of Trent and the Isle of Axholme, while to the east lies the Lincolnshire Marsh fringed by a sand-dune and salt marsh coastline. This basically simple pattern of a belted lowland hides a great deal of interesting detail, not a little of it the result of glaciation and its aftermath. Systems of regions for the county have been devised for various purposes, the latest of which is by J. W. Blackwood (1972) proposing 13 natural regions, based on solid geology and surface deposits, as a basis for ecological surveys and wildlife conservation. However, for the purposes of this chapter it will be convenient to consider the county as a series of north-south zones, starting in the west. (Figs. 2 and 3).

The **Isle of Axholme** — really an island within an island — is composed largely of hard red Keuper Marl which gives a series of low flat-topped hills of 50 — 130 ft. Banked against the west side are blown sands and beyond them peat extends over the county boundary. East are the Carr lands of the River Trent. Before drainage the Isle was an area of vast swamps and moorlands with some woods; of these only fragments of former turf-diggings — the turbaries — remain. Drainage started by Vermuyden in the 17th century was completed by Rennie in the 19th century. Warping drains leading off from the tidal Trent were used to spread silt on the surrounding land.

The meandering section of the **Trent** through Lincolnshire forms the eastern boundary of the Isle and flows in a broad vale of alluvium. Fronting these wet lowlands (which are covered in part by sands on the east in the form of inland dunes) are a series of low hills (again with blown sand) caused by the tougher Hydraulic Limestone in the Lower Lias. Afforestation has taken place at Laughton, Scotton and the hills culminate in the impressive 200 ft. Burton cliff which is partly undercut by the Trent and overlooks the Trent Falls where the Trent joins the Humber estuary. Between this cliff and the limestone hills is a narrow clay vale drained north by the Winterton Beck. South of Gainsborough is a series of low 50 — 100 ft. hills, with a bluff of Keuper Marl overlooking the Vale of Trent, and the wide Till valley, on Lias clays, draining south-east.

South of the Roman Fossdyke and west of Lincoln are expanses of sands and gravels, resulting from former glacial meltwater drainage through the Witham gap, which are relatively well wooded. Southwards again is the upper course of the Witham with the vale of its tributary the Brant parallel to the Heath escarpment. Both of these rivers flow on Lias clays and probably contain the largest area of what remains of Lincolnshire grassland.

The Lincoln **Heath** is a narrow north-south line of hills capped by oolitic limestone of the middle Jurassic, varying up to nearly 250 ft. in height north of Lincoln (known as the "Cliff"), to over 350 ft. between Lincoln and Ancaster (the "Heath") and to between 400 and 500 ft. on the south Kesteven Plateau. The hills narrow from 15 miles wide in the south of the county to less than 3 miles wide in places north of Lincoln. They have a steep escarpment slope facing west, and an opposite slope, the dip slope, which declines gently eastwards. The hills are cut by the distinctive Lincoln gap and, at a higher level, the Ancaster Gap with its in-fill of river gravels.

Underlying the pervious oolitic limestone, which does not exceed 100 ft. in thickness, are a series of sandy and clay deposits which form the lower part of the scarp slope. In the north these are in turn underlain by the Frodingham Ironstone which is opencast quarried extensively east and north of Scunthorpe. This is also an area of Cover Sands — blown sand which gives rise to sandy warrens and miniature dune systems, to grass and heather heaths and to open pine, oak and birch woodland. Small pools and areas of marsh and bog add to the variety of habitat. When Abraham de la Pryme saw these heaths in 1695 they reminded him of "the sandy deserts of Egypt and Arabia, which I had a most clear idea of when I beheld these sandy plains". Until the third quarter of the last century this was a remote district, much of it in its natural condition, but the rapid growth of Scunthorpe and the ironstone mining encroached upon it and disturbed it. Even so, large tracts such as Manton and Twigmoor Warrens remained largely untouched until the early 1920s. Since then, however, the encroachment of ironstone mining on Crosby and Brumby Warrens has continued, whilst Manton Warren and many smaller patches of heath were ploughed up after 1950. Now only one or two small pieces like Twigmoor Warren remain in a natural state.

The hills between Kirton and Lincoln, and as they broaden south from the gap to Ancaster, are dry and sparsely wooded. There is an extensive pattern of dry valleys in the dip slope between Lincoln and Ancaster. Apart from the Roman Ermine Street the pattern of straight roads and hedges tells the story of late 18th and early 19th century enclosure, before which much of it was trackless heath and rabbit warrens. South of the Ancaster gap, which is floored with sands and gravels, the limestone ridge broadens rapidly into the Kesteven Plateau, has an extensive covering of Boulder Clay and is heavily wooded. The ridge here is dissected by the north-flowing Witham and south-flowing Glen and Eden. The latter's distinctive meandering course cuts through the Cornbrash, and Great Oolite clay and limestone into the Estuarine series. The Marlstone Ironstone forms a secondary escarpment to the west of Grantham where it has been extensively quarried, as has also the Northants Ironstone (Middle Jurassic) near Colsterworth. East of the latter, sink holes occur in the limestone.

Between the Cliff and the Wolds is the broad **Clay Vale** of mid-Lincolnshire. Although it is floored by thick deposits of Oxford and Kimmeridge Clay these seldom appear at the surface because of extensive covering of glacial boulder clay, gravels and blown sand. Where the vale narrows markedly in the north it is occupied by the River Ancholme which was straightened and the adjoining peaty Carr lands drained in the 17th and 19th centuries. The low 50 — 100 ft. hills on which stand villages like Howsham and the Kelseys are capped by Boulder Clay. Cover Sands are banked up against the Wolds just north of Caistor and at Wrawby, and spread south along the foot of the Wolds giving rise to the woodlands in the Market Rasen area. South of the headwaters of the Ancholme and south-east from the valley of the Langworth Beck is an extensive, undulating, well-wooded Boulder Clay cover. Most of this is derived from the underlying Upper Jurassic clays, but nearer the Wolds there is a greater chalk content. The woods and heathy moors of the Woodhall-Coningsby area are on sands and flinty gravels washed down from the Wolds during the Last Glaciation.

The extension of **the Fens** towards Lincoln, bounded by the Witham and the Roman Car Dyke at the foot of the Boulder Clay covered dip slope of the Heath, is predominantly of peat. Another major area of peat occurs in Deeping Fen and continues over the county boundary. A third area of peat is in the East Fen. A line of Fen edge gravels from Potterhanworth to Timberland is wooded. Gravels also border the Fens south of Swaton and encroach upon it south of Bourne where they are extensively quarried at Tallington, Langtoft and Deeping St. James. The other area of gravels is in the northern part of Wildmore Fen. The rest of the Fens is divided into alluvial and finer silt soils, the latter occupying the eastern part nearer the Wash, particularly the zone of 'Townlands' — the silt bank of varying width on which the older settlements stand and which separates lands reclaimed from the fen from those reclaimed from the sea. The most extensive reclamations of salt marsh in the last three centuries have been between Skegness and Wrangle, and round the former Bicker Haven and the Welland and Nene estuaries.

The Fens have undergone many changes in post-glacial times; they were not all "foule and flabby quavemires", nor were they always "overflowed by the spreading waters of the rivers". Certainly the Fens were drier in Roman times. Although piecemeal reclamations continued, the turning point was the 13th century after which winter flooding occurred in most years and summer grazing was held in common. The two main phases of drainage were in the 17th and 19th centuries. The Holland and Witham Fens were finally drained during the second half of the 18th century, but it was not until the first decade of the 19th that the vast expanse of the East, West and Wildmore Fens was completely drained and brought under cultivation. Arthur Young, writing in 1799, described a visit to the East Fen; "Sir Joseph Banks had the goodness to order a boat, and accompanied me into the heart of this fen, which in this wet season had the appearance of a chain of lakes, bordered by great crops of reeds". There was also doubtless much carr of willow and alder and the higher and drier parts probably had pine, oak and birch scrub. Today nothing is left of the natural fenland; the reclamation has been complete. Even the former 'Wash' land at Cowbit was drained in 1968. At first much of the land was put down to grass, but now almost every field is under the plough.

The **Wolds,** capped by chalk, rise to a plateau of about 400 feet and converge towards the Cliff in the northern part of the country. The simple west-facing escarpment of the Wolds is only apparent in the north, and the east margin is not a simple dip slope but has been subject to a variety of erosional agencies during the Pleistocene period. During this period sea cliffs were produced which were masked with glacial deposits of the Last Glaciation. The latter also modified the stream valley systems.

Underlying the White Chalk, which does not exceed about 150 feet in thickness in the Wolds, is the Red Chalk producing a vivid splash on some steep hillsides. Beneath this is the Carstone, a friable khaki-coloured sandstone, forming the middle part of the scarp slope and lower down is the Tealby series of clays, limestones and ironstones. The upper ironstone — the Roach Stone — forms noticeable ledge features in the south. Also in the south the Spilsby Sandstone, formerly important as a building stone but giving sour soil, occurs at the surface over a wide area, particularly round the Lymn valley. Except in the

THE FLORA OF LINCOLNSHIRE

north many streams have cut through all these strata to expose the Kimmeridge Clay which occupies the wet valley floors as a contrast to the dry sandstone above.

Because of glaciation not all strata produce a directly related soil. During the Penultimate Glaciation Lincolnshire was completely covered by ice moving roughly from north to south. Thus only thin soils have developed on the tops, but the south-west Wolds have a thick cover of Boulder Clay between the valleys. During the Last Glaciation the ice penetrated little further west than the eastern edge of the Wolds and most valleys are floored with gravel and silt deposits.

Physiographically the Wolds can be divided into north, centre and south. North of a line from Caistor to Laceby there is a steep, simple, west-facing escarpment, some of it uncultivated to this day. Because of the greater thickness of Chalk few of the valleys are occupied by streams. The extensive Brocklesby estate woodlands occur on the dip slope. The central section of the Wolds, from the Caistor-Laceby line to that of the A.631 from Market Rasent to Louth, exhibits a double escarpment to the west, and contains the highest area of the Wolds, 550 feet, near Normanby le Wold. The headwaters of the Waithe Beck have cut through to the Lower Cretaceous rocks and a series of steep-sided valleys on the eastern margins have resulted from glacial stream diversions. The Nettleton and Usselby Becks and the river Rase have cut deeply into the western escarpment.

The southern Wolds are a complex area. The true escarpment (Donington — Red Hill — Tetford — Sutterby — Langton) is obscured by the Lower Cretaceous ridge with its capping of glacial deposits to the south-west, or it overlooks the Lymn valley. The chalk becomes an increasingly narrow tongue towards Candlesby. There are a number of different and interesting valleys in this part of the Wolds. The River Bain rises near Ludford and flows almost parallel to the strike of the rocks, and along its upper and central course there are a number of deposits of gravel. Towards Horncastle the valley is wider, flatter and is floored by Kimmeridge Clay. Another strike stream, but with different valley form, is the Lymn. Rising in a coombe near Belchford, it tumbles through the spectacular New England gorge into the main valley which is surrounded by a wide ledge of Spilsby Sandstone and floored again by Kimmeridge Clay. A number of its tributaries notably Snipe Dales (and Sow Dale occupied by a stream flowing south to the Fens) have steep, dry valley sides of sandstone and a flatter floor of clay. The valley emerging at Keal Carr is of similar form.

Until the 18th century much of the Wolds was in a natural or semi-natural condition. When Arthur Young travelled through the country in 1799 some enclosure of the Wolds had been made in the previous fifty years. "Forty years ago," he wrote, "it was all warren for thirty miles from Spilsby to beyond Caistor", and elsewhere he spoke of "the bleak wolds and heaths being almost enclosed and planted within twenty or thirty years". In 1799, however, there were still large tracts of open grassland with thorn and gorse scrub on the chalk tops. Thirty years later when Cobbett travelled over these hills the transformation of the landscape had been completed and it was "a very fine country, large fields, fine pastures, flocks of those great sheep". Writing in 1852, Clarke confirms that "all the open fields have disappeared, a great part having been enclosed within the last thirty years. The gorse has been grubbed, the rough sward burned and all the warrens, with one or two exceptions, have been brought into cultivation ... the highest parts are all in tillage and the whole length of the Wolds is intersected by neat whitethorn hedges, the solitary furze bush appearing only where a roadside or plantation border offers an uncultivated space." The hedges are fewer and smaller at the present day, but otherwise the landscape of the higher Wold tops is that described by Clarke. The great arable fields are even more intensively cultivated and the rare fields of permanent pasture are confined to the valleys where there are streams. Here and there a hillside too steep for cultivation or the wide grass verges and occasional beech clumps beside an ancient trackway like the Bluestone Heath Road remind one of the open heath.

East of the Pleistocene sea cliff edge of the Wolds is the **Lincolnshire Marsh.** Its main divisions are the Middle Marsh — of hummocky, undulating Boulder Clay, the result of the Last Glaciation — with outwash gravels in the Thornton Abbey-Keelby and Alford-Bonthorpe areas; and the Outmarsh. The latter is of flat marine silts at about 10 ft. OD. Occasional hummocks of Boulder Clay project through the silt cover near Hogsthorpe and the Middle Marsh reaches the coast at Cleethorpes. In the northern part of the Middle Marsh 'blow wells' occur where sand and gravel lenses allow water from the chalk to reach the surface and produce a different habitat. In the southern part there are woods in the lee of the Wolds. The Outmarsh was traditionally important for grazing but an increasing area is coming under the plough. The remains of numerous brickworks are to be found on the Outmarsh clays, those at Barton, Goxhill, Mablethorpe and Skegness still being worked. The Outmarsh north-west of Saltfleet has been built up by reclamations from the sea, whereas that between Mablethorpe and Skegness has been subject to erosion by the sea.

Before the stormy 13th century the Lincolnshire **Coast** was protected by a line of off-shore boulder clay islands, from Holderness to Norfolk. The final destruction of this barrier gave rise to a storm beach which became the basis for the present sand dune system and is best seen in the North Somercotes Warren. Between Mablethorpe and Skegness the coastline has retreated and since the Storm-Flood of 1953 is now protected by varying forms of sea wall; the sand dunes are impoverished and non-existent in places. At Saltfleetby-Theddlethorpe a strip of fresh-water marsh occurs in the dunes; north-west of Saltfleet accretion has been considerable and the dunes and reclamation banks are fronted by a wide beach and extensive 'fitties' or saltmarshes. South of Skegness accretion by longshore drift and by material derived from

THE FLORA OF LINCOLNSHIRE

offshore has extended the dune line to Gibraltar Point and built the coast outwards in a complex of dune ridges and strip saltings. The Wash coast is fringed by saltings over half a mile wide which are subject to reclamation from time to time. They are most extensive between the estuaries of the Witham and Welland, and at Holbeach and Gedney.

Selected references

ALVEY, R. C. : Post-glacial fauna and flora from inter-tidal exposures in the Ingoldmells area, Lincolnshire. *Mercian Geologist* **3**, 1969, 137-142.

BARNES, F. A. and KING, C. A. M. : A preliminary survey at Gibraltar Point, Lincolnshire. *Bird Obs. and Field Station Report*, Lincs. Trust for Nat. Cons. 1951, 41-59.

BARNES, F. A. and KING, C. A. M. : The Lincolnshire coastline and the 1953 Storm-Flood. *Geography* **38**, 1953, 141-160.

BARNES, F. A. and KING, C. A.M. : Salt marsh development at Gibraltar Point, Lincolnshire. *East Midland Geogr.* **2**, No. 15, 1961, 20-31.

BLACKWOOD, J. W. : The distribution of nature reserves in the natural regions of Lincolnshire. *Trans. L.N.U.* **XVIII**, 1972, 1-6.

BOYLAN, J. P. : The Pleistocene deposits of Kirmington, Lincolnshire. *Mercian Geologist* **1** (4), 1966, 339-350.

COOPER, C. : Tetney Blow Wells. *Trans. L.N.U.* **XVII**, 15-18.

HINDLEY, R. : Sink-holes on the Lincolnshire Limestone between Grantham and Stamford. *East Midland Geogr.* **3**, No. 24, 1965, 454-460.

KESTNER, F. J. T. : The old coastline of the Wash. *Geogr. Journ.* **128** (4), 1962, 457-478.

LINTON, D. L. : Landforms of Lincolnshire. *Geography* **39**, 1954, 67-78.

MUSKETT, P. J. : Periglacial gulls in the upper Witham valley. *Trans. L.N.U.* **XVII**, 1971, 210-216.

OWEN, A. E. B. : Coastal erosion in east Lincolnshire. *Lincs. Historian* **9**, 1952, 330-341.

ROBINSON, D. N. : Coastal evolution in north-east Lincolnshire. *East Midland Geogr.* **5**, Nos. 33-34, 1970, 62-70.

ROBINSON, D. N. : Geology and scenery of the Lincolnshire Wolds. Excursion report in *Mercian Geologist* **4** (1), 1971, 63-68.

RUDKIN, E. H. and OWEN, D. M. : The medieval salt industry in the Lindsey Marshland. *Lincs. Archit. and Archaeol. Soc.* **8**, 1959-60, 76-84.

SMITH, A. G. : Post-glacial deposits in south Yorkshire and north Lincolnshire. *New Phytologist* **57**, 1958, 19-49.

STRAW, A. : Some glacial features of east Lincolnshire. *East Midland Geogr.* **1** (7), 1957, 41-48.

STRAW, A. : The glacial sequence in Lincolnshire. *East Midland Geogr.* **2** (9), 1958, 29-40.

STRAW, A. : Drifts, meltwater channels and ice margins in the Lincolnshire Wolds. *Trans. and Papers Inst. Brit. Geogr.* **29,** 1961. 115-128.

STRAW, A. : The Quaternary evolution of the lower and middle Trent. *East Midland Geogr.* **3** (20), 1963, 171-189.

STRAW, A. : Some observations on the Cover Sands of north Lincolnshire. *Trans. L.N.U.* **XV,** 1963, 260-269.

STRAW, A. : The development of the middle and lower Bain valley, Lincolnshire. *Trans. and Papers. Inst. Brit. Geogr.* **40,** 1966, 145-154.

STRAW, A. : Lincolnshire Soils. *Lincs. Nats. Union*, Lincoln, 1969.

STRAW, A. : Pleistocene events in Lincolnshire: a survey and revised nomenclature. *Trans. L.N.U.*, **XVII,** 1969, 85-98.

SWINNERTON, H. H. : The physical history of east Lincolnshire. *Trans. L.N.U.* **IX,** 1936, 91-100.

SWINNERTON, H. H. and KENT, P. E. : The Geology of Lincolnshire. *Lincs. Nat. Union*, Lincoln, 1949.

WATTS, W. A. : Pollen spectra from the interglacial deposits at Kirmington, Lincolnshire. *Proc. Yorks. Geol. Soc.* **32** (2), 1969, 145-151.

CHAPTER 3

WEATHER AND CLIMATE

F. A. Barnes, B.Sc.

The essential insularity of the British climate, under the influence of prevailing westerly maritime air-streams, is modified in Lincolnshire by the eastern location, bringing free exposure to eastern continental influences, and by some degree of shelter afforded by the southern Pennines. The low relative relief of the county is reffected in a corresponding uniformity of climate, such that variations from area to area are often obscured by the effects of quite local site differences. Only in a rather narrow coastal strip east of the Wolds does the maritime influence of the North Sea become evident. Much of Lindsey, especially the north-east, is affected by wintry showers in unstable north to north-east air-streams off the North Sea, and layers of low stratus cloud in summer often penetrate freely at night westwards over the county and beyond, and may persist by day in the east. The sea fog which often drifts over the coast and the Outmarsh rarely extends inland as far as the Wolds.

The low but somewhat variable annual rainfall indicates a climate for Lincolnshire drier than the average for England and Wales, and the rainfall tends to be concentrated in the second half of the year over most of the county, though a weak summer maximum is evident in the Fens. Local configuration, notably the disposition of the uplands of the Wolds, Heath and Kesteven Plateau and of the major valleys of the Trent and the lower Witham (including the Fens) largely determines the patterns of distribution of rainfall, length of snow-lie, the incidence of low cloud, fog and mist. But altitude as such influences temperature less than very local site conditions of slope, aspect, soil type and land use.

Inland from the coast the daily temperature range increases, while day time humidity, wind strength and sunshine all decrease; but day to day the influence of the sea varies greatly. While maritime influence penetrates further inland over flat county as a rule, this hardly applies in the case of the Fens because of their distance from the open sea, and the special character of the Wash, though the Fens are markedly exposed to strong east winds.

The coastal and eastern location of Lincolnshire has obvious implications when the wind has an easterly component, and especially when anticyclonic development over northern Europe brings characteristically raw, dull weather in the winter and spring, and sheets of very low stratus cloud (haar) shroud even the low hills during night and morning in summer, often persisting throughout the day in coastal districts when the weather becomes warm and sunny inland. East winds in winter bring colder air into Lincolnshire than they do into eastern Scotland because of the shorter sea track.

Wind

Long records indicate south-west as the most prominent surface wind direction in the East Midlands, especially in winter, but the balance of directions has varied periodically as general weather patterns, and with them rainfall and temperature conditions, have fluctuated. In the relatively warm, dry decades from the 1890s to the 1920s south-west to south winds were more prevalent, especially in winter, than more recently, and the frequency of west winds has since increased.

A prominent feature for all inland stations is a secondary maximum of north-east winds, masked in the annual frequencies, in spring and early summer, with a peak in May. These cold, often dry vernal easterlies are especially important for their association with the hazards of late killing frosts and retarded spring growth. Near the coast sea breezes occur on quite sunny days.

Winds of different directions are associated with characteristic temperature and moisture conditions. The average dew point temperature of a north-east wind in February is 6°C or more below that of a south-east wind at inland stations, but in July the north-east wind is likely to be the moister. Of particular note are mean dew points between 0°C and 2°C with north to east winds in early spring. With dry bulb temperature of 4°C or less and overcast skies, these are representative of the raw weather which is such an unpleasant though intermittent feature of that season in Lincolnshire.

At sheltered stations on low ground well inland the annual average wind speed is no more than 6 or 7 kts, but at the coast the median wind is about twice as strong, and winds of more than 15 kts are more than twice as frequent. This increases the 'cooling power' and emphasises the bleakness of easterlies in winter, and the 'bracing' qualities reputedly enjoyed by the coast in summer.

In certain circumstances severe erosion of soil by wind occurs, especially in the Black Fens and on light soil in other areas. The danger to spring cultivations occurs especially in March and April after a period of low rainfall and frequent frost. Serious 'blows' occurred in the Fens in 1956, in the Isle of Axholme in 1943 and were widespread in March, 1968. The most damaging winds were those from between northwest and south-west, mainly 25-30 kts. Local fetch is important, especially in the Fens, though other open areas lacking windbreaks, like the Wolds, are susceptible, and the removal of hedgerows and ploughing of old pastures in recent years has probably aggravated the problem.

Rainfall

In the 35 year period 1916-1950 the mean annual rainfall in Lincolnshire ranged from about 29-30 inches at a few places in the higher parts of the central Wolds, to 20-21 inches in southern Holland. But only over the Wolds and their margins, the higher parts of the Heath north of Lincoln and a few square miles of the Kesteven Plateau east of Grantham was 25 inches exceeded. Only southern Holland, the Outmarsh east of the Wolds and parts of the Carrlands of the lower Trent valley near Crowle, all areas within a few feet of sea level, averaged less than 23 inches. Thus well over half the county had a mean annual rainfall of between 23 and 25 inches, that of the Witham valley below Lincoln and the Trent valley being characteristically 23-24 inches and that of the Kesteven Plateau 24-25 inches.

A relationship with altitude is the most obvious feature of the annual rainfall pattern, though some anomalies occur. The only regional gradient of note is that implied by the rather higher falls, height for height, of the north compared with the south, probably related in part to its greater exposure to North Sea showers in winter, and nearer proximity to the Pennines. In summer sea breeze convergence may tend to favour conventional storms in the general area of the Wolds. The steepest gradient of rainfall occurs along the steep eastern margin of the Wolds, where the fall increases westwards at about 1 inch per mile for several miles west of Louth.

Rainfall graphs for **1936-60 at Cranwell**, Skegness and Spalding (Pode Hole) show the annual regimes. November is the wettest month, December markedly drier and January wetter again (the second wettest at Spalding). Rainfall amounts then decline to reach the year's minimum in April, and thereafter increase to a secondary maximum in July/August, one or other of which is generally the second wettest month. Everywhere September provides a drier interlude before the rise to the November peak, but at Skegness October is still drier than September.

Corresponding graphs for earlier periods present a different picture. In 1866-90, which includes the wet 1870-mid 1880s, July was the wettest month in eastern Lincolnshire and September was almost as wet, while October was also wetter than in recent decades. There was no significant break between the rainfall peaks of summer and autumn. March was the driest month. Graphs reveal that September was wettest in the rainy years around 1870-80, but fell off sharply from about 1890 to only half its earlier figure until, about 1920, the monthly fall increased to an intermediate level which has since been maintained.

Such changes in rainfall seasonality emphasise that quite minor fluctuations in annual rainfall figures may be associated with alterations

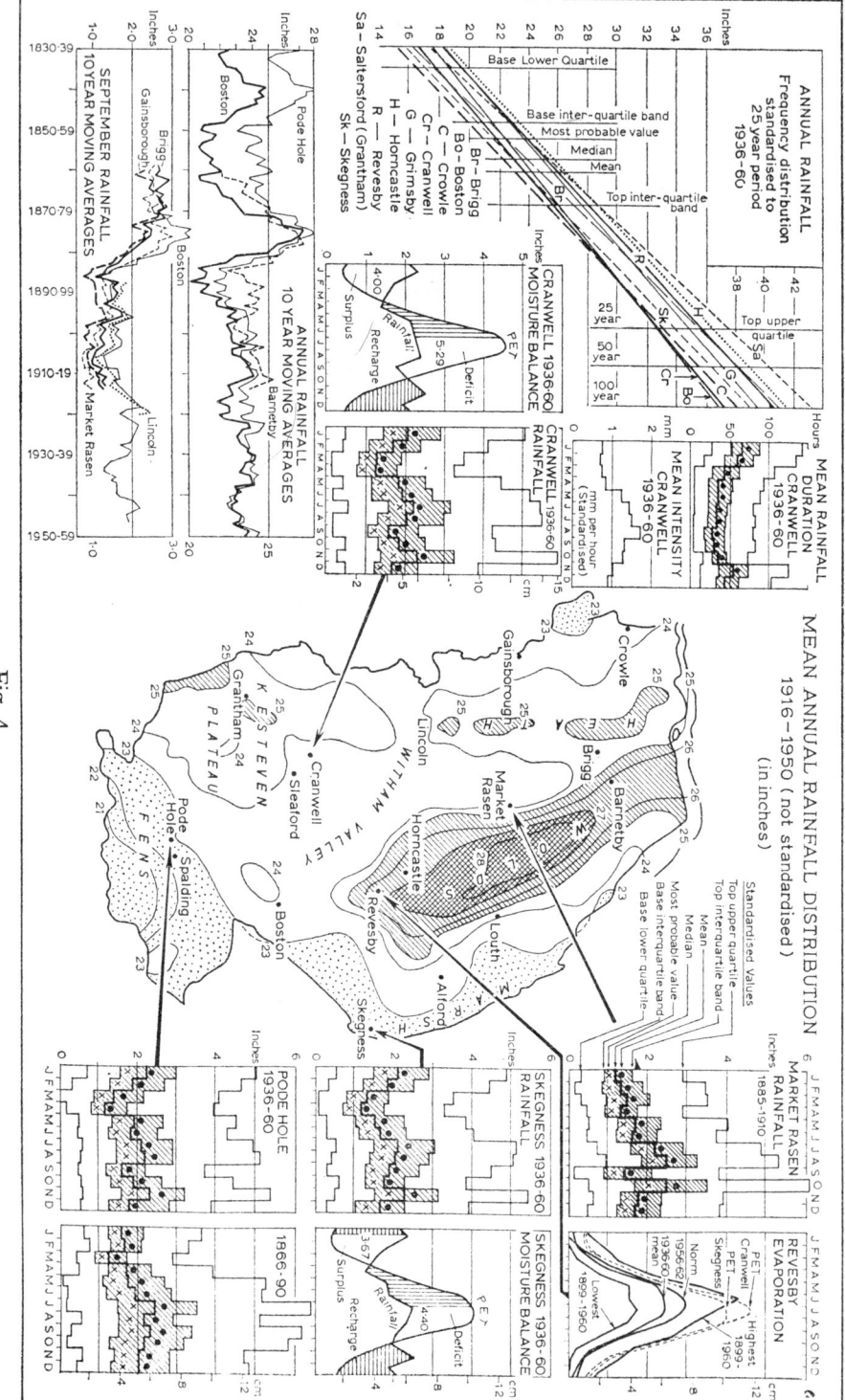

Fig. 4

of seasonal water balance which may be very important from a biological and agricultural standpoint. For example, the wetness of late summer and early autumn, and the high autumn water tables which must have hampered autumn ploughing on many soils, may go far to explain the preoccupation with field drainage in the decade before 1890, if not, indeed, the decline in 'high farming' in that period.

The wettest year of the last century and a half was 1872 over much of Lincolnshire and adjoining counties, such very wet years usually having high and unevenly distributed summer rainfalls. In the present century 1960 was the wettest year at many places, including Cranwell, Skellingthorpe, Brigg and Boston, but 1937 was wetter in some places and 1931 or 1912 in others.

The driest years vary more from place to place because of the greater influence of individual storms on the totals. Records indicate that while 1887 was the driest in some areas, 1921 was drier in many others. 1887 had a rainfall as low as 12.94 inches at Boston, while totals below 14 inches have been recorded at Gainsborough and Keadby. Annual totals above 40 inches are uncommon, though 41.37 inches was reported from Louth in 1872 and 39.83 inches from Gainsborough in the same year.

The number of rain days (days with 0.01 inches of rain or more) in a year is more readily correlated with rainfall amount, which is mainly expressive of altitude, than with regional location. Typical average frequency rises from about 160 days in the lowland to 180 days or more on high ground.

A proper evaluation of the moisture conditions requires that rainfall should be considered in relation to actual or potential evaporation, for rainfall is, of itself, a poor indicator of its biological effectiveness. In particular 'official' drought, arbitarily defined by spells without measurable rain, is a less useful concept than 'biological' or 'agricultural' drought. The latter occurs when precipitation has been insufficient to meet 'water need', or potential evapotranspiration, and soil water reserves have been depleted to the point at which plants suffer physiologically from lack of water. Unlike official drought it is essentially a summer phenomenon.

It can be shown that the whole of Lincolnshire is subject to risk of agricultural drought in more than five years out of ten, and more in the areas of lowest rainfall. There is an average soil water deficit decreasing north and north-west from about 6 inches in the Fens (Spalding) to about 3.5 inches in Kesteven and north-west Lincolnshire. Run-off figures suggest that over a large part of the county annual run-off will be the equivalent of about 7 inches of rainfall. Absolute droughts (15 days or more with no daily measurement of 0.01 inches of rain) occur with an average of about one a year (1919-38) in south-east Lincolnshire and 0.7 times per year in the north-west. They occur mainly between mid-July and mid-September.

Snow

The frequency of snowfall and the length of snow-lie respond sensitively to differences of altitude and exposure, and the latter also to aspect. In Lincolnshire the degree of exposure to showers in the north to north-west airstreams is important. Manby, four miles east of the Wolds and fully exposed to North Sea showers, records a frequency of days with snow fall even higher than that on high ground south of Lincoln. The distribution pattern tends to show rather sharp concentrations with the northern Wolds probably the snowiest part of the county. Over the past 20 years the average annual number of days with snow fall on the higher parts of the Wolds must have been considerably above 30, and probably about 35. Days with snow lying at 9 a.m. do not much exceed 20 on average anywhere, except on the Wolds, and are fewer than 15 on the coast.

Snow is most frequent in February, but nearly as frequent in January, and the proportions in December and March are considerable, but more variable. Snow is infrequent in April, May, October and November. The amount of snow tends to be more concentrated into the later winter months than the frequency of snow fall, with 85-90% falling after the turn of the year.

Air temperature

From the figures for mean monthly and mean annual temperatures, no great variation is apparent within the county, and the chief difference is between the coastal stations which have mean annual temperatures 0.2°C or 0.3° higher than inland stations. Under the influence of periodic onshore winds, in turn influenced by sea surface temperature, the monthly average temperatures at the coast are higher than those inland in autumn and early winter, but lower in spring, while in high summer they differ little, for lower daily maximum temperatures, moderated by sea breezes, offset higher night minima.

Inland a monthly mean temperature of about 3.0°C (37.5°F) in January rises to about 16.3°C (61.3°F) in July, the warmest month, and the annual regime varies little from place to place. At the coast the variation is from 3.3-3.5°C in January to 16.2-16.3°C in July. Comparatively the coast is warmest in autumn and coldest in late spring, in accord with the sea temperature regime. In February a wide sea area off Lincolnshire has a surface water temperature as low as 4-5°C on average, and has little warming effect on continental air blowing onshore especially that with a short sea track from east or north-east. One consequence is a small area round the Wash with a mean daily January temperature of less than 3.5°C (39°F), the lowest in England and Wales except for the northern Pennines. In August when the wind is from the sea the afternoon maximum temperature is limited near the coast by the sea surface temperature, and a narrow coastal strip of Lindsey has a mean maximum temperature a degree or two below the general level.

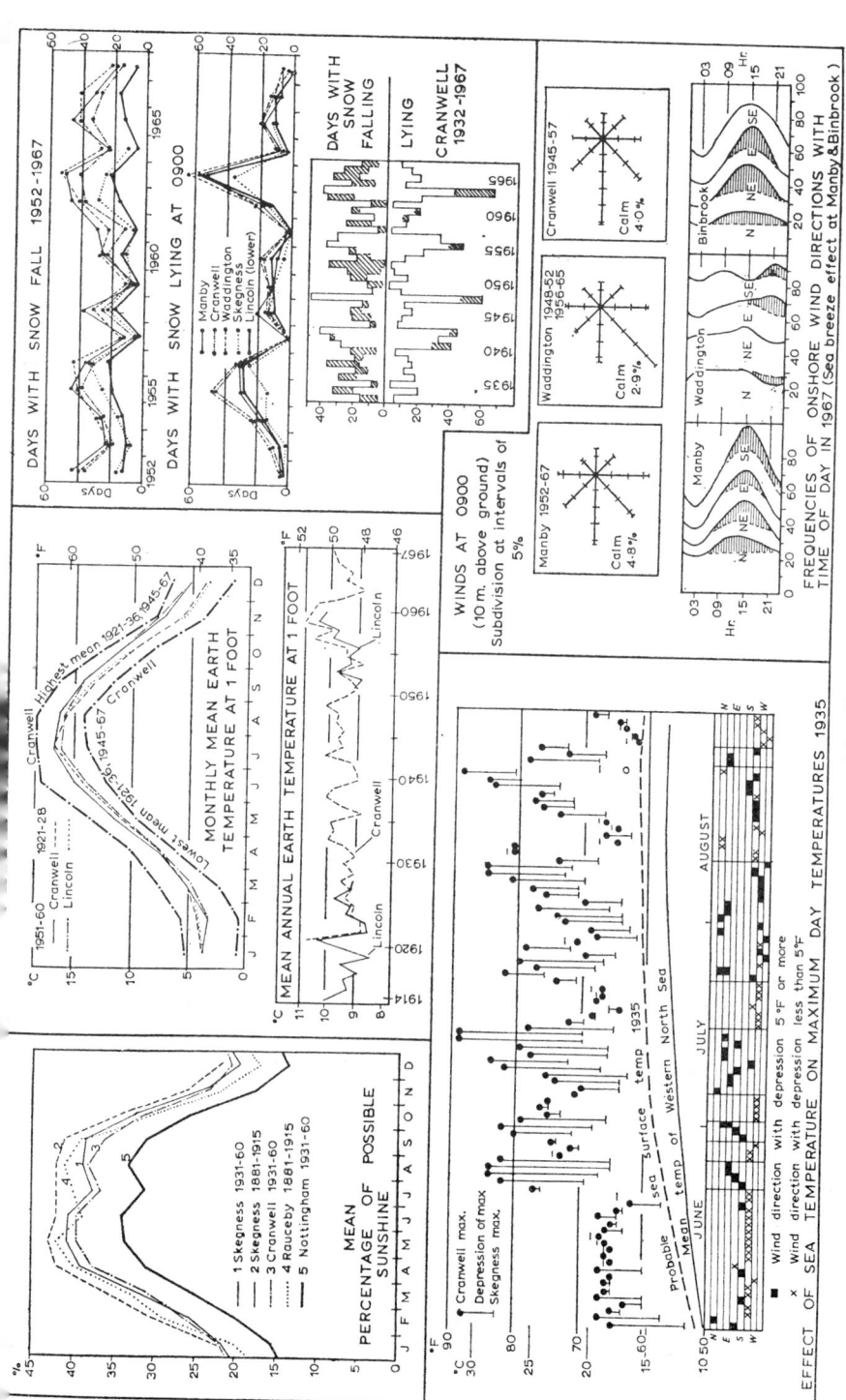

Fig. 5

It is rare for any year's annual mean temperature to differ from the average by more than 1.5°C. Year to year variations are larger in the winter months, when cyclonic and anticyclonic spells give wide temperature differences, than in summer, when insolation is more decisive. The temperatures of individual days vary much more widely, site characteristics strongly influencing extreme minima, and concave sites may be many degrees colder than ridges and slopes on the same occasions. The lowest temperatures always occur with snow cover.

Except near the coast maximum temperatures reach 27°C on five to ten days a year on average. Although July is the warmest month taken as a whole, the highest temperatures of individual days are recorded in August and September. A typical station in Lincolnshire probably has an **absolute** temperature range of about 55°C (100°F), between about −23°C and + 35°C, this being reduced by 11 to 17°C at the coast.

Sea breeze effect

A sea breeze develops strongly in spring and summer, when temperature maxima in warm weather are commonly 6-8°C and occasionally 11-12°C cooler at the coast than inland. Temperatures which fluctuate with the onset and retreat of the sea breeze are not uncommon in summer, especially on the southern part of the Lindsey coast, and they may be associated with the advance and retreat of sea fog along the coast. On hot days the sea breeze may penetrate as far as the Wolds, but the temperature amelioration diminishes rapidly inland, so that Alford and Louth, for example, are little affected.

Frosts

Since frost occurrence depends on favourable synoptic circumstances, the annual average number of air frosts varies less extremely from place to place than it does with time, though their severity is influenced greatly by site factors. Valley sites may experience killing frosts when nearby slopes and ridges remain above freezing point, for on quite clear nights a temperature difference of 5°C or more can develop rapidly after sunset as a result of cold air drainage. The annual frequency of screen frosts does not vary much from 55 to 70 days on average in Lincolnshire, but is about 40 days in the coastal strip.

Frequency of ground frosts varies more widely, and is more influenced by site factors than frequency of air frosts. Over much of Lincolnshire the annual average is about 70 to 80, but fewer on favourable ridge sites and near the coast, though well above 100 at 'cold' concave sites, where they may occur in any month. Soil differences also have an effect. For example as between silt and peat soils in the Fens, where crop distribution is influenced by the degree of susceptibility to ground frost. At most places ground frosts are normally absent from early June to mid-September, and few air frosts are recorded between early May and early October. The absolute frost-free period is nearly 110 days for most places, and much less in frost hollows, though often greater near the coast and on favoured sites inland.

Low air temperature of itself is serious only when prolonged, so that frost penetrates deeply into the ground as it did in January-February, 1963, and in frost susceptible soils, that is those with a considerable water content, when plant roots are disturbed by the heaving associated with ice segregation.

Soil temperature

At one foot deep the soil is warmest in late July, normally reaching a maximum of about 16.7°C (62°F), and falling to a minimum of about 3°C (37°F) in January-February. Values of 21°C (70°F) are attained only at shallower depths. Freezing point is reached at one foot deep only in rare cold spells like that of early 1963. The soil temperature at six inches deep on a south-facing slope of 1 in 4 will be up to 2°C warmer than that on a similar slope facing north in summer, and on sunny days the difference is greater, especially in spring while the sun is relatively low. As a consequence a south aspect, inducing early flowering, brings increased danger from late frost, especially on the lower parts of slopes.

Soil temperatures are also influenced by other factors, and on gentle slopes the character and conditions of the soil, especially its texture and water content, are likely to be more important in leading to local variations of soil temperature, and may significantly influence the effective length of the growing season. Using a 'growing threshold' temperature of 42°F the growing season is 250 days in western Lincolnshire and 260 days east of the Wolds and in the silt fen.

Sunshine and cloud

There is an annual average of about 1,550 hours of bright sunshine at the coast, decreasing westwards and northwards to less than 1,350 hours in the extreme north-west. Recent figures have indicated a decrease in hours of sunshine compared with 1881-1915 of one or two per cent. The most recent data show that the mean annual daily sunshine decreases from about 4.3 hours at the Lindsey coast to about 4.0 hours well inland, and 3.8 hours locally (Lincoln) where smtoke pollution affects it. Skegness and the extreme east of Holland is he sunniest part of the county.

Fog and Air Pollution

During 1923 to 1950 the frequency of fog (visibility less than 1,100 yards) decreased from about 40 days in Kesteven to about 15 days on the coast at Skegness. In the northern part of the coastal area, with greater degree of air pollution, the total approached 20 days a year. Fogs are fairly evenly distributed through the winter months.

Low-level smoke is responsible for many instances of visibility below the fog level in towns, but pollution from distant sources depresses the visibility almost universally. Pollution from local sources is most significant in the case of smoke, and background pollution from more distant sources is especially important for sulphur dioxide. In the latter case the general background pollution derives from the industrial areas of the Midlands and northern England and is up to three times greater in concentration in winter than in summer. Sulphur dioxide concentrations, at least in the Trent valley area of western Lindsey, appear to have decreased rather substantially since the early-middle 1960s, and there is no reason to suppose that damage to plant life will occur from this cause anywhere in the Lincolnshire countryside.

Thunderstorms

More than 15 thunderstorms a year occur in the western half of Lincolnshire, and 20 along the Nottinghamshire border, falling to 8 in southern Lincolnshire. The Humberside area and southern Kesteven have less than 12 a year. Experience suggests that the area of the central and southern Wolds is especially prone to occasional heavy thunderstorms yielding exceptionally heavy rainfall, and resulting in flash floods. It is possible that the explanation lies in convergence between the east coast sea breeze and the general wind inland from the coast, combined with the high-level heat sources provided by the elevated Wolds, a combination which provides especially favourable local circumstances for thunderstorm development in deep, moist, conditionally unstable air masses in summer.

Selected references

BARNES, F. A. : Weather and Climate in *Nottingham and its Region* (ed. K. C. Edwards), British Association, Nottingham 1966, 60-102.
BARNES, F. A. : The intense thunder rains of 1st July 1952 in the northern Midlands. *East Midland Geographer* **14**, 1960, 11-26.
CORDEAUX, J. : The great ice-storm of July 3rd 1883 in north Lincolnshire. *Meteorological Record* for Quarter ending 30th September 1883, 37-39.
HOWE, G. M. : The moisture balance in England and Wales, based on the concept of potential evapotranspiration. *Weather* **11**, 1956, 74-82.
LAMB, H. H. : *The English Climate.*
MANLEY, G. : Topographical features and the climate of Britain. *Geographical Journal* **103**, 1944, 241.
MANLEY, G. : Snow cover in the British Isles. *Meteorological Magazine* **75**, 1940, 41.
MILLER, R. C. and STARRETT, L. G. : Thunderstorms in Great Britain. *Meteorological Magazine* **91**, 1962, 247-255.
PENMAN, H. L. : Evaporation in the British Isles. *Quart. Journ. Royal Meteorological Society* **76**, 1950, 372-383.
ROBINSON, D. N. and BARNES, F. A. : The Horncastle and Bain valley flood. *East Midland Geographer* **15**, 1961, 52-55.
ROBINSON, D. N. : Soil erosion by wind in Lincolnshire, March 1968. *East Midland Geographer* **30**, 1968, 351-362.
SKERTCHLEY, S. J. B. and MILLER, S. H. : *The Fenland, Past and Present* (Wisbech 1878). Reviewed in *Meteorological Magazine* 1878, 135. (Contains a summary of 15 year's observations at Fenland stations.)
SPENCE, M. T. : Soil blowing in the Fens in 1956. *Meterological Magazine* **86**, 1957, 21-22.
TINN, A. B. : Sea breezes on the Lincolnshire coast. *Meteorological Magazine* **57**, 1922, 11.

CHAPTER 4

THE FLORA OF
THE MAJOR HABITATS

WOODLANDS

Lincolnshire is less wooded now than many parts of England. The deciduous woodland is very limited. Apart from the Forestry Commission plantings on the sands and gravels in the last 40 years, the total woodland acreage is very low. The 1924 Forestry Census gave a figure of 2.4% woodland cover for Lincolnshire. After afforestation the 1966 figure was still less than 3% compared with national average of 8%, showing 47.4 thousand acres of woodland in the County — two thirds in Lindsey, one third in Kesteven and less than 100 acres in Holland.

Fig. 7 shows the present distribution of mainly deciduous and mainly coniferous woodland in the county. In the north-west at Burton Stather and Brumby, woods occur on the Lias escarpment, and on limestone with patches of blown sand at Broughton, Scawby and Appleby. The old oak woods round Gainsborough are of small extent now. The new woods planted from the 1750s at Brocklesby, Limber, Cabourne and Swallow are on thin chalk soils, much in the form of a huge shelter belt, and Forestry Commission plantations at Scotter and Laughton, round Market Rasen (Willingham Forest), Osgodby, Walesby and Tealby, North Willingham and Linwood are largely on cover sands.

The Wragby area of clay woodlands is chiefly taken over by the Forestry Commission (Bardney Forest) and was formerly Oak and Small-leaved Lime running from Linwood and Legsby to Horncastle and Woodhall Spa. Another line of woods on the east of the Wolds from Bradley to Welton is chiefly on boulder clay. Some also occur on the eastern slopes which together with those on the edge of the Marsh are chiefly of Oak and Ash. Small woods occur in the Wold valleys; Claxby Wood resembles the Lower Cretaceous woods of the Spilsby area.

South-west of Lincoln Oak woods occur with Small-leaved Lime and Birch at Skellingthorpe, Eagle and Norton Disney, and on river

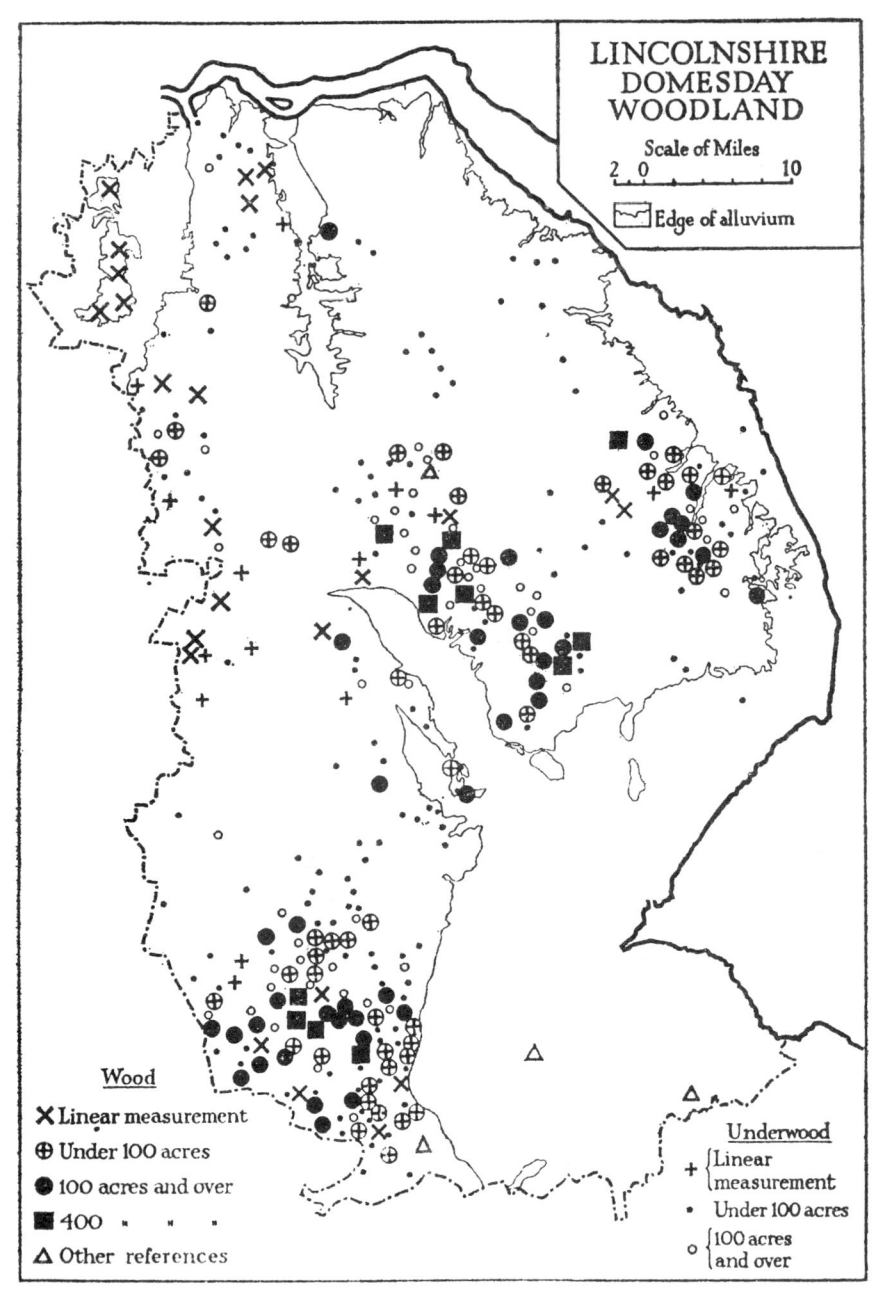

Fig. 6
(*Reproduced by kind permission of Professor H. C. Darby*)

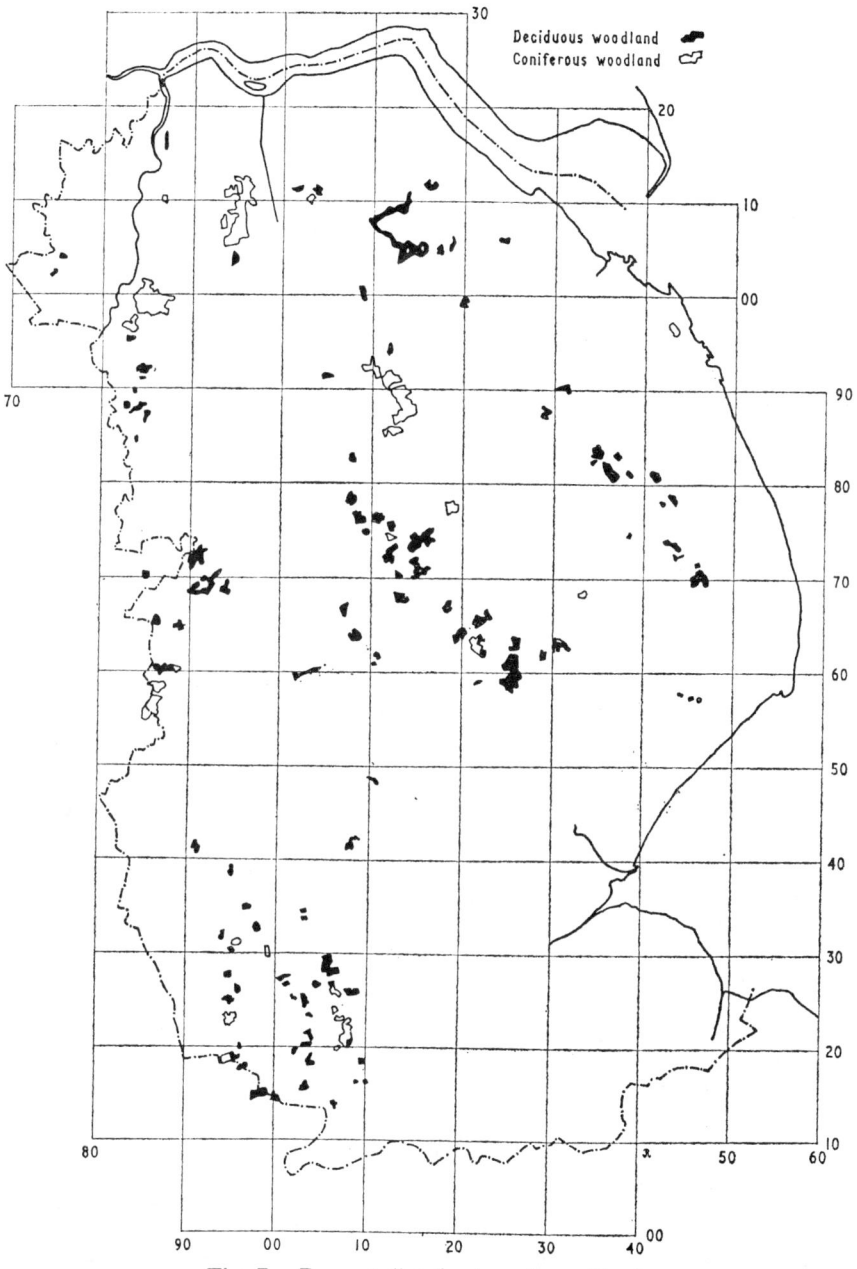

Fig. 7. Present distribution of woodland
(Based on information supplied by the Forestry Commission)

gravels with conifer plantations at Doddington and Stapleford. A line of woods on a gravel ridge from Branston to Martin on the edge of the Witham valley includes Potterhanworth, Blankney and Nocton. Another line occurs on the Fen-edge gravels from Tumby to Revesby.

The name Kesteven is derived from the Celtic *coed* meaning wood, and part of it was a royal forest from Norman times until the 13th century. The broad limestone plateau south of the Ancaster gap has a fair covering of boulder clay and there is much woodland to be found between Sleaford and Bourne and on the western slopes.

In Domesday Book it is interesting to find woods mentioned in various places where no woods exist now. They were mainly recorded as 'woodland for pannage' (pigs) or 'underwood' (scrub). There was a line of woods from Sturton-by-Stow to Saxilby all of which have vanished, including the one still named as High Wood. Metheringham (180 acres) South Kyme (300 acres) South Hykeham, South Carlton with Riseholme and many more, had large areas of which no trace remains. Near Louth, Cockerington had 130 acres and Stewton 260 acres. Woodlands have also gone from Wood Enderby (458 acres) and Mareham-le-Fen (300 acres). Timberland has no woods now but justifies its name in having had 50 acres of underwood. Further south Gelston near Caythorpe had 200 acres, Welby 230 acres, Pointon 100 acres and Corby 1,130 acres.

Some of the old coppiced Bardney Forest, between Bardney and Wragby, was brought under cultivation between 1824 and 1880, particularly at Langton-by-Wragby, Apley and Newball. Branston and Mere had a big area of woodland north of Potterhanworth torn up and cultivated in 1859 when the Mere Charity of St. John's Hospital was reorganised. Other woodlands at Fiskerton and Langton-by-Wragby were reduced to remnants.

How much of the Fenland was wooded it is difficult to know, but there were woodlands near Spalding with a different flora for which we have no records. In Domesday Book Spalding had a wood of Alders 'rendering eight shillings'. Fleet and Moulton had small woods in 1824 but these have now gone (Elloe-stone Wood and Primrose Holt and a few others round Moulton and Holbeach). There was a 10 acre wood of Oak, Elm and Hazel near Fleet before 1930 where Bluebells Primroses, Violets and Early Purple Orchids were indicative of relic woodland. A few small decoy woods remain round Wainfleet and Friskney, one of which is now a nature reserve.

In the north-west of the county the Carrs (watery thickets) had prehistoric woods of Oak, Birch, Yew and Pine but these have all gone, though remains of bog oak and other species are found when drainage and deep ploughing reveal them in river valleys. The Isle of Axholme had large ancient woods, as place names such as Belwood, Belshaw, Melwood indicate. A few yards of woodland plants (Yellow Archangel, Dog Mercury, Hairy St. John's Wort) in a hedge at Melwood and in

THE FLORA OF THE MAJOR HABITATS

two or three spinneys such as Swithin's Thick are all that is left. Burton Stather and Brumby Wood have interesting woodland plants and also Broughton Lily Woods which were skilfully replanted by the Earls of Yarborough up to 1950 (now owned by the Duke of Westminster).

Many woods have been carefully replanted by far-seeing landlords but others have been neglected or grubbed up. After felling some were sold to the Forestry Commission. The Commission now has a huge acreage of old valuable deciduous woodlands and new ground chiefly planted with young conifers not yet come to maturity — a county total of about 18,000 acres. Nearly 100 separate woods are owned or leased by the Commission varying from three to 1,800 acres on a variety of soils. Of these 14,010 acres are under plantations — 10,340 acres conifers, 3,670 acres broad leaved (pure or mixed with conifers) — and the rest scrub and coppice for replanting (Tilney-Bassett, 1971). The earliest Commission acquisitions were at Laughton and Bourne in 1926. Corsican and Scots Pine were planted on light soils: on cover sands at Scotton, Laughton, Willingham and Osgodby, on sands and gravels at Woodhall, on sand dunes at North Somercotes and on river gravels at Stapleford. Oak and Ash were planted on heavier soils in the late 1920s and later mixed with conifers. Sometimes the conifers were later removed and the Oak stands left. Ash and European Larch were used at Bourne. After World War II Norway Spruce was used on the clays either pure or with Oak. In the 1960s Oak planting was replaced by pine. Conifers alone tend to destroy the ground flora due to the dense canopy and pine needle litter. Nothing replaces it but Rose-bay Willowherb and a tangle of grasses and Brambles, although the rides have relics of a more interesting edge flora.

Plantings by private owners have increased the mixed deciduous woodlands. Starting from 1787 successive Earls of Yarborough have planted large acreages of timber, including exotics, on the former open calcareous heath of the Wolds at Brocklesby, Limber, Cabourne and Swallow. Pelham's Pillar was built between 1840 and 1849 to commemorate the planting of 12½ million trees. Old woods at Broughton, Claxby, Roxton and Limber have also been replanted on the same estate. Fox coverts were planted on and west of the Heath north and south of Lincoln but they have a very limited flora. Most have been felled in this century and have grown up with scrub.

In a triangle in the mid-Lincolnshire clay vale, between Market Rasen, Fiskerton and Revesby but mainly in the Wragby-Bardney area, are a number of woods with mainly Oak/Ash and Small-leaved Lime with Hazel and Ash. The Small-leaved Lime woods are a feature of Lincolnshire and names such as Linwood (near Market Rasen), Lindebergh (Limber), Linwood (near Blankney) and Basswood (near Gainsborough) are derived from it. The Small-leaved Lime was more widespread once than it is now. Ray said it was abundant round Wragby and Horncastle but strangely enough none of the later botanists came to see, or to notice that the Wild Service Tree grew with it. Bass from the outside was used for matting or "basses" (satchels for carrying dinners or tools). The timber was used for carving, perhaps spoons and platters and for coach building. (See Fig. 8, p. 35 and Plate 2).

These old woodlands had been coppiced regularly but the composition of the coppice varies. Few have remained completely under coppice management. The ground flora in coppice is rich and varied — Primroses, Early Purple Orchid, Greater Butterfly Orchid, Herb Paris, Wood Sorrel, Wood Anemone, Woodruff, Bugle, Water Avens and Wood Avens. The ground flora is very luxuriant after felling and for the first few years if replanting has been with conifers. This was seen at Rand, Wickenby and Horsington in the early 1960s, when Greater Butterfly Orchid was quite a feature and Fiskerton had large quantities of Yellow Archangel. Coppice ground flora varies within a wood according to soil. At Newball Wood associations of Small-leaved Lime, Hazel, Hawthorn and Ash carry regions of Woodruff, Dog Mercury, Wood Anemone, Cow-wheat, Lilies of the Valley, Bluebells, Bracken and ferns, with area dominants. Some gravelly tracts are found in these woods. The coppice at Newball has Bird's-nest Orchid always under Hazel; it is also recorded from Stainton growing with Dog Mercury, Woodruff and Wood Anemone and from other small woods nearby.

At the southern end of the Wolds, where streams have cut through the Spilsby Sandstone to the Kimmeridge Clay, the valleys are wooded. The canopy on the drier slopes is Beech and Ash with a very attractive flora of Wood Forget-me-not, Red Campion, Wood Anemone, Garlic or Ramsons and Bluebells. On the waterlogged clay base under Alder and Willow, carrs develop with Marsh Marigolds and Golden Saxifrage, for example in Keal Carr and the New England valley between Somersby and Salmonby. Some of these woods have been replanted with poplar. Similar conditions occur on the western edge of the Wolds at Claxby and Hainton.

On the Boulder Clays along the eastern edge of the Wolds from Louth through Burwell and Muckton to Welton, is a series of large and small woods. They are mainly Oak/Ash standards with Ash/Hazel coppice and a good rich ground flora.

The Fen-edge sands and gravels of the Woodhall region, which are lake-edge deposits from the Last Glaciation, give podzol soils. The woods here and at Kirkby Moor are pure Birch, or Birch/Oak/Pine with Ash in places. There has been much afforestation here and at Roughton Moor. In Fulsby and Tumby (one-time property of Sir Joseph Banks) the old woods have a rich flora. At Tumby Lilies of the Valley are profuse, Ferns, Climbing Fumitory and Bracken are plentiful. The Fen-edge decoy woods mentioned above also have Birch, Oak, Alder, a profusion of Climbing Fumitory, fine ferns and Bracken. At Friskney Decoy exotics have been planted.

The river and glacial sands and gravels which overlie the Lias in

THE FLORA OF THE MAJOR HABITATS

the Trent Vale support large woodland areas at Skellingthorpe, Doddington, Eagle, Norton Disney and Stapleford. They are mostly replanted but Oak/Birch remains. There is Small-leaved Lime at Skellingthorpe and areas of Oak/Birch with Royal Fern and Hard Fern. In the deciduous areas are Oak/Hazel, Aspen and Alder Buckthorn, particularly on the edges of plantations, with Giant Campanula, Tuberous Bitter Vetch, Millet and Yellow Pimpernel. The woodland at Norton Disney is very mixed with Bluebells, plantings of Daffodils now naturalised, two species of Birch, Rhododendron, Bracken and Lilies of the Valley, the latter now nearly all gone. Stapleford Moor heathland is now mostly planted with conifers but formerly had Marsh Gentian and Sundew.

On the more acid soils Foxgloves are very scarce and, except in the Somersby area, are found mostly in the small woods between Newark and Gainsborough. Burton has a small but fine Foxglove population in a pine wood beside the willow carrs and woodland.

The Boulder Clay woodlands of southern Kesteven are extensive with Oak/Ash standards and Hazel coppice, but also with much conifer replanting. Ground flora includes Wood Vetch, Everlasting Pea, Primrose, Bluebells, Dog Mercury, Greater Butterfly Orchid and one of the only localities in the County for Wood Spurge. Tortoiseshell Wood, North Witham, with Oak and Wild Service Tree and a ground flora including Herb Paris, Greater Butterfly Orchid and Wood Melick Grass, is now a nature reserve. Grimsthorpe has mixed deciduous woodland where the introduced Mistletoe is well established.

Woodland Flora

Native trees are Oak (*Quercus robur* chiefly, *Q. petraea* scarce or local) Ash, various Elms, Hawthorn (*Crataegus oxyacanthoides* abundant in the Lime woods on clay), Small-leaved Lime, (*Tilia cordata*), Rowan (*Sorbus aucuparia*) chiefly in acid woods, widely distributed but never abundant, Wild Service Tree (*Sorbus torminalis*) sparingly in Lime woods, Birch (*Betula pendula* and *B. pubescens*), Alder, Aspen, Willow sp., Hazel, and Dogwood. Holly does not appear to be indigenous in the County but is listed in the Dragonby peat remains; when seen in hedges and copses it is probably bird sown.

In the deciduous woods the flora is best seen in May and June. There are significant differences in the composition of the ground flora. Marsh Marigold (*Caltha palustris*) is locally abundant in a few Wold woods, and Columbine (*Aquilegia vulgaris*) flowers profusely in Broughton Woods after felling. Climbing Fumitory (*Corydalis claviculata*) occurs in a number of peaty woods, particularly on the fen edge.

Sweet and Hairy Violet (*Viola odorata* and *V. hirta*) occur on calcareous soils and Hairy St. John's Wort (*Hypericum hirsutum*) in most of the deciduous woods. Red Campion (*Silene dioica*) is locally abundant on the Wolds as a striking feature with Wood Forget-me-not. Wood

Sorrel (*Oxalis acetosella*) is rather rare and in small quantity amongst leaf mould. Bush and Wood Vetch (*Vicia sepium* and *V. sylvatica*) are uncommon but locally abundant in a few woods. Water Avens (*Geum rivale*) is a feature of the Wold and adjoining woods, but less common elsewhere.

Stone Bramble (*Rubus saxatilis*) has to be searched for in calcareous woods, but has been found at Broughton in recent years and in the Gainsborough district in the 1940s. Three species of *Alchemilla* occur but are rare, and local in calcareous woods. Golden Saxifrage (*Chrysosplenium oppositifolium*) is found in the Wold woods and was formerly in the Fens. Wood Sanicle (*Sanicula europaea*) is in calcareous woods on the Wolds and the limestone Heath but is not common.

Bog Myrtle or Sweet Gale (*Myrica gale*) was formerly much more abundant on peat but occurs now only in three Divisions. Wood Spurge (*Euphorbia amygdaloides*) is on the edge of its distribution in a few old woods in the extreme south. Dr. Lees said that Common Wintergreen (*Pyrola minor*) was one of the most salient features of the fir woods round Rasen. Now it is very rare and dies out after dry seasons. Ling (*Calluna vulgaris*) is uncommon in woods and rare except on the acid commons.

Primrose (*Primula vulgaris*) occurs rather sparingly in clay woods but flourishes after coppicing or felling. Yellow Pimpernel (*Lysimachia nemorum*) is not common occurring chiefly in woods east of Wolds. Wood Forget-me-not (*Myosotis sylvatica*) is a feature of Wold woods though scarce elsewhere. Gromwell (*Lithospermum officinale*) only occurs rather scarcely in the north and west with a very little in the south. Wood Speedwell (*Veronica montana*) is in most of the Wold woods, but scarce elsewhere.

Cow-wheat (*Melampyrum pratense*) is scarce and in small quantity in a small number of woods. Yellow Archangel (*Lamiastrum galeobdolon*) is local, but not everywhere. Giant Bellflower (*Campanula latifolia*) is a northern plant found here and there on the Wolds but sparingly in south Lincolnshire. Nettle-leaved Bellflower (*C. trachelium*) is a southern plant growing chiefly on the limestone in south Lincolnshire but not on the chalk Wolds. A hybrid between the two was found at Broughton.

Woodruff (*Galium odoratum*) is uncommon and in small quantity in many woods. Moschatel (*Adoxa moschatellina*) is chiefly found in the north, east of the Wolds and a little in the south. Small Teasel (*Dipsacus pilosus*) is scattered and in small quantity but not in the north-west. Great Woodrush (*Luzula sylvatica*) is local and in small quantity. Lily of the Valley (*Convallaria majalis*) occurs at Broughton, Legsby and Tumby in great quantity but is dying out under new conifer plantations; only small patches elsewhere. Bluebells (*Endymion non-scriptus*) are rather uncommon but can be found in most woodland districts.

Herb Paris (*Paris quadrifolia*) is scarce in calcareous woods, often found with Dog Mercury (*Mercurialis perennis*) which is common in

THE FLORA OF THE MAJOR HABITATS 35

many woods. Ramsons (*Allium ursinum*) is dominant over large areas in old woods in the north and east, but only in small quantity in the south. Broad Helleborine (*Epipactis helleborine*) is rare occurring chiefly east of Wolds and in the south-west. Greater Butterfly Orchid (*Platanthera chlorantha*) is local in a number of woods where clearing has been done but it needs some shade. Fly Orchid (*Ophrys insectifera*) is very rare and needs searching for, particularly among Dog Mercury.

Smooth Sedge (*Carex laevigata*), Pendulous Sedge (*C. pendula*), Pale Sedge (*C. pallescens*) and Thin-spiked Sedge (*C. strigosa*) are found in a few woods but are rare. Remote Sedge (*C. remota*) and Wood Sedge (*C. sylvatica*) are common. Hairy Brome (*Bromus ramosus*), Tall Brome (*Festuca gigantea*) and *Brachypodium sylvaticum* are common, but Wood Melick (*Melica uniflora*) is patchy and Mountain Melick (*M. nutans*) very rare. Wood Barley (*Hordelymus europaeus*) is rare being found in four woods. Bush-grass (*Calamagrostis epigejos*) is locally dominant and Purple Small-reed (*C. canescens*) not uncommon in clay woods. Wood Millet (*Milium effusum*) is not common but is a feature of some woods, while Tufted Hair-grass (*Deschampsia cespitosa*) is often found in damp rides and is common. Fibrous Twitch (*Agropyron caninum*) is rare. In secondary woodland Herb Bennet (*Geum urbanum*), Enchanter's Nightshade (*Circaea lutetiana*), Bugle (*Ajuga reptans*) and Twayblade (*Listera ovata*) are noteworthy.

Fig. 8. Small-leaved Lime from brass in
St. Cornelius Church, Linwood

GRASSLANDS

Heathland and Low Moorland

The grass heathlands, a feature of the County and once quite extensive are found on the cover sands and on river and glacial sands and gravels. Much of the land of this type unprofitable for agriculture — rabbit warrens and commons — has been cultivated and afforested with conifers, and others have been affected by the ironstone industry. Utilisation of these lands began in the late 18th Century and continued during and after the last war.

Wet and dry sand and gravel areas usually support a calcifuge flora, but where the cover is thin over chalk, limestone or highly calcareous clays, a mixture of species occurs. Variety of habitat provided by peaty flashes and pools give these areas an interesting flora.

The main areas of heathland are found on **1.** the cover sands over clays and limestones in the north-west — Commons, Heather and Warrens from Gainsborough to Appleby and Moorland from Elsham to Linwood Warren; pockets of blown sand and gravels also occur north of Caistor and at Panton; **2.** the river and glacial sands and gravels of the Trent Vale from Stapleford to Lincoln and in the Upper and Middle Witham Valley; **3.** the river and glacial deposits of sand and gravels in the Woodhall Spa — Kirkby Moor area; and **4.** glacial sand and gravels on the limestone and in areas round Ancaster — Sleaford.

1. In the north-west at Laughton and Scotton Commons (cover sands on clays) the heathland has been extensively afforested. Small areas of wet and dry acid grassland and heather remain, but few pools, although some of the land is low lying and very wet. There is a natural succession to birch scrub and woodland with oak and pine. At **Scotton** on the nature reserve, wet and dry heaths with Purple Moorgrass (*Molinia caerulea*) and Wavy Hair-grass (*Deschampsia flexuosa*) respectively, combined with a mixed *Calluna* association provide a variety of dominants over the area. At **Laughton** which is much wetter are Marsh Dock, Marsh Cinquefoil (*Potentilla palustris*), Cotton-grass and Bristle Clubrush. Many sedges have been recorded and *Teesdalia* on the drier areas. On the new forestry rides Petty Whin and other sandy plants — Birdsfoot Vetch (*Ornithopus*), Smooth Cat's Ear (*Hypochoeris glabra*) and Hoary Cinquefoil (*Potentilla argentea*) are still found.

In the Scunthorpe and Brigg areas, cover sands on the limestone ridge supported sandy rabbit warrens and commons (warreners obtained their livelihood from rabbit skins) and on the escarpment and low lying land are grass and heather heaths. Again small marshy pools survive with a succession to oak/birch. **Crosby Warren** was one of the most striking ecological regions of the County in the early 19th century with the richest flora. Due to ironstone quarrying and new small

THE FLORA OF THE MAJOR HABITATS

plantations the best plants dried out and have since disappeared. **Santon Warren** has also partly gone and **Brumby Warren** has been destroyed.

Risby Warren is a higher exposed and drier heathland with extensive tracts of blown sand of varying depth. Marram Grass (*Ammophila arenaria*) was planted here in 1910 to prevent further blow out and erosion. There is a mixture of acid and calcareous grassland heath (due to the varying depths of sand on limestone) in marked regions. Where the sand is thin Tor Grass (*Brachypodium pinnatum*) is the dominant grass with swards of Purple Milk Vetch. On the thicker sand Sand Sedge (*Carex arenaria*) is extensive. *Festuca ovina/tenuifolia* grassland is extensive. Stands of *Calluna* and areas of pine with *Dryopteris borreri* are present. Sand plants occurring are Field Mouse-ear Chickweed, *Myosotis ramosissima* and *M. discolor*, *Aphanes microcarpa*, Dove's-foot Cranesbill (*Geranium molle*) and Wall Speedwell (*Veronica arvensis*). (See Plate 3).

Appleby and **Manton Warrens** have extensive plantations. A young conifer plantation at Staniwells Wood still has its grassy areas, with Field Mouse-ear, Small Bugloss, Flea Sedge, Purple Milk Vetch, Twayblades and extensive areas of Marsh Orchids occur in the wet flushes. Manton Warren has a very sandy heath with a show of Field Mouse-ear Chickweed and Sand Sedge. At **Manton Common** there is wet *Molinia* heath on the low lying land with Marsh Gentian and Bog Myrtle. Bilberry was found in a sand warren in Broughton, close to the wood. **Twigmoor** is a very fine locality with mixed heath habitats. There is sand over peat in places, and wet *Molinia* heath.

The low lying Messingham Common has both wet and dry sandy habitats together with peat. *Molinia*, *Calluna*, Marsh Cinquefoil, Bog Asphodel, Bell Heather and Cross-leaved Heath occur and there is a small area of Marsh Gentian. The dry sandy Wavy Hair-grass heath has damper tracts with Creeping Willow (*Salix repens*) and wet flashes have Cotton-grass and Bogbean.

From Elsham and Wrawby Moor to Linwood Warren blown sand occurs over upper Jurassic clays, with gravels in places. **Linwood Warren** now a nature reserve, which has cover sands over Kimmeridge Clay, is again partly afforested but varied heathland habitats are left. One of the finest dry Wavy Hair-grass heaths, it has Mat-grass (*Nardus stricta*), Heath Bedstraw and *Ornithopus*, and the wet *Molinia* heath has Cotton-grass and on drier areas of *Calluna* are *Ericas* and Heath Rush. Sundew (*Drosera rotundifolia*) is much diminished. Marsh Gentian (*Gentiana pneumonanthe*) has been recorded in the past — "perhaps fifty flowering specimens" in 1895 — and is still present on the nearby golf course. Marsh Cinquefoil occurs in wet flashes.

Osgodby Moor, Holton le Moor, Moortown and **Nettleton Moor** have a very acid grassland flora, although most is cultivated or planted. Plants occurring are *Agrostis tenius*, Bog Pimpernel, Heath

Rush, Slender Cudweed (*Filago minima*), and Wavy Hair-grass. *Aphanes microcarpa* and *Rumex tenuifolius* are plentiful there. Marsh Gentian was first recorded in Britain at Nettleton Moor. **Elsham** and **Wrawby Moors** were visited by John Ray in 1700 and James Britten in 1862.

2. and 3. The sands and gravels of the Trent and Witham valleys and around Woodhall are not so acid as the cover sands. Although the flora is predominantly calcifuge other species do occur for example *Acinos*, Salad Burnet and Hoary Cinquefoil. At **Laughterton** on Naylor's Hills Sheep's-bit (*Jasione montana*) occurs and at Broom Hills there is much Broom among the *Deschampsia*, together with Greater Broomrape (*Orobanche rapum-genistae*). Broom is common on the gravels, and Sand Sedge (*Carex arenaria*) is common in vast tracts on the sands east of the Trent.

The Trent Vale has light gravels laid down by the river, and glacial gravels on the rises. **Stapleford Moor** was an ancient heath of river gravel now planted with conifers. Areas of dry Wavy Hair-grass heath occur with Common Bent-grass (*Agrostis tenius*) and *Calluna*. On patches of wet *Molinia* are Mat-grass, Heath Rush (*Juncus squarrosus*), *Ericas*, *Potamogeton polygonifolius*, *Rumex acetosa* and *R. acetosella*.

The Woodhall Spa sands and gravels are of glacial lacustrine origin. At **Kirkby Moor**, now a nature reserve, there are surface deposits of sand and scattered gravel pockets, with Wavy Hair-grass heathland in varying assorted communities. *Euphrasia anglica* is near its north-east limit in the grassland. *Calluna* heath, Bracken, Wood Sage, Creeping Willow, Bog Pimpernel (*Anagallis tenella*), Early Marsh Orchids and Marsh Violet (*Viola palustris*) occur. At **Roughton Moor** are again mixed habitats with Lousewort, Sundew, Bog Pimpernel and Gorse, with calcicoles also present. The golf course at Woodhall also has very assorted heathland communities with Wavy Hair-grass and Common Bent-grass (*Agrostis tenuis*).

At Tattershall the gravel pits are colonising with many heath and sandy plants and aquatics including Slender Cudweed (*Filago minima*), Small Bugloss, Purple and Yellow Loosestrife, Great Water Dock, Frogbit, Marsh Speedwell, Great Hairy Willowherb, Celery-leaved Crowfoot and Red Goosefoot.

4. Sand over limestone near Ancaster gives grass heaths, once much more extensive. At **Rauceby Warren**, now a nature reserve, and on the golf course Purple Milk Vetch occurs with Heath and Lady's Bedstraw, Field Mouse-ear Chickweed, Sand Sedge and the rare Glabrous Rupture-wort (*Herniaria*).

In the County as a whole the three Sundews are nearly gone. Petty Whin and Dwarf Furze have become very rare. Both *Ulex gallii* and *U. nanus* were recorded by Dr. Lees at Linwood, Osgodby Moor and Marton in 1877. *U. gallii* was at North Hykeham and on Fossway, Swinderby about 1930 and at Swinderby in 1963. There are no records of *U. nanus* since 1877. It was recorded by Miss M. E. Dixon in 1858 at Holton le Moor, F. A. Lees saw it in 1877 but it has certainly not grown there since 1907. There has been no later record for it in the Gainsborough district. Butterwort has become very scarce. *Selaginella* is perhaps extinct and the *Lycopodiums* nearly absent. *Osmunda* and *Myrica* only flourish in one or two spots, but *Teesdalia*, Smooth Cat's Ear and Heath Milkwort can still be found. Field Mouse-ear Chickweed (*Cerastium arvense*) is present in quantity and is a feature of some of the sandy regions. Corn Spurrey (Pickpurse or Dother) has been abundant also Parsley Piert (*Aphanes microcarpa*) and *Rumex tenuifolius* particularly on the cover sands denoting extreme acidity. In the Holton le Moor area Corn Spurrey and Parsley Piert have decreased through spraying. Blue Fleabane is common all over the Scunthorpe area and at Panton.

Leached sand over limestone is occasionally met with in south Lincolnshire. Sheep Sorrel and Lesser Stitchwort are often indicators of leaching acidity on more alkaline soils. Hoary Cinquefoil occurs on sand and gravel, and Basil Thyme grows chiefly on calcareous soil but is on gravel at Woodhall.

Chalk and Limestone Grassland

Chalk grassland is scarce due chiefly to the fact that it is largely overlain with glacial boulder clay and drift, but also because of the thinness of the chalk. The hill slopes are often of Red Chalk and Lower Cretaceous strata which do not bear calcareous plants. Limestone grassland is also in short supply mainly owing to cultivation. There is some leaching on hill tops in the south-west. Small Scabious and Rockrose are indicators of good chalk pasture; *Centaurea nemoralis* also needs lime as well as *C. scabiosa*.

Much land on the Wolds and Heath was brought under the plough after 1750, and since 1940 old pastures have been disappearing rapidly. Before the enclosures most of the limestone Heath was rough pasture, together with some of the Wolds — the rest being rabbit warrens. Another indicator of calcareous soil is Burnet Saxifrage (*Pimpinella saxifraga*) which turned up recently on a steep bank near Epworth on Keuper Marl. It is well distributed over the whole County. Another calcicole, Salad Burnet (*Poterium sanguisorba*) was found in Greetwell cutting during a railway strike but is otherwise inaccessible.

Purple Milk Vetch seems to require some sand mixed with the chalk or limestone, and is rare in the north except on Risby Warren which has some of the largest remaining expanse of limestone grassland. Peacock said that it was absent from the Wolds but it has been found in recent years north and east of Caistor. Burton Cliffs, steeply rising from the River Trent, also produces calcareous grassland.

The boulder clay on and to the south-west of the Wolds varies in its chalk content, but is largely under the plough. In some valleys the chalk becomes like sugar limestone and supports hardly any vegetation, notably in the Swaby valley and at Walmsgate and Ravendale. Horse-shoe Vetch, Basil Thyme, Wild Basil (*Clinopodium vulgare*) and Wild Calamint are all uncommon and decreasing. Marjoram is very patchy, being locally abundant in some districts but scarce or absent in others.

There are old records for Fly Orchid, Man Orchid and Lady's Tresses on the chalk but none of these has been found for at least thirty years. Some of the records for the Wolds are unconfirmed, notably Pasque Flower (*Pulsatilla*) and Horse-shoe Vetch (*Hippocrepis*) in the Louth area. On the limestone, Fly Orchid and Spider Orchid at Ancaster and Holywell were only rumours before 1920. H. Fisher recorded in 1930 that Field Fleawort (*Senecio integrifolus*) was almost extinct but he picked and pressed three specimens. Pyramid Orchid occurs sparingly in a few Wold valleys, but is commoner on the limestone.

Autumn Gentian is found sparingly on loose soil on both the chalk Wolds and limestone Heath; also the Early Felwort (*Gentianella anglica*), but they are a complex genus and time of flowering is the chief method of identification. Limestone grassland in the County is rich in species — Pasque Flower, Bee Orchid, Man Orchid, Pyramidal Orchid and Horseshoe Vetch — but scarce in quantity, much of the typical flora being found on roadside verges and in quarries. (See Plates 5 and 6).

Clay Grassland

Of the Lincolnshire grasslands, the richest were on the clays of the Marsh, central clay vale and parts of the Trent vale. The fall in acreage of permanent grassland in the county — from 37% to 16% in the thirty year period up to 1966 — affected most those on clay. The "improvement" of grassland by applications of fertilisers has also reduced the number of flowers. Before the 1950s old pasture fields at Huttoft, Snarford and Scopwick, now ploughed up, contained a colourful collection of flowers now rarely seen in such profusion. They included Dyer's Greenweed (*Genista tinctoria*), Spiny Restharrow (*Ononis spinosa*), Zigzag clover (*Trifolium medium*), Tufted Vetch (*Vicia cracca*), Dropwort (*Filipendula vulgaris*), Great Burnet (*Sanguisorba officinalis*), Water Avens (*Geum rivale*), Pepper Saxifrage (*Silaum silaus*), Parsley Water Dropwort (*Oenanthe lachenalii*), Cowslip (*Primula veris*), Yellow Rattle (*Rhinanthus minor*), Devil's-bit Seabious (*Succisa pratensis*), Sneezewort (*Achillaea ptarmica*), Meadow Thistle (*Cirsium dissectum*) and Saw-wort (*Serratula tinctoria*). The fine meadows at Bratoft, now a nature reserve, have a similar wealth including Cowslip, the now rare Green-winged Orchid (*Orchis morio*), Spotted Orchid, Adder's Tongue Fern, Great Burnet, Saw-wort, Dyer's Greenweed and Devil's-bit Seabious.

NOTE

Out of many lost habitats, two of the last small calcareous bogs are worthy of mention. **Waddingham Common** was swept away in the interest of agriculture in spite of protests to preserve its unique flora. Grass of Parnassus was present in some quantity, also Fragrant Orchid (*Gymnadenia conopsea* var *densiflora*), Black Bog-rush (*Schoenus nigricans*), Few-flowered Spike-rush (*Eleocharis quinqueflora*), Tall Yellow Sedge (*Carex lepidocarpa*), Hemp Agrimony (*Eupatorium cannabinum*), Wild Mignonette (*Reseda lutea*), Devil's bit Scabious (*Succisa pratensis*), Butterwort (*Pinguicula vulgaris*), Bog Pimpernel (*Anagallis tenella*), *Calluna* and Cross-leaved Heath (*Erica tetralix*) and Saw-wort (*Serratula tinctoria*). **High Toynton** had a smaller bog also with Grass of Parnassus, Devil's-bit Scabious, Brookweed, Red Rattle and Tall Yellow Sedge — now drained and planted with poplars.

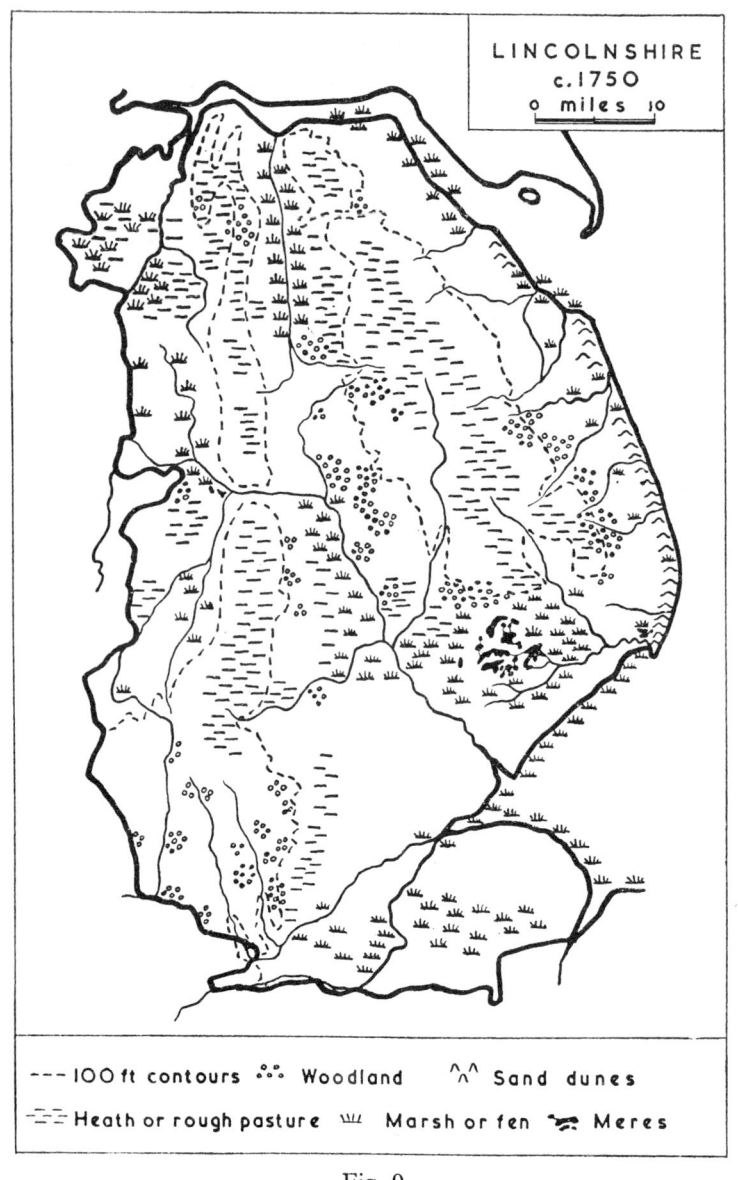

Fig. 9

THE FLORA OF THE MAJOR HABITATS

FENLANDS

Wash and Witham Fens

These include the whole of south-east Lincolnshire, south of the Wolds and east of the limestone plateau with a branch extending north-westward along the Witham to Lincoln. Following earlier attempts, the Holland, Deeping, Kesteven and Witham Fens were finally drained in the late 18th Century, and the East, West and Wildmore Fens in Lindsey in the first decade of the 19th Century. Today there is no natural fenland. At first the land was grassed but it is now cultivated. Drainage by windmill, followed by steampump and now diesel have altered the whole area and the fens have become extremely rich agricultural land. The areas of peat fen are now well inland, much of the rest being silt and the result of the reclamation of salt marshes from post-Roman time. Bicker Haven and Surfleet Lows were once estuaries and a few estuarine plants remain there and also beside inland creeks, for example, Wild Celery, Sea Milkwort, Sea Club-rush and other less saline plants like Hairy Buttercup, Strawberry Clover and Brookweed.

To some it is still a shock to find that there is no undrained land left in the Lincolnshire fenland and that Norfolk, Cambridgeshire and Huntingdonshire have many more fenland relics.

Arthur Young and several other authors gave very vivid pictures of the varying habitats existing in the East Fen before 1800. It was an area with great variations from deep peat moss with chains of lakes bordered by crops of reeds to Oak Pine and Birch on higher ground. The 18th century plant list for East Fen is extensive and includes *Cladium mariscus*, *Sonchus palustris*, *Senecio paludosus* and *S. palustris* and many others. Certain acidic areas which were present in the fens producing Cranberries can be better compared to areas such as Dersingham Fen and Royden Common which produce an acid bog flora and where there are still many of the plants formerly growing in the East Fen of Lincolnshire, for example Cranberries, the three heathers and *Cladium*. The East Fen list of plants in Arthur Young's *General View* (1799) does not mention the two *Ericas*, Bog Asphodel or Grass of Parnassus but surely they were there?

The fens east, south and west of Spalding have been long under cultivation though often only for grazing. H. E. Hallam in *Settlement and Society* (C.U.P. 1965) traces reclamations north and south of the main road from Spalding to Long Sutton before 1300. No list of the flora of this part of the fens has ever been given, not even to the Spalding Gentlemen's Society. Such a list for about 1700 would have been most interesting.

Henry of Huntingdon and William of Malmesbury delighted in the scenery of the apple orchards, but no plants were given. The former (*c.* 1134) described this "fennie countrie" as "passing rich and plenteous, yea and beatuiful to behold watered with many rivers running down to it, garnished with a number of meres, both great and small, which abound in fish and fowl: and it is finely adorned with woods and islands." William of Malmesbury (c. 1140) recorded that the fens "were a very paradise and seemed a heaven for the delight and beauty thereof; in the very marshes bearing goodly trees, which for tallness, as also without knots, strived to reach up to the stars. It is a plain countrie and as level as the sea, which with greene grass allureth the eye. There is not the least portion of ground that lieth waste and void there; here you shall find the earth rising somewhere for apple trees; there you shall have a field set with vines, which either creep upon the ground or mount on high upon poles to support them."

Other glimpses of past floras are given in Liber Eliensis The Chronicles of Crowland, Camden's *Britannia* (Gough's edition 1789), and by the naturalists Ray and Montague; also in verses by Michael Drayton in *Polyolbion* (1662).

Areas liable to flood in winter would have had only an aquatic flora and there must have been much of this. Now with mechanical clearing of dykes and drains these plants have gone. Flowering Rush, Great Spearwort, Frogbit, Great Water Parsnip, Purple Loosestrife, Valerian and Hemp Agrimony were abundant along drain sides up to 1950 but are now disappearing. Marsh Mallow is only in a few patches. Rev. J. Dodsworth of Bourne had a record for *Damasonium* which has not been recorded before or since 1836. (Recently a record unconfirmed turned up for Willoughby c. 1920). Mrs. Arthur Wherry was shown *Pinguicula* beside a drain near Billingborough about 1920 but that again is the only record for the fens. In Peacock's Cambridge manuscript there are 43 records for it over nine Divisions of the County; it is now all but gone. Baston Fen (now a nature reserve) and Cowbit Wash have kept a few remains of indigenous plants but they are only a fraction of what must have been there once. There are no records for Marsh Orchid in fen Divisions 17 and 18, though recently Early Marsh Orchid was found in Baston Fen and near Deeping (Division 16). Marsh Helleborine has only one record for south Lincolnshire (Ancaster 1930), but surely this must have been in the fen lands.

The Witham Fens east and west of Lincoln have had fen conditions near the River Witham and Fossdyke. *Viola stagnina* was found at Boultham, almost in the city, in 1833 by Miss Cautley, a Suffolk botanist. In 1836 it was sent by Dr. John Nicholson to Sir W. Hooker and it was recorded as a species new to Britain. Since then it has been found in various places by the River Witham up to 1936 when it was recorded at Fiskerton. Habitats are now scarce. It is also recorded by W. Bunting (1969) in the Isle of Axholme near the Yorkshire border. *Ranunculus lingua, Teucrium scordium, Lathyrus palustris* and other fen plants have been recorded close to the City of Lincoln. Burton gravel pits (now a nature reserve) and parts of Skellingthorpe have some relics of fen flora.

THE FLORA OF THE MAJOR HABITATS

ISLE OF AXHOLME

In north Lincolnshire the term 'fen' is not used. However, 'Carr' means a watery thicket, so we may presume that the low-lying land on either side of the River Ancholme was chiefly water-logged, growing willows, alders and, in early times, oak. Remains of bog Oak, Ash, Yew and Birch have been found in peat in ploughing and draining.

Final drainage here was not achieved until the late 18th and early 19th centuries. A low plateau of Keuper marl is surrounded by water-logged peat and sand. It was formerly much wooded and parts of it are similar to Thorne Waste and Hatfield Chase. Cultivation after drainage has left little of the original flora. 'Warping' of the poor soils was carried out on a large scale on the low lying marshes of the Isle and to the east of the Trent, raising the ground level and improving the soil by depositing silts. Turbaries or turf pits occur in various places where commoners had rights of turbary, that is to dig turf or peat for fuel. These areas of raised bog partially destroyed by peat cutting, which remain on the west side at Crowle Waste, Epworth and Haxey are now nature reserves and have a range of fen and wet heath conditions. *Andromeda polifolia*, usually found in the north-west, is present. It is found in quantity on neighbouring Thorne Waste, but not on Crowle. *Eriophorum vaginatum* is present, together with *Cladium mariscus*, Star Thack or Fen Sedge. This was once more widespread in the County, being abundant in southern Fens, but Epworth and Haxey are now its only localities. ('Thatch' fen Potterhanworth, Starholme Farm, North Kelsey and Star Carr, Wrawby, also indicate possible former distribution.) On the sandy areas on peat *Teesdalia*, *Rhinanthus serotinus* and *Apera spica-venti* are found.

COASTLANDS

The Lincolnshire coast, including the tidal Humber and Trent banks, is about three-fifths of the County boundary. It presents a large variety of coastal habitat for maritime plants. Sand dunes are extensive (15 miles) and the mud flats and saltings of the Humber and Wash are of prime interest. There are no cliffs of chalk or limestone in the usual pattern, but Burton Cliffs near the mouth of the River Trent have steep limestone pasture with saltmarsh plants growing at tide level. A very small area of low chalk bank exists at South Ferriby.

The coast line between the Humber mouth and the boundary of Norfolk has altered a good deal during the last 600 years. Much land has been reclaimed round the Wash and to a lesser degree north of Saltfleet. As reclamation takes place there is a change in vegetation to the grassland type as the salt leaches out and the land dries. Remnants of coastal and estuarine vegetation can be seen inland on occasion for example Sea Milkwort (*Glaux maritima*), *Ranunculus sardous* and Distant Sedge (*Carex distans*) at Surfleet Lows. At Freiston the vegetation observed in 1919 has considerably altered and new areas of salt marsh have built up. Erosion is taking place between Mablethorpe and Skegness, but there is considerable accretion north of Saltfleet and at Gibraltar Point. The storm surge and floods of January 31st — February 1st, 1953 caused great inroads of the sea and washed away so much of the dunes that a concrete wall was built for about 15 miles south of Mablethorpe. This has changed the flora and the vast caravan sites of Mablethorpe, Ingoldmells and even the quieter spots have also caused the disappearance of some of the old records. So too has the South Humberside development from Immingham to Grimsby. The remaining Humber bank areas of salt marsh are rich in Sea Aster (*Aster tripolium*), Sea Lavender (*Limonium vulgare*) and typical mud flat flora. Cord-grass (*Spartina*) has been planted in several areas recently at Goxhill, and more mature stands occur at Barrow Haven.

There are many interesting old records of Lincolnshire coastal plants: Sea Pea (*Lathyrus japonicus*), Sea Purslane (*Halimione pedunculata*), Oyster Plant (*Mertensia maritima*), an amusing record of Portland Spurge (*Euphorbia portlandica*), Matted Sea Lavender (*Limonium bellidifolium*) and Golden Samphire (*Inula crithmoides*). Also Yellow Horned-poppy (*Glaucium flavum*) and Rock Samphire (*Crithmum maritimum*) which have come and gone in the last 20 years. The Horned-poppy is very fluctuating and came up after the 1953 floods then disappeared reappearing in 1973 at Gibraltar Point, while the *Crithmum*, an old Thompsonian record, was found in 1964 also at Gibraltar Point but its habitat has since been eroded away. Shingle is very scarce which also accounts for the scarcity of Sea Campion (*Silene maritima*), Sea Holly (*Eryngium maritimum*) and Sea Bindweed (*Calystegia soldanella*), the latter being much commoner 30 years ago. These plants are all now present at Gibraltar Point, the Sea Holly and Sea Bindweed increasing. The Sea Campion is found on the inland side of the dunes. Sea Heath (*Frankenia laevis*) has appeared in 1973. Asparagus (*Asparagus officinalis*) was recorded between 1690 and 1940 though possibly it could still be found as a native at Mablethorpe. However many recent searches have failed to do this, and the Moulton coastline is so altered from Gerarde's date that it is unlikely to turn up again.

Ray's Knotgrass (*Polygonum raii*) had not been recorded at Gibraltar Point from Streatfeild's time (1870) until 1968, merely because no one was interested to search for it. Other rarities might still turn up. Hare's-ear (*Bupleurum tenuissimum*) appeared in 1967 but perhaps had come by seed being washed up. Crow Garlic (*Allium vineale*), Pyramidal Orchid (*Anacamptis pyramidalis*), Parsley Water Dropwort (*Oenanthe lachenalii*) and Clematis (*Clematis vitalba*) occur but are not abundant. There is no record for Alexanders (*Smyrnium olusatrum*) though it grows in plenty near Kings Lynn.

Creeping Willow (*Salix repens*) so often a feature of the west coast dunes, is absent from our dunes except for two small plants discovered in 1965 on the old dunes near Somercotes. Restharrow (*Ononis repens*)

THE FLORA OF THE MAJOR HABITATS 47

is only present in small quantities from Grimsby to Gibraltar Point and Squinancy Wort (*Asperula cynanchica*) is not recorded. Sea Lyme-grass (*Elymus*) is also sparsely distributed at Mablethorpe and Gibraltar Point. One patch of Wood Sage (*Teucrium scorodonia*) is present at Gibraltar Point.

Several coastal plants are at their northern limit in Lincolnshire· Shrubby Seablite (*Suaeda fruticosa*) has come and gone and returned several times. One old plant was recorded at Gibraltar Point in 1927, but it is now increasing on the new dunes. Marsh Mallow (*Althaea*) is decreasing at its northern limit. There are two unconfirmed Cleethorpes records before 1930 and a good dyke of it at Ingoldmells has gone due to road widening since 1960. Sea Clover is also very uncertain and might turn up again. Bulbous Poa (*Poa bulbosa*) and Tuberous Foxtail (*Alopecurus bulbosus*) are rarities in the north, and *Parapholis incurva* and the native Cord-grass (*Spartina maritima*) are rarities around the Wash. Perhaps someone will find these elsewhere. Search should also be made for Darnel Poa (*Catapodium marinum*) and Grey Hair-grass (*Corynephorus canescens*) for which there are no recent records for Division 11, though both were recorded many years ago.

The Saltfleetby-Theddlethorpe dunes and the extensive area of sand dunes and marshes south of Skegness between Seacroft and Gibraltar Point have been designated as nature reserves. The former is the County's only National Nature Reserve, administered by the Nature Conservancy and the latter is a Local Nature Reserve established by Lindsey County Council and Skegness Urban District Council and administered by the Lincolnshire Trust for Nature Conservation. Both present great diversity of coastal habitat — dune scrub, dune, saltmarsh and mud flat, and have a very rich flora. (See Plate 11).

The Saltfleetby-Theddlethorpe area has changed over the years for a variety of reasons. Before the 1930's there was regular grazing by cows and rabbits and the unique freshwater slacks (the only ones in the County) were kept lawn-like, with huge patches of Bog Pimpernel, Creeping Buttercup and Silverweed. Now they are rank and grown over with long grass and rushes (*Juncus maritimus*) but many interesting plants remain. Marsh Helleborine (*Epipactis palustris*) must have been very scarce and it is said to have been overwhelmed by the 1953 floods and not been seen since. A recent reintroduction has not been successful. Before, during and after World War II the area was largely out of bounds due to the presence of a bombing range which afforded protection from other forms of interference. Present in the dune slacks are populations of *Dactylorchis praetermissa* and *D. incarnata* hybridising happily with huge trusses of flowers. Pyramidal Orchid occurs throughout the dunes. *Thalictrum minus* grows on the north side of the dunes at Saltfleetby. (See Plate 9).

At Gibraltar Point there is a large area dominated by Sea Buckthorn (*Hippophae rhamnoides*). It is a very prominent feature of the dunes. A strange population of this species exists in a sand quarry 10 miles inland at South Thoresby. The mature salt marsh has a fine stand of Sea Lavender. Clematis, never recorded previously, is now established in several places on *Hippophae* from Theddlethorpe to Gibraltar Point. Lizard Orchid has been recorded at Gibraltar Point.

Spartina anglica is present in quantity in the Wash and on the new marshes at Gibraltar Point. It follows the coast up to Cleethorpes and is present along the Humber Bank.

Salicornia known as Samphire is gathered for pickling and cartloads of it have been sold in Boston market. There are very varied species: *S. dolichostachya* below high tide level, then *S. europaea* among *Halimione portulacoides* with *S. ramosissima* on the higher levels. *S. pusilla* has been found in two very different forms — a prostrate form at Gibraltar Point and a bushy form at Horse Shoe Point and Grainthorpe Haven. *S. perennis* is in small quantity north and south of Boston.

KEY TO NATURAL REGIONS

1. Coast. 2. Cover sands. 3. Outmarsh clays and silts. 4. Ancholme, Trent valley and Isle of Axholme clays, silts and peat. 5. Fenland clays, silts and peat. 6. Fen Edge clays and gravels. 7. Lindsey Middle Marsh boulder clay. 8. Lindsey Wold and Vale boulder clay. 9. Kesteven Plateau boulder clay. 10. Wold chalk. 11. Lower Cretaceous strata and Kimmeridge Clay. 12. Limestone. 13. Trent Vale clays and gravels.

NATURE RESERVES

1. Gibraltar Point—Skegness. 2. Saltfleetby—Theddlethorpe Dunes. 3. Friskney Decoy Wood. 4. Isle of Axholme Woods. 5. Hoplands Wood. 6. Tortoiseshell Wood. 7. Baptist Cemetery, Boston. 8. Surfleet Lows. 9. Baston Fen. 10. Heath's Meadows, Bratoft. 11. Moor Closes, Ancaster. 12. The Shrubberies, Long Sutton. 13. The Yews, Donington. 14. Snipe Dales. 15. Little Scrubbs Meadow. 16. Mill Hill, Claxby. 17. Candlesby Hill. 18. Red Hill, Goulceby. 19. Wilsford Heath Quarry. 20. Ancaster Valley. 21. Rauceby Warren. 22. Scotton Common. 23. Linwood Warren. 24. Kirkby Moor. 25. Moor Farm. 26. Epworth and Haxey Turbaries. 27. Crowle Waste. 28. Sea Bank Clay Pits. 29. Gosberton Pits. 30. Barrow Blow Wells. 31. Burton Old Gravel Pits. 32. Hubbert's Bridge Claypit. 33. Surfleet Seas End Reed Bed. 34. Tetney Blow Wells. 35. Covenham Reservoir. 36. Barton Reed Beds. 37. Dole Wood.

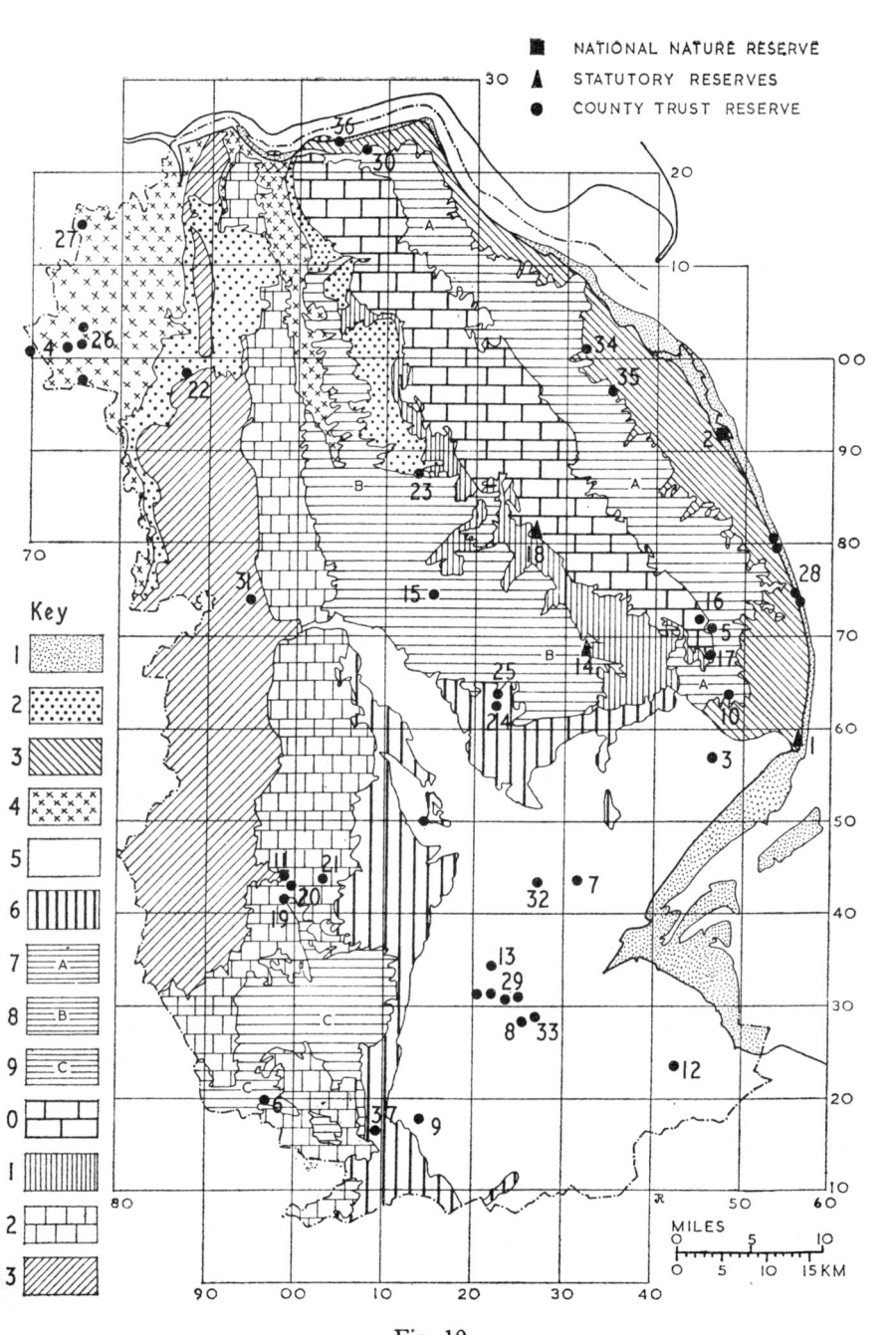

Fig. 10
Natural Regions and the distribution of
Nature Reserves in Lincolnshire
(after J. W. Blackwood)

NOTES ON FERNS AND SEDGES

Ferns

Several old records were made of club mosses before most of their habitats had dwindled or dried out. They are inconspicuous and it is always possible that one or two may be refound, as occurred in 1967. *Selaginella* is at its extreme south east limit and has not been seen recently. The Royal Fern has been almost exterminated by Victorian market gardeners and cultivation. Habitats for *Aspleniums* have gone recently when railway platforms were demolished at small stations. Churches near the coast were searched by Rev. W. W. Mason and Miss Susan Allett and most of them still retain the same ferns, but a few have gone with restoration work. Inland walls and churches seldom have any ferns on them, whatever the type of stone.

In the acid moors and woods the Buckler Ferns are found. Male Fern objects least to lime in the soil. There are often big stands of *Dryopteris filix-mas*, *D. borreri* and *D. dilatata* but *D. carthusiana* is never very plentiful. *Blechnum* is scarce and *Polypodium vulgare* is more widespread in a variety of habitat; ssp. *vulgare* is not found on branches of trees as it grows in the west of England but it is found on walls as well as on the ground; ssp. *interjectum* is on walls in the south. The Shield-ferns are uncommon and are found on the east side of the Wolds almost entirely. Mountain Fern is rare and in small quantity. Marsh Fern has only been recorded once in recent years at Sleaford (1962) and was thought to be extinct, but was found at Crowle last year. Where bracken grows on chalk or limestone it is an indication of sand or leached soil over the rock.

Sedges

In Scotland and the moorlands of England and Wales sedges are everywhere, but in this County they have little chance of survival, though in the undrained past they would have had many more habitats. Some botanists have been sedge minded, others have not. Canon Fowler was really keen on finding sedges but many others have passed them by. Rev. J. Dodsworth was also a capable observer in south Lincolnshire where there are now fewer habitats.

Carex pendula is surely large enough to escape being unnoticed but no one had recorded it before 1956 (though Mrs. Stewart had recorded it in her Bentham and Hooker in the 1920's). It is certainly uncommon now but is in five Divisions. *Carex strigosa* also eluded many botanists until recent years, but is in fair quantity in the woods where it has been found.

John Cragg of Threckingham in his agricultural account of Metheringham remarks on "poor pry meadows". This means sedgy ground and a tithe map of Bigby, 1840 has two "pry" closes. These sedges must remain unidentified.

THE FLORA OF THE MAJOR HABITATS

Carex lasiocarpa was thought to be extinct since 1900 until refound by Dr. Sledge 1946. Canon Fowler would surely have noticed *Carex ericetorum* had it been in his Hooker's Students' Flora, but not being so, it remained unnoticed until 1951. It is darker in colour and earlier in flower than *C. caryophyllea* which it resembles.

Linwood Warren was toothcombed by Dr. Lees, but Canon Fowler was overjoyed to find *Carex curta* there in 1911 — not recorded before. In 1950 seventeen species of *Carex* were found there in one afternoon and there are another six making 23 altogether. *Carex curta* was there in quantity, but since 1955 it seems to have disappeared.

Freshney Bog was also a good sedgy bog up to 1950. Scotter and Laughton Commons have lost much of their boggy ground through drainage, afforestation and spreading birches, but some older records might still be found. There are still many species there. Recently *Carex extensa* was found in quantity at Saltfleetby. Flea Sedge (*Carex pulicaris*) has always been rare and in small quantity and has to be searched for.

Eriophorum angustifolium needs study. Two forms are found with a different flowering period in one Wold valley. There is a slender form flowering earlier with fewer branches and a stouter form with broader leaves and many heads flowering later. Both are found elsewhere in Lincolnshire and in other counties. *Eriophorum vaginatum* exists in a very small quantity in a few spots, although there is plenty on Crowle Waste. *Blysmus compressus* has been overlooked generally but has turned up in calcareous pasture in fresh localities; a second site in Broughton and in other places on the Wolds and Cliff, usually near springs. *B. rufus* on the other hand has decreased, having been found at Skegness and Gainsborough 100 years ago. Recently it has been found in quantity at Cleethorpes, but "improvers" there are destroying its habitat, and has also been recorded recently at Saltfleetby. *Schoenus nigricans* and *Rhyncospora alba* have reached the point of extinction. *Cladium mariscus*, the Star Thack or Fen Sedge, still exists in the Isle of Axholme but has not been seen elsewhere recently.

Plate 1. Oak woodland, Kirkby Moor (TF 225629)
(*Photo by R. V. Collier*)

Plate 2. Lime woodland, Ivy Wood, Bardney (TF 145737)
(*Photo by R. V. Collier*)

Plate 3. Risby Warren (SE 925136)
(*Photo by D. N. Robinson*)

Plate 4. Martin Moor (TF 215648)
(*Photo by D. N. Robinson*)

Plate 5. Limestone grassland, Ancaster Valley (SK 988430)
(*Photo by R. V. Collier*)

Plate 6. Chalk grassland, Red Hill, Goulceby (TF 264807)
(*Photo by D. N. Robinson*)

Plate 7. Crowle Waste (SE 759145)
(*Photo by D. N. Robinson*)

Plate 8. Baston Fen (TF 140173)
(*Photo by D. N. Robinson*)

Plate 9. Saltfleetby–Theddlethorpe National Nature Reserve (TF 467917)
(*Photo by D. N. Robinson*)

Plate 10. Saltmarshes of the Welland Estuary
(*Cambridge University Collection: Copyright reserved*)

Plate 11. Gibraltar Point Local Nature Reserve

Plate 12
Rev. E. A. Woodruffe-Peacock
(1858—1922)

Plate 13
Canon W. Fowler
(1835 — 1912)

Plate 14
Rev. W. Wright Mason
(1853—1932)

Plate 15
Miss S. C. Stow
(1870—1956)

CHAPTER 5

BOTANICAL HISTORY AND BOTANISTS

Much has been written about the remarkable **Rev. John Ray,** so it only needs a very grateful note for his third itinerary in Lincolnshire in 1670 when he passed through on a journey to Hull. He recorded the abundance of Small-leaved Lime about Wragby and Horncastle, and Marsh Gentian at Tattershall Moor and Wrawby. **Dr. Martin Lister,** his friend, rode to Burwell, where his father had lately come to live, in the summer of 1666, to get away from the plague which was raging in Cambridge. His records from Burwell are preserved in the Bodleian Library. Some of his plants are still found in Burwell woods, including Lady's Mantle (*Alchemilla vulgaris*) and the Hawkweed (*Hieracium perpropinquum*), though there is now no chance of finding Yellow-wort or Lady's Tresses as the park is ploughed up. He also listed the two Fluellens. His *Geum* at Tetford "with the larger flowers" must have been the monstrous form often seen in the Wolds with a double flower "hose in hose" with a stalk with seedhead, coming through the middle. He found Sea Buckthorn and Asparagus on the dunes and wrote to Ray about them and other local plants. Mason saw the Asparagus in 1922 — can it still survive at Mablethorpe? It is not the Cornish prostrate form, but evidently an early record of *Asparagus officinalis* ssp. *officinalis* var *altilis* and why not native? The cultivated Asparagus must have arisen from a wild plant in dim ages. C.T.W. says native distribution "obscure owing to frequent cultivation", but why should not Gerarde 1597 and Lister 1698 have found indigenous colonies of it in Lincolnshire?

Many early 18th century records from Boston list the same plants found by various eminent botanists. **Dr. Christopher Merrett, M.D.,** Surveyor, and **Dr. Patrick Blair,** physician to the Port of Boston, found a safe place to live at Boston when they had been in trouble. **Dr. Leonard Plukenet** and **Dr. Richard Pulteney** also visited south Lincolnshire and left records of their finds. **Dr. Vincent Bacon** was brought up near Grantham where his brother became headmaster of the King's School. He was apprenticed to an apothecary in London and became doctor to the Hugenot silk weavers at Spitalfields. He belonged to a Botanical Society founded by Dr. T. Martyn and became an F.R.S. He has been well written up by D. E. Allen (*Proc. B.S.B.I.,*

6, 310, 1967). **Dr. W. Stukeley** of Holbeach and Stamford also left records of rare plants and **Professor John Sibthorp** from Lincoln and Easton gives us the best list for the fens in 1780, which includes Cranberry but with no exact locality.

The Rev. Abraham de la Pryme (1672-1704) was curate of Broughton and his experiences have been printed by the Surtees Society. He also wrote about Hatfield Chase in Yorkshire in the *Philosophical Transactions*.

Arthur Young visited Sir Joseph Banks in 1797 and explored the last part of the East Fen, shortly to be drained. In his *"General View of the Agriculture of the County of Lincoln"* 1799, he writes "There are about 300 acres of land in the East Fen where cranberries grow in such abundance as to furnish a supply for several different counties. The land is chiefly common... *Empetrum* and several other mountain plants being found upon the cranberry ground and in no other part of the fens. They are so plentiful that one man has got nine score pecks in a season... Sir Joseph Banks had the goodness to order a boat, and accompanied me into the heart of this fen, which in this wet season (1797) had the appearance of a chain of lakes bordered by great crops of reeds, *Arundo phragmites*."

Sir Joseph Banks should be the most important botanist of the county but unfortunately his records are not available. Many of his possessions are in Australia and America and some in Paris. A few of his herbarium specimens are at the British Museum but Kew has none. The *Botanists Guide* 1805 gives about twenty of his best finds but a full list of his Lincolnshire plants would have been of great value. As a boy he would have enjoyed discovering the exciting parts of the fens where the rarities grew. The Matted Sea Lavender he notes was "where the Sheep bite close" meaning the drier grassy rises on the saltmarsh. He also describes the Water Germander growing "where the geese frequently land", Cowbane (now extinct) — "in the East Fen chiefly on the edges of the narrow channels, called Rows, which communicate the Deeps with each other" and Milk Parsley "in the East Fen in vast plenty". These give a very vivid picture of the habitats. It is possible that some records in Gough's *Camden* (1789) were from Banks, as they are given under his name in Turner and Dillwyn's *Botanists Guide* 1805.

One interesting entry concerns *Crambe maritima* "among the sand-hills, on the coast, in abundance" with a footnote: "This plant has been seen in the Orkneys, on sand incapable of producing any other crop and has proved to furnish some food for sheep". This suggests the plant was formerly present in Lincolnshire, though it has also been suggested that the original manuscript had a slip-of-the-pen error and that *Cakile* was intended.

Alternatively, the reference makes sense if, knowing it was used for sheep grazing in the Orkneys, he encouraged Lincolnshire sheep farmers to send their flocks to graze on it, thus causing its extermination.

BOTANICAL HISTORY AND BOTANISTS

Crambe has not been recorded in the County since. Sir Joseph Banks with Mr. Correia da Serra visited the submerged forest at Sutton-on-Sea in 1796 and found peat containing holly leaves, and tree stumps of oak and birch.

John Cragg (the great grandfather of Captain W. A. Cragg of Threekingham, who attended L.N.U. meetings up to the 1930s) was a keen botanist at the time of Arthur Young's visit. He kept a commonplace book with a list of Threekingham plants between 1790 and 1820. He found a Lizard Orchid 'beside the Turnpike near 3 feet high, by Dunsby Wood'. He was puzzled by the Small-leaved Lime and confused it with aspen. Another list of plants at Metheringham, compiled by him, was unfortunately lost. He had a set of the first edition of Sowerby's *English Botany* which his great grandson sold. He was a landowner and surveyor of various enclosures.

Edmund Oldfield in the *History of Wainfleet* (1829) also commented on the East Fen plants. "A principal part of the East Fen which appertained to this parish of Friskney was denominated the Mossberry or Cranberry Fen from the quantities of cranberries which grew upon it in its wild and uncultivated state."

Rev. R. J. Bunch, a Lincoln boy, became a Fellow of Emmanuel College, Cambridge. His herbarium, with many plants collected near Lincoln *c*. 1830, is at Leicester University. His friend **Rev. J. F. Wray,** son of the Vicar of Bardney, was also at Cambridge and his herbarium, chiefly of plants from Bardney 1820-1830 was found at Ipswich together with some collected by **Miss C. Cautley.** Many of the records in E. B. Drury's *Guide to Lincoln* (1831) were clearly found by Bunch. Drury was also a friend of John Clare the Poet of Helpston and moved from Stamford to Lincoln.

Dr. John Nicholson of Lincoln was taught by Sir W. Hooker at Glasgow in 1836, and he is the first recorder for England of *Viola stagnina,* though Miss Cautley had found it in 1833. Apart from a herbarium belonging to a friend of his, called the Simpson collection, containing two or three of his specimens, no account of his finds has survived. He left Lincoln in 1856 for London, where he died *c*. 1880.

Rev. J. Dodsworth, Vicar of Bourne, made a herbarium in 1836 and marked finds in his Smith's *Compendium* which Peacock gave to the Herbarium of the British Museum. The 'ruin of his herbarium' came into Peacock's possession in 1893, and a number of surviving specimens are in the Lincoln Herbarium. He was an excellent and painstaking botanist.

Dr. R. G. Latham (of Dictionary fame) was a friend of the Rev. J. Dodsworth. A few of his specimens are at the Cambridge Botany School, but Peacock says his records are lost. **Rev. T. V. Wollaston,** son of the Rector of Scotton, made a valuable list of plants which he sent to *The Phytologist* (1843, p. 522). His father found the Great Fen

Ragwort "on the banks of ditches near Brayford Water, half a mile from Lincoln".

Rev. J. K. Miller, Vicar of Walkeringham (Nottinghamshire), near Gainsborough, made a careful list of Lincolnshire and Nottinghamshire plants 1847, which was published by Sir C. Anderson in his *Short Guide.* **Dr. B. Carrington, M.D.** of Lincoln might have left us more records, but his herbarium has perished and only his paper on plants near Lincoln (1849) has survived.

H. C. Watson of Kew took advantage of the new railways to visit Lincoln, Grimsby, Louth and Boston in 1851, and was the first to put a full list into print, in *CYBELE Britannica* (1868). His marked London Catalogues are at Kew. **Rev. R. E. G. Cole** made a careful herbarium at Doddington in 1856 which he gave to the Lincoln Herbarium.

At Louth there was an active group of botanists (or a botanical class) in 1855-6, under **Rev. J. T. Barker** (Congregational Minister). **Drs. E. B. and T. W. Bogg** made a collection which was obtained from another brother, J. Stuart Bogg of Altrincham by Dr. F. A. Lees in 1892 and is incorporated in the Lincoln Herbarium. Their uncle, Edward Bogg of Benniworth House near Donington-on-Bain wrote on the local geology in Weir's *History of Horncastle* (1820).

Dr. John Lowe of Sleaford wrote a paper for the Botanical Society of Edinburgh (1856) on Lincolnshire plants near Sleaford. Later he moved to King's Lynn and continued as a botanist there. **Miss M. E. Dixon** of Caistor kept a record of plants from 1850-1870 in her 7th edition of Withering's *British Plants* and a few are first records. Her friend **Mr. J. Daubney,** a Caistor solicitor, brought her *Drosera intermedia* out of Caistor Moor, and she found Bog Asphodel, Marsh Gentian and Butterwort near Nettleton Lodge.

Rev. Canon William Fowler of Winterton and Liversidge, Yorkshire, was one of our best pioneers. As a youth he was tutored by the Rector of Saltfleetby St. Clements and recorded the plants of the salt marshes. He wrote some papers for *The Phytologist* (1856-58) and others later for *The Naturalist* (1878-1890). He became friendly with Dr. F. A. Lees, and in 1877 he went to south Lincolnshire to record there for the Botanical Locality Record Club. He must have been a great walker and probably had friends he could stay with in the south. He arranged his time and his travelling well and he added large numbers of records to the two vice-counties. As a very old man he retired to his birthplace where his forebears had been builders of skill and intelligence. He came to two or three L.N.U. meetings, was President in 1898, and his benign countenance and friendliness showed up. His joy at finding *Carex curta* at Linwood Warren as a new record there in 1911 is still remembered. He was Vicar of Liversidge 1866-1910 and a member of the Y.N.U. Canon Wm. Fowler is confused in Britten and Boulger's *Biographies of Botanists* with Canon Wm. Weekes Fowler,

Headmaster of Lincoln Grammar School who was an entomologist. Both were Presidents of the L.N.U. (See Plate 13).

James Britten of the B.M. visited the Caistor district in 1862, at the age of 16, and published a list of 23 plants in the *Journal of Botany* and *The Phytologist*. In 1872 he compiled a full list of Lincolnshire plants for White's *Directory of Lincolnshire* which was repeated in 1882 with a few additions.

Dr. F. A. Lees, a Yorkshireman, became interested in Lincolnshire plants due to **Dr. R. M. Bowstead** of Caistor who sent up a record of *Pyrola minor* to the Botanical Locality Record Club in 1876. North Lincolnshire, except for Canon William Fowler's records of Winterton and Saltfleet, was little known botanically. Dr. Lees took a medical practice at Market Rasen 1876-78 in order to study the flora of that district. He published his numerous outstanding finds in the *Reports of the Botanical Locality Record Club*, of which he was editor, together with those of Canon Fowler. He took an immense interest in Lincolnshire Botany, particularly in Linwood Warren, and later paid several visits to Market Rasen, Louth and Spilsby, meticulously recording everything in notebooks which are now at Cambridge and with Dr. Sledge. He and Canon Fowler added a large number of County Records to Watson's *Topographical Botany* (1873).

Lees' interleaved *Outline Flora of Lincolnshire* (1893) (now at the Cambridge Botany School) gives many first records for both vice-counties. It consists of a separate of the original Outline Flora from White's *Directory of Lincolnshire* (1892) annotated and leather-bound into a larger manuscript volume. It includes a medley of correspondence, photographs, reviews, cuttings and herbarium specimens. Also "Additional Notes for the Flora 1893", a rough typescript list of loose pages, with the comment "*Senecio paludosus* — Wollaston found the last specimen ever found in England and that was near Lincoln!" Letters copied into the manuscript include one from F. W. Burbidge of Trinity College, Dublin (1896) to Peacock concerning a flowering specimen of *Liparis loeselii* from Lincoln. Also a loose postcard from Peacock to Lees recording the *Vaccinium myrtillus* found by Claye in 1917. Lees also records with relief, the finding in the county, at long last, of *Cardamine amara* from a bog at Aylesby near Grimsby in 1910. He also notes details of the disappearing *Teucrium scordium* found at Washingborough and Cowbit (G.C.D. 1876, Bot. Soc. & Exch. Club Rep. 1911) and amongst the many additional species for the Flora points out especially *Thesium* and *Pilularia*. There are herbarium specimens of *Iris spuria* together with a cutting of the first printed notice of the species at an L.N.U. meeting and *Nardurus* which was a new British grass. (Other separates of the 1892 White's *Outline* are at Kew and the Lincoln Museum.) Lees' herbarium was purchased by Bradford Corporation in 1905, and his library is housed in Bradford City Library.

Between 1850 and 1890 many clerics, doctors and ladies became keen botanists and natural history societies were formed in several centres. At Alford **Joseph Burtt Davy** was secretary and **J. W. Chandler** and **Susan Allett** recorded plants. **Miss Susan Skipworth** of South Kelsey in 1835-1850 (afterwards Mrs. J. L. ffytche), **Miss E. J. Nicholson** of Wootton and **Dr. T. P. J. Grantham** of Burgh-le-Marsh later lent their plants to Peacock in the 1890's. **Rev. T. V. Wollaston, Rev. J. K. Miller** and **Rev. T. Owston** botanised in West Lindsey before 1840.

Joseph Burtt Davy was a keen young botanist coming to Alford in 1890 for six months. Unfortunately he was there chiefly in winter but his diary survives. He copied out lists of Lincolnshire plants at H.B.M. He became secretary of the Alford Naturalists' Society. When Dr. Lees was compiling his *Outline Flora of Lincolnshire* for White's 1892 Directory, Burtt Davy was to have been co-editor but his health broke down and Rev. E. A. Woodruffe-Peacock (lately returned to his native county at Cadney cum Howsham) took his place. Many notes and much of the correspondence between Lees and Woodruffe-Peacock have survived — not always friendly. Both were living on a shoe string and post cards at ½d. were chiefly used. Burtt Davy went abroad and recovered but never returned to Lincolnshire, though he later lived at Oxford. When at Alford he once walked 14 miles to Horncastle, took a train to Woodhall Spa, searched in vain for a piece of un-reclaimed fen near Dogdyke and returned to Alford by train via Boston — cost four shillings! It was a new experience for him to see the heather and moorland plants round Woodhall.

Miss Susan Allett of Strubby was a remarkable finder of rare plants. She was a member of the Alford Naturalists' Society before 1890 when cook at Tothill Rectory; about 1900 she was cook at Nettleton House near Caistor. When at Tothill she found ferns particularly, but also *Menyanthes* and *Parnassia* in the 'carr marsh'. At Sandbraes, Caistor, she found *Vicia lathyroides* and *Teesdalia*, and at Nettleton and Moortown five kinds of *Equisetum*, two being rarities. She wrote letters to Dr. Lees and Peacock, one of which survives, showing her to be of good intelligence but poor scholarship. She was also a cook near Bath and in London. After Peacock's death no more was heard of her. She sent him a large packet of ferns from Burwell wood in 1909 and a letter in 1920.

Rev. E. A. Woodruffe-Peacock published a most interesting *Critical Catalogue of Lincolnshire Plants* in fourteen parts in *The Naturalist* (1894-1900) from his knowledge of the Lincolnshire flora at that date. It was limited because of both lack of transport and of his knowledge of plants which was growing all the time. During this period he lost his first wife and had a serious illness but the L.N.U. which started in 1893 continued to flourish under his secretaryship. He became President in 1905. Field Meetings were held on Bank Holidays and

BOTANICAL HISTORY AND BOTANISTS

cheap tickets were useful to get to the coast. Horse drawn brakes were used to convey members from railway stations and a few others arrived on bicycles. Peacock recorded the findings of the field meetings in great detail. His card index and locality register have unfortunately been lost, but the notes for his *Rock-Soil Flora*, though incomplete, were left to Cambridge University. He was in advance of his time as an ecologist and his *Rock-Soil Flora* leaflet shows how he worked his card index of soils and habitats and frequencies.

In his later years he made a study of hybrids and also took much interest in pollination and animal, bird and man carriage of plants. Sir Arthur Tansley was much impressed and offered to publish his flora, but this came to nothing and his unfinished manuscript is at Cambridge. He compiled a section on the Lincolnshire Flora for the Victoria County History which has never been published. He was to blame for putting in localities for plants of later finders if the original earliest record lacked one. He confessed to this but did not mark which they were. His *Check List of Lincolnshire Plants* (1909) is a very valuable one but has a few mistakes. His *Drosera anglica* and *D. intermedia* and the *Cuscutas* are confused and he allows that he did not know *Avena pratensis* till a later period. He attributes the "Simpson Collection" to Dr. John Nicholson but only two or three plants in it were given by Dr. Nicholson (see under *Verbascum, Nat.*, 1896). He formulated the Divisions of the County on which the Check List is based. (See Plate 12).

William Higginbottom and **Mrs. Jarvis** of Hatton were some of Peacock's early recorders in the 1890s. Rev. A. E. and Mrs. Jarvis had a natural history society in Hatton and and W. Higginbottom, the village carpenter, was a keen member. They made several good records. Mrs. Jarvis provided Peacock with a list of Hatton plants, but unfortunately left Lincolnshire, returning once or twice to Woodhall Spa.

Rev. W. W. Mason, "the most consistent of botanists" as Peacock called him, was born at Wainfleet in 1853. He was Rector of Leverton 1878-94 and then moved to industrial Bootle, but corresponded with Peacock and occasionally visited his native county. After 10 years at Melmerby, Cumberland (where he added *Myosotis brevifolia* to the British Flora) he returned in 1924 to Salmonby and attended several L.N.U. meetings. He would not allow himself to become President of the L.N.U. but he did much towards making a full list of Lincolnshire plants for the eighteen divisions. His handwriting is a joy to read — the very opposite of Peacock's which is large and written with a broad quill pen. Mason was a close friend of **Dr. G. C. Druce** who named a form of hawthorn from Melmerby after him — *Crataegus monogyna* var *masonii*. He was said to have "a good eye for micro-species and varieties". After retiring from Lincolnshire he lived at Holton near Oxford for a time and died at Louth, 1932. (See Plate 14).

Miss E. M. Lane-Claypon was a keen young L.N.U. member in the 1890s, painting flowers and fungi with great skill, and contributing many records from Wyberton and L.N.U. field meetings for the Check List. In 1905 she married Rev. J. P. Cheales of Friskney when she dropped botany. After his death in 1948 she gave her best flower paintings to the L.N.U. and came to two or three L.N.U. meetings. She was still pleased to talk of wild flowers up to her death, at the age of 91, in 1964, and remembered where she had found them.

Harry Fisher (1860-1935) of Newark, Nottingham and Grantham, could have been a great help to Lincolnshire but he was unable to work with Peacock. His vast herbarium is at Wollaton Hall Natural History Museum, Nottingham. He sent Peacock his Lincolnshire records in the 1890s before he went to Franz Josephland in the Arctic. He did not return to this county till 1911, when he became director of the *Grantham Journal*. About 1930 his interest in botany was revived by Mr. H. Pugsley, an old friend of his London days, and he collected masses of critical plants, *Rosa*, *Rubus*, *Euphrasia*, etc., chiefly round Grantham. Miss Stow said he taught her much. Judging from his correspondence with Pugsley and from Mr. J. Chambers (chemist) of Grantham, I learnt that he was of an uncertain temper and had a very high opinion of his own powers.

Miss S. C. Stow, another very keen L.N.U. member, was the very best of Peacock's helpers. She supplied him with countless soil notes with her records from the Grantham district and at L.N.U. meetings did much recording for him. At one time she became interested in mosses and plant galls. She was the first lady president of the L.N.U. (1914) and wrote the sectional reports 1922-32. *Nardurus* was her outstanding find for the British Flora. Her early Herbarium was at Grantham, but was destroyed after 1960. (See Plate 15).

Mrs. Newman of Lincoln and **Miss Mackinder** of Belleau were early helpers of Peacock's. They contributed specimens to the County Museum. **J. S. Sneath** was one of the L.N.U. stalwarts. His specimens in the Lincolnshire Herbarium show his care and love of plants. His happy face with a bunch of flowers he had placed in his square bowler hat, his grey beard and his Victorian manners remain in my memory.

Arthur Smith, who became curator of Lincoln Museum in 1905, was not a botanist first and foremost, but he collected dock aliens at Grimsby, and did much work as L.N.U. Secretary (1904-1933), was President in 1934 and tidied up the Herbarium. **Rev. F. S. Alston** was a member of the L.N.U. for many years and became President in 1921. He recorded 190 plants from the Horncastle district in a copy of Hayward's *Botanist's Pocket Book*, which his son Hugh gave to the Herbarium of the British Museum, where he was Assistant Keeper of Cryptogamic Botany. They recorded a large number of aliens at Woodhall Spa, 1917-19, some of which were new to Britain. **Rev. A. N. Claye** was a keen member of the Wild Flower Society as well as of the L.N.U. (President 1922). He lived at Brigg from 1893 to 1918, moving to Hagworthingham for two years and then settling at Stockport for

BOTANICAL HISTORY AND BOTANISTS

the rest of his life. He kept up his interest in recording plants until he died in 1956, and sent a list of his Lincolnshire finds after he had retired.

Samuel Hurst, M.P.S. with his friends **Fred Kime** (fishmonger) and **Frank Waite** (corn merchant) sent in many records for the Boston district, but most are unlocalised. They were a friendly helpful trio who attended L.N.U. meetings regularly. Mr. Hurst had a square cut red beard and was always smiling, Mr. Kime with a dark grey beard, less cheerful but always keen on many sides of natural history, Mr. Waite clean shaven and rather quiet.

When Rev. W. W. Mason returned to Lincolnshire after an exile of 30 years he compiled a manuscript register from the Check List and his own finds (1925-30). **F. T. Baker,** Secretary of the L.N.U. from 1934 to 1961 and President 1961-2 (former Director of Lincoln Libraries, Museum and Art Gallery), added a large number of printed records to the register, with herbarium specimens and new finds from field meetings (1930-36). He also searched out many old records and added his own new records from the Lincoln district and from the Lincoln Museum Wild Flower Table. **Dr. H. B. Willoughby-Smith** (President of the L.N.U. 1924), **Miss C. D. Marsden, G. H. Allison, C. S. Carter** (President, 1928 and 1929), **F. L. Kirk** (President, 1946-7), **A. Roebuck** (President, 1932), **S. A. Cox, Reg. May** (President, 1956), **Mrs. M. E. Stewart** and **Miss Janet Cook** all contributed to this and made their own records.

In recent years a new generation of botanists has arisen who have added many new localities in the grid square plant distribution scheme 1954-60.

Mrs. Eva Dunn née **Wright** was a member of the Wild Flower Society in 1920 and again in 1953-59. She knew and remembered every plant she had seen in her girlhood at Roxby and later at Elsham, and was the greatest help in the grid square survey. She urged all her neighbours to bring her wild flowers in plastic bags from definite spots and she then identified them and put them on record cards. In this way she recorded plants in fifteen squares and made a number of divisional records. Her triumph was *Gagea lutea* a new vice-county record for north Lincolnshire found by Mr. and Mrs. Stones. As an active minded invalid she was seldom able to look for plants herself except in a wheelchair or from a car.

Miss M. N. Read has been a very active member at L.N.U. meetings and has walked many miles round Horncastle to make records. **J. H. Chandler** (President 1965-6) was similarly active round Stamford, and **Mr. and Mrs. R. C. L. Howitt** undertook recording in the Fenland squares of south Lincolnshire. These four were untiring in their efforts for the *Atlas of the British Flora* covering at least six 10Km squares.

C. J. Allerton, Mrs. K. Wherry, the late **Miss R. E. Taylor, Shirley Derry, Stephanie Monk, Roger Hull, Alan Gray** and **Alan Lowe** (the latter five starting as students) sent in records. Since 1960 **Miss W. Heath, Mrs. Z. Porter** and **Mrs. I. Weston** (President 1967-8) have been particularly active at L.N.U. meetings and elsewhere in the county.

The B.S.B.I. Grid Square recording, 1954-60, has been an enormous help in levelling up records all over the county. Soils and types of habitat make a great difference to the number of plants in a 10 Km. square. Some squares have no woods and most of the fields are arable; others have large woods and a variety of soils. The most notable squares are Stamford (not all Lincolnshire species) with 790, Woodhall Spa and Horncastle 621 and Broughton, and Twigmoor and Manton 604. The lowest numbers are three reclaimed fenland squares, with under 250, where it is difficult to find any indigenous plants and weeds are scarce. (See Fig. 11, p. 288).

CHAPTER 6

LINCOLNSHIRE RECORDERS

Key

fl. — flourishing at that date
FRS — Fellow of Royal Society
FLS — Fellow of Linnaean Society of London.
B & B — Britten & Boulger, Biographies of Botanists.
BSBI — Botanical Society of the British Isles.
DNB — Dictionary of National Biography.
LNU — Lincolnshire Naturalists' Union.
WFS — Wild Flower Society.

Allen and Saunders : (published 1834). List of plants in *History of Lincs.* Vol. I, p. 117.

Allen, W. : fl. 1877. BRC (not W. B. Allen, 1875, B & B); Market Rasen, later Bridgnorth.

Allerton, C. J. : fl. 1961. Wigtoft schoolmaster; LNU.

Allett, Miss Susan : fl. 1888-1920. Alford Nat. Hist. Soc.

Allison, G. H. : 1862-1956. Grimsby and Ashby by Partney; LNU; bryologist.

Alston, Rev. F. S. : 1863-*c*.1930. Scrivelsby etc.; LNU; father of following.

Alston, A. H. G. : 1902-1958. Assistant Keeper, Dept. Botany BM; FLS; b. West Ashby.

Amner, Miss S. : fl. 1954-55. Spalding school teacher; LNU.

Anderson, Miss M. : fl. 1894. Lea; article in *Nat.* 1903. 189.

Bacon, Miss G. : c. 1875-1949. WFS; BSBI; visited Newark 1927.

Bacon, Dr. Vincent, F.R.S. : fl. 1726 (d. 1739). Grantham; *Nat.* 1898, 177; B & B.

Baker, F. T., O.B.E. : fl. 1929 . . . Lincoln; LNU Secretary 1934-61.

Ball, M. E.: fl. 1965 . . . Assistant Regional Officer, Lincs., Nature Conservancy.

Banks, Sir Joseph, Pres. Royal Soc. : 1743-1820. Revesby and Kew; B & B; DNB.

Bates, Mrs. E. : fl. 1891. Brigg.

Batters, R. : fl. 1931 . . . Lincoln; LNU.

Barker, Rev. J. T. : fl. 1856. Congregational minister, Louth.

Bayldon, Rev. J. : 1842-1912. Partney and Low Toynton; friend of Dr. J. Burgess.

Bayley, Rev. R. S. : fl. 1834. *Notitiae Ludae;* list not localised; Nonconformist minister.

Beeby, W. H. : 1849-1910. FLS, BM, B & B; visited Deeping Fen 1881.

Bell, J. : fl. 1960-61. Wigtoft schoolboy; WFS.

Blackstone, J. : fl. 1746 (d. 1753). Apothecary; B & B; DNB.

Blackwood, J. W. : fl. 1970 . . . Assistant Regional Officer, Lincs., Nature Conservancy.

Blair, Dr. Patrick, F.R.S. : 1666-1728. Physician to port of Boston 1723-28; B & B; DNB.

Bloxam, Rev. Andrew : 1801-1878. Rector of Twycross, Leics.; B & B; DNB; records in *Top Bot.*

Bogg, Dr. E. B. : fl. 1852. Louth (Herb.) Records to H. C. Watson.

Bogg, Dr. T. W. : fl. 1856. Louth (Herb.) Records to H. C. Watson.

Bond, L. H. : fl. 1925 . . . Grantham LNU, FRIBA.

Bond, Miss S. : fl. 1954 . . . Grantham High School; daughter of L. Bond, WFS.

Booth, W. : c. 1895. Howsham schoolmaster.

Bowstead, R. M., M.D. : 1834-1898. Caistor.

Bratt, D. : fl. 1893. Nocton schoolmaster.

Bray, B. : fl. 1852 (d. 1870). King's Lynn.

Brewster, Rev. H. C. : 1832-1915. Rector of South Kelsey 1865-1915; MS. *History of South Kelsey* in Cathedral Library, Lincoln.

Britten, James : 1846-1924. Visit to Caistor 1862; *Nat.* 1864, 84; BM; FLS; B & B.

Brooks, Miss S. C. : fl. 1860. Great Ponton Rectory; herbarium at Grantham Museum.

Brown, John : fl. 1944. Sheffield, BSBI.

Browne, E. M. : fl. 1879-1900. Master at King's School, Grantham; BRC.

Browning, G. W. : fl. 1837. Stamford; paintings lost; d. Cape Town.

Buddle, Rev. Adam : 1660-1715. b. Deeping St. James; DNB.

Bullock, A. : fl. 1907. Grimsby schoolmaster.

LINCOLNSHIRE RECORDERS

Bunch, Rev. R. J. : 1804-1870. b. Lincoln, Fellow of Emmanuel Cambridge.

Burgess, Dr. J. T. : fl. 1879-1915. Spilsby; paintings Lincoln Museum; *Nat.* 1893, 325.

Burchnall, H. W. : fl. 1908-1953. Butterwick schoolmaster.

Burton, F. M. : 1829-1912. Gainsborough; geologist; LNU; FLS.

Burton, W. A. : fl. 1943 . . . Marston schoolmaster.

Burtt, G. W. : fl. 1865. Leadenham; herbarium.

Butcher, Dr. R. W. : fl. 1943-52. Nottingham; LNU; FLS.

Cammack, Dr. T. A. : fl. 1856. Boston; list in P. Thompson.

Carlton, H. : fl. 1920 . . . Horncastle chemist; LNU.

Carr, Amos : fl. 1880 (d. 1884). Sheffield; Yorks. Nats. Un.; BRC.

Carrington, Dr. B. : 1827-1893. b. Lincoln; *Bot. Gaz.* 1849; MD Edin 1852; FLS; d. Eccles.

Carter, C. S. : 1865-1933. Louth; LNU.

Cautley, Miss C. : fl. 1833. Visited Scotter, Doddington; herbarium.

Cave, Mrs. L. : fl. 1958 . . . Grantham schoolmistress; LNU.

Chandler, J. H. : fl. 1936 . . . Stamford; LNU; BSBI; herbarium.

Chandler, J. W. : fl. 1890. Alford Nat. Hist. Soc.

Charters, Rev. R. H. : fl. 1868. Gainsborough Grammar School.

Claye, Rev. A. N. : 1863-1956. Brigg. (1893-1919) and Stockport; LNU; WFS.

Clayton, W. J. : fl. 1920 . . . Croxby, Holton le Moor.

Claypon, Miss E. M. Lane : fl. 1886-1904. Wyberton; paintings; LNU (later Mrs. Cheales).

Clifton, E. H. : fl. 1961. Lincoln; LNU.

Coates, E. : fl. 1856. Friend of W. Fowler.

Cockin, Rev. C. E. : fl. 1885. Rector of Lea 1874-94.

Cole, Rev. R. E. G. : 1831-1921. Doddington 1856 and 1861-1921; herbarium.

Coles, Dr. Susan M. : fl. 1974. Institute of Terrestrial Ecology.

Collier, R. V. : fl. 1972 . . . Conservation Officer, Lincolnshire Trust for Nature Conservation.

Cooke, Miss J. : fl. 1930. Spalding; LNU; WFS.

Cordeaux, J. : 1831-99. Great Coates; LNU; ornithologist.

Correia da Serra, Rev. J. F., F.R.S. : 1750-1823. *Phil. Trans.* 86-89, 1796-99 (in P. Thompson); B & B.

Cox, S. A. : fl. 1930 . . . Grimsby; LNU; ornithologist.

Crabbe, Rev. G. : 1754-1832. Plant list Belvoir district in Nichols *History of Leicester*, Vol. 1; *Bot. Guide;* B & B; DNB.

Cragg, John : 1761-1832. Threekingham, Surveyor.

Craster, Miss : fl. 1895. Denton rectory.

Crow, B. : fl. 1893. Sec. of Mechanics' Inst. Louth; LNU.

Cullum, Sir T.G., F.R.S. : 1741-1831. FLS; B & B; DNB.

Dalton, Rev. J. : 1764-1843. FLS; Stayed at Fillingham castle c. 1800; *Bot. Guide;* herbarium at York; B & B.

Daubney, Rev. W. H. : 1852-1927. Rector of Leasingham 1894; LNU.

Davy, J. Burtt, F.R.S. : 1870-1940. Ph.D. Cambridge; Oxford; Alford Nat. Hist. Soc.; FLS.

Deakin, Richard, M.D. : 1809-1873. Sheffield; *Floragraphia Br;* B & B; DNB.

De la Pryme, Rev. Abraham, F.R.S. : 1672-1704. Curate of Broughton, 1695-97; DNB.

Derry, Miss S. : fl. 1960. Lincoln; LNU.

Dixon, Miss M. E. : 1838-75. Caistor.

Dodsworth, Rev. J. : 1799-1877. Rector of Bourne; patron of Ingoldmells 1859; B & B.

Druce, G. C. : 1850-1932. Oxford; visit to Spalding 1911; FLS.

Drury, E. B. : 1831. List of plants in *Guide to Lincoln* not localised; possibly by Rev. R. J. Bunch; friend of John Clare.

Dunn, Mrs. Eva : 1902-59. Elsham and Roxby; WFS.

Edees, E. S. : Visited Lincolnshire 1965; bramble specialist, Staffordshire; BSBI; FLS.

Elwes, Miss Sophia : fl. 1840. Brigg; herbarium at Elsham Hall.

Firbank, Miss K. G. : fl. 1891. Barton-on-Humber girls' school.

Fitter, R. S. : fl. 1957. BSBI.

Fisher, H. : 1860-1935. Newark and Grantham.

Foggitt, Mrs. G. : (See Bacon, G.).

Fowler, Miss Ethel : d. 1966. Winterton; daughter of the following.

Fowler, Rev. Wm. : 1835-1912. Winterton; vicar of Liversedge, Yorks. 1864-1910; Yorks. Nats. Un.; B & B; Bot. Loc. Rec. Club; LNU.

(Refs. under Fowler are to Wm. except where otherwise stated).

Frankish, Mrs. Anne : fl. 1972. Scunthorpe.

Frith, Dr. W. S. : fl. 1920. Brigg; LNU; WFS.

Gerarde, J. : Herbal 1597.

Gibbons, Miss E. J. : 1920 . . . Holton le Moor; FLS; LNU; WFS; BSBI; herbarium.
(Refs. under Gibbons or J. Gibbons are to E.J. unless otherwise stated).
Gibbons, Miss M. E. : 1907 . . . Holton le Moor; LNU; WFS.
Gibbs, L. : fl. 1884. Boston; LNU.
Gibson, Rev. Edmund : 1669-1748. Bishop of Lincoln and London; translated Camden 1695 and used Ray's list; DNB.
Gilham, Miss L. : . . . Gainsborough; *Trans. LNU* 1914.
Gilson, H. : fl. 1856. Boston; list in P. Thompson.
Gillett, Rev. E. A. : fl. 1893 (1842-1927). Woolsthorpe 1879-1903; LNU.
Goodall, Rev. R. W. : fl. 1892. Curate Lincoln; LNU.
Gough, R. : 1735-1809. Translated Camden 1789; includes Ray's list and Banks' records.
Goulding, R. W. : 1868-1929. Louth; librarian to Duke of Portland; LNU.
Grantham, Dr. T. P. J. : fl. 1840. Burgh le Marsh; list of plants in Anderson's *Guide to Lincs.;* herbarium lost.
Gray, A. J. : fl. 1950 . . . Tetney; LNU; BSBI.
Grierson, Dr. G. A. : 1864-1958. Lincoln and Grimsby; LNU.
Hailstone, S. : 1768-1851. FLS; list in *Proc. Linn. Soc. Lond.* II, 189, 1852; *E. Bot.* 1035, 2737; herbarium list in York Museum; B & B; DNB.
Hampson, Rev. W. S. : fl. 1867. Rector of Stubton 1857-68.
Haslam, Miss M. M. : fl. 1948 . . . Boston High School; LNU.
Hawkins, J. : fl. 1842-90 (1820-1917). Grantham.
Hawley, Sir Henry C. W. : 1876-1923. Tumby; mycologist; LNU.
Heath, Miss Winifred : fl. 1960 . . . Nettleham; WFS; LNU.
Healey, H. A. H. : fl. 1911. Son of Dawsmere schoolmaster.
Hexham, H. : ?1585-?1650. Edited Mercator's Atlas 1636; DNB.
Higginbottom, W. : fl. 1890 . . . Hatton; wheelwright.
Hill, Sir J., M.D. : 1716-75. Reputed to have been first Supt. Kew 1763; b. ?Spalding; DNB; B & B.
Hinchliff, Rev. H. M. W. : fl. 1899. Lincoln; LNU; Nonconformist.
Hind, F. : fl. 1930-40. Skegness; LNU; ornithologist.
Holden, Henry : 1662-1710. Birmingham; visited Kirton 1688; herbarium.
Hope-Simpson, J. : fl. 1953. Univ. Bristol.
Hopkins, Miss B. : fl. 1950. Boston.

Hopwood, S. F. : d. 1955. Louth; LNU; ornithologist.
How, W. : 1620-56. Author of *Phytologia Brit.*
Howard, Dr. T. : fl. 1897. Sibsey.
Howitt, Godfrey, M.D. : 1800-1873. Visited Frieston shore 1826; Notts. flora.
Howitt, R. C. L. and Mrs. B. M. : 1950 . . . Southwell and Newark; LNU; BSBI; WFS; Notts. flora.
Hudson, S. : fl. 1894. Epworth.
Hudson, W., F.R.S. : 1730-93. Apothecary; *Flora Anglica;* herbarium mostly burnt, some at BM; FLS; B & B.
Hull, Roger : fl. 1949 . . . Hackthorn; WFS; BSBI.
Hull, Miss J. : fl. 1948 . . . Hackthorn; WFS.
Hurst, S. J. : fl. 1908-1936. Boston (with F. Waite and F. Kime); LNU; MPS.
Hutchinson, R. : fl. 1890 (d. 1950). Willoughton.
Irvine, Alexander : 1793-1873. (Flora of London, 1830); plants from Scotter and Hemswell; DNB; B & B.
Jarvis, Mrs. A. E. : fl. 1890. Hatton rectory; LNU; list of Hatton.
Jennings, Rev. W. T. : fl. 1900. Huttoft 1893-1902.
Johnson, T. : ?1600-1644. Edited *Gerarde's Herbal* 1630; B & B and DNB; visited Lincolnshire; apothecary.
Kew, H. W. : 1868-1948. Louth; LNU; grandson of T. W. Wallis.
Kirk, F. L. : 1940 (d. 1959). Schoolmaster Donington & Alford G.S.; entomologist; LNU.
Larder, J. : fl. 1893 (d. 1923). Louth; LNU.
Latham, Dr. R. G. : 1812-1888. b. Billingborough; herbarium and notes lost; some specimens at Camb.; B & B; DNB; friend of Rev. J. Dodsworth.
Lees, Dr. F. A. : 1847-1921. Leeds and Market Rasen; FLS, Yorks. Nats. Un.; Bot. Loc. Rec. Club; B & B.
Lewin, Miss E. F. : fl. 1897. Tealby vicarage.
Ley, Rev. Augustin : 1842-1911. Hereford; Brambles — *J. Bot.* 1871; B & B.
Lister, Dr. Martin, F.R.S. : 1638-1712. Visited Burwell 1666; see *Trans. LNU* 1927; DNB; B & B.
Lousley, J. E. : fl. 1955 . . . Visited Lincs. 1955-56 & 1970; BSBI.
Lowe, J. Alan : fl. 1960-61. Cleethorpes; WFS.
Lowe, Dr. John : 1830-1920. Flora of Sleaford district — *Trans. Bot. Soc. Edinb.* 1856: 13; later of King's Lynn; B & B.
Lowe, Mrs. M. fl. 1958 . . . Denton; Stainby schoolmistress; WFS.

LINCOLNSHIRE RECORDERS

McClintock, D. : fl. 1942 ... Skegness, Cleethorpes etc.; visited Lincs. 1942, 1956 and 1963; BSBI; WFS.
Mackinder, Miss : fl. 1880. Belleau; herbarium; friend of Mrs. Newman.
Malmesbury, William of : 1200. Quoted by P. Thompson of Boston.
Mathews, Rev. T. : fl. 1856. Boston (Baptist); list in P. Thompson.
Mapletoft, Rev. R. : 1609-1677. Sub-dean of Lincoln 1660.
Marsden, Miss C. D. : 1873-1960. Louth; LNU; WFS.
Martyn, J., F.R.S. : 1699-1768. Prof. Bot. Cambridge 1733-61; B & B; DNB.
Martyn, Rev. T., F.R.S. : 1735-1825. Prof. Bot. Cambridge 1762; B & B; DNB.
Mason, J. L. : fl. 1964. Nature Conservancy; BSBI.
Mason, Rev. W. Wright : 1853-1932. b. Wainfleet; Rector of Bootle, Melmerby and Salmonby.
May, Reg : fl. 1920 ... Limber; ornithologist; LNU.
Melvill, J. Cosmo : fl. 1882. FLS; Prestwich, Lancs.; Visited Woodhall Spa.
Melville, Dr. R. : fl. 1965. FLS; Kew; rose and elm specialist; visited Lincs. 1965.
Mercator : 1638. Atlas.
Merrett, Christopher, M.D., F.R.S. : 1614-1695. Surveyor to port of Boston 1666; B & B; DNB.
Miller, Rev. J. K. : c. 1790-1855. Vicar of Walkeringham, Notts.; B & B; list in Anderson's *Guide* 1897; *Nat.* 1893.
Miller, Thomas : 1807-1874. b. Gainsborough; wrote *Common Wayside Flowers*, 1860 (no records); DNB; B & B.
Mills, F. H. : fl. 1893-4. Torksey schoolmaster.
Monk, Miss S. : fl. 1961. Roughton; LNU (now Mrs. Tyler).
Morris, Mrs. M. : fl. 1928 ... (d. 1953). Grimsby; LNU.
Moss, C. E. : 1872-1931. Cambridge; FLS.
Mossop, Rev. J. : fl. 1830-70. Covenham rectory (paintings).
Nash, The Misses H. M. : fl. 1900 ... Lincoln; LNU.
Newman, L. F. : fl. 1919. Wash coast survey with G. Walworth — *J. Ecol.* 1919.
Newman, Mrs. M. : fl. 1886 ... Lincoln; herbarium; LNU.
Newton, Gordon : fl. 1963. Cleethorpes schoolboy; LNU.
Nicholson, Miss E. J. : fl. 1835. Wootton; herbarium.
Nicholson, John, M.D. : fl. 1836 (d. 1880). b. Lincoln, d. London.

Nicholson, Miss M. : fl. 1829 . . . Grayingham.
Nicholson, Guy : b. 1860. Hibaldstow.
Noel, Miss E. F. : fl. 1930 . . . (d. 1950). Louth; FLS.
Oldfield, Edmund : fl. 1829. *History of Wainfleet.*
Orchard, E. E. : fl. 1930 . . . Ancaster; son of schoolmaster; herbarium.
Ordoyno, T. : fl. 1807. Newark; short *Flora of Notts.*
Owston, Miss R. J. : fl. 1895 (d. c. 1950). Lincoln; LNU.
Owston, Rev. T. : fl. 1840 (1809-1895). Gringley and Dalby; few herbarium specimens.
Parker, C. : fl. 1899. Grimsby; LNU.
Parker, S. W. : fl. 1958 . . . Moortown; WFS.
Parkinson, John : 1567-1650. Notts.; DNB; B & B.
Parsons, Dr. H. F. : 1876. Goole; Yorks. Nats. Un.; B & B.
Peacock, Edward : fl. 1853. Bottesford; father of E.A.W.P.
Peacock, Rev. E. A. Woodruffe- : 1858-1922. Cadney; *Check List of Lincs. Plants* 1909; FLS, LNU, B & B.
Pears, Miss A. : fl. 1895. Mere; related to W. W. Mason; LNU.
Peck, W. : fl. 1815. *History of Isle of Axholme;* list of plants.
Peel, Mrs. : 1847. Willingham by Stow rectory.
Peet, W. M. : fl. 1959 . . . Sleaford; LNU.
Perring, Dr. F. H. : 1953 . . . Cambridge; Nature Conservancy; BSBI.
Peterken, G. F. : fl. 1970 . . . Nature Conservancy.
Phillips, G. S. and Mrs. : 1965 . . . Cleethorpes; BSBI; LNU.
Plukenet, Leonard, M.D. : 1641-1706. The Queen's Botanist; visited Boston; DNB; B & B.
Porter, Mrs. Zoe : fl. 1961 . . . Tattershall and Bourne; LNU; WFS.
Preston, H. : fl. 1890-1939. Grantham; LNU; geologist.
Proctor, H. : fl. 1954 . . . Market Rasen schoolmaster.
Pulteney, Dr. R., F.R.S. : 1730-1801. b. Loughborough; FLS; DNB; B & B.
Rainey, J. J. : fl. 1890. Spilsby; LNU.
Rake, Dr. T. Bevan : fl. 1853. Newark, Fulbeck and Fordingbridge 1891.
Rasell, Miss Anne : fl. 1961. Stamford High School.
Rawnsley, Miss S. F. : fl. 1895. Halton Holegate rectory.
Ray, Rev. John, F.R.S. : 1627-1705. Visited Lincolnshire 1661 and perhaps later; DNB; B & B.
Read, Miss M. N. : fl. 1948 . . . Horncastle; BSBI; LNU; WFS.

Redshaw, E. J. : fl. 1970 . . . Pinchbeck; Conchology Secretary LNU; FLS.

Reid, Clement, F.R.S. : 1853-1916. Geological Survey; interglacial plants; FLS.

Relhan, Rev. R. : 1754-1823. Fellow of King's College, Camb.; Rector of Hemingby; friend of T. Martyn; DNB; B & B; *Bot. Guide*, 1805; herbarium lost.

Reynolds, Alfred : fl. 1895. Owston Ferry; specimens sent to Hull and lost by bombing.

Reynolds, Bernard : fl. 1908. Boston; records *J. Bot.* 1910, 57.

Richardson, Dr. R., F.R.S. : 1663-1741. Ref. to subterraneous trees at Goole in *Phil. Trans.* XIX, 526; DNB; B& B .

Ridlington, Miss : fl. 1930. Spalding; WFS.

Roebuck, A. : fl. 1908 (d. 1971). Caistor and Sutton Bonington; LNU; entomologist.

Roebuck, W. D. : 1851-1919. Leeds; conchologist; LNU; FLS.

Rose, Dr. F. : fl. 1948-50. Kent; BSBI.

Salt, Jonathan : 1759-1810. Sheffield; FLS; herbarium.

Sandwith, Mrs. C. I. : fl. 1905 . . . (d. 1960). Harworth, Notts.; FLS; WFS.

Sandwith, N. Y. : 1901-1965. Kew; son of above; FLS; WFS.

Searle, H. : fl. 1881. Ashton-under-Lyne; Cleethorpes; Bot. Loc. Rec. Club.

Seaward, Dr. M. R. D. : fl. 1960 . . . Lincoln; LNU Secretary 1961-5; herbarium; bryologist and lichenologist; FLS.

Seppings, Dr. E. : fl. 1960. Boston.

Shaw, H. K. Airy : fl. 1947 . . . Kew.

Sibthorp, H. : 1713-1797. b. Canwick; Prof. Bot. Oxford; B & B.

Sibthorp, J., F.R.S. : 1758-1796. FLS; Prof. Bot. at Oxford; list of plants in *J. Bot.* 1910; DNB.

Simpson, N. D. : fl. 1945. Visited Boston etc. 1945; BSBI.

Sinclair, George. : 1786-1834. *Hortus Gramineus Woburnensis*, 1824; G. Whitworth of Normanby le Wold corresponded with him; DNB; B & B.

Skipworth, Miss Susan : 1824-1912. South Kelsey; married John Lewis ffytche of Thorpe Hall, Louth; herbarium 1835.

Sledge, Dr. W. A. : fl. 1948 . . . Leeds Univ.; visited Lincs.

Smith, Albert Malins : 1880-1964. Stallingborough and Saltaire, Yorks.

Smith, Arthur : 1869-1947. Grimsby and Lincoln; FLS; LNU Secretary 1903-33.

Smith, A. E., O.B.E. and Mrs. M. : fl. 1938... Alford and Willoughby; LNU.

Smith, Dr. H. B. Willoughby : 1879-1948. Gainsborough; LNU; BSBI.

Sneath, J. S. : 1840-1924. Lincoln; LNU.

Stanwell, Miss Mary : fl. 1864-90. Gainsborough; herbarium; list Lees' MS.

Stark, Adam : 1784-1867. *History of Gainsborough* 1847; DNB.

Steele, E. E. : fl. 1936... Fiskerton; LNU.

Stewart, Mrs. M. E. : fl. 1926-30. Partney; LNU; WFS.

Stollery, Mrs. D. : fl. 1957... Lincoln; LNU.

Stones, T. and Mrs. : fl. 1959. Brigg; friends of Mrs. E. Dunn.

Stovin, Mrs. and Miss : fl. 1814. Sheffield; visited Nocton.

Stow, Miss S. C. : c. 1870-1956. Caythorpe and Grantham; LNU; herbarium at Grantham.

Streatfeild, Rev. G. S. : fl. 1873. Curate of Louth and Boston; *Lincs. and the Danes* 1884; B & B.

Strickland, E. : fl. 1822. Appleby.

Stukeley, Rev. Wm. : 1687-1765. b. Holbeach; FRS; DNB.

Sutton, C. V. : fl. 1959. Lincoln School.

Taylor, Miss R. E. : fl. 1946 (d. 1957). Lincoln; LNU; herbarium.

Taylor, S. A. : fl. 1932... Leicester Univ.

Thompson, Rev. J. H. : 1823-1864. Louth area; Cradley 1856-80; records in *Top. Bot.;* B & B.

Thompson, Pishey. : fl. 1820-56. *History of Boston* 1856; plants of Skirbeck Hundred; DNB.

Tryon, Mrs. K. : 1853-1932. Middle Rasen vicarage.

Tuckwell, Rev. W. : 1829-1919. Waltham 1893-1905.

Turner, Dawson and Dillwyn, L. W. : fl. 1805. *The Botanist's Guide.*

Walcott, Rev. M. E. C. : fl. 1861. *Guide to coast of Lincs. and Yorks.;* plant list from P. Thompson.

Walker, Mrs. : fl. 1894. Billinghay vicarage; localities unreliable; specimens in county herbarium.

Walker, C. : fl. 1973... Assistant Regional Officer, Lincs., Nature Conservancy.

Wallace, E. C. : fl. 1942. Skegness; BSBI.

Wallis, T. W. : 1821-1900. Louth woodcarver; herbarium lost.

LINCOLNSHIRE RECORDERS

Walters, Dr. S. M. : fl. 1950 . . . Camb. Botany Sch.; BSBI.
Walworth, G. : fl. 1919. Wash coast survey with L. F. Newman — *J. Ecol.* 1919.
Ward, Dr. John, M.D. : fl. 1820. Horncastle; plant list in Weir *History of Horncastle* 1820.
Waterfall, C. C. : 1857-1938. Visited Lincs. 1895.
Watkins, A. J. : fl. 1920-50. Gainsborough; herbarium at Gainsborough.
Watkinson, Mrs. B. : fl. 1958 . . . Cleethorpes; LNU.
Watson, H. C. : 1804-1881. Visited Lincs. 1851; FLS; DNB; B & B.
Webb, Rev. R. H. : b. 1805-1880. Records in *Top. Bot.;* Herts. Flora; B & B.
Welburn, Mrs. : fl. 1892 . . . Orby vicarage.
Weston, Mrs. Irene : fl. 1961 . . . Riseholme; LNU; BSBI; WFS.
Wherry, Mrs. K. : fl. 1916 . . . Brigg. and Bourne; LNU; WFS.
Whitehead, J. : fl. 1888. Ashton under Lyne; B & B.
Whitelegge, J. : fl. 1878. Ashton under Lyne; specimens in Liverpool Univ. Herb.
Whitworth, G. : fl. 1824 (?1770-?1844). Normanby-le-Wold (see Sinclair, G.).
Wilkinson, Rev. C. : fl. 1889. Toft; record for Sleaford.
Winch, N. J. : 1768-1838. Barton-on-Humber 1803; *Bot. Guide*, 1805; FLS; DNB; B & B.
Wollaston, Rev. H. : 1770-1823. Rector of Scotton.
Wollaston, Rev. T. V. : 1822-1878. Son of above; *Phytologist*, 1843, 522; F.L.S.; DNB.
Woods, E. V. : fl. 1897. Grimsby; LNU.
Woodward, T. J. : 1745-1820. b. Huntingdon; FLS; DNB; B & B.
Woolward, Miss F. : fl. 1890 (d. 1923). Belton rectory; paintings lost; LNU.
Wray, E. Verdun : fl. 1950 . . . Humberstone; FLS; LNU; BSBI.
Wray, Rev. J. F. : 1801-1859. Bardney and Stixwould; herbarium.
Wylie, Misses M. and J. : 1893. Horkstow vicarage; LNU.
Young, Arthur, F.R.S. : 1741-1830. List of plants in East Fen, 1799; DNB.

THE FLORA OF LINCOLNSHIRE

Location of Lincolnshire Herbaria for Vice Counties 53 and 54

Alston, A. H. G. : Herb. Brit. Mus.
Banks, Sir Joseph : Herb. Brit. Mus.
Blackstone, J. : Herb. Sloane.
Bogg, Drs. E. B. and T. W. : Lincoln City and County Mus.
Brookes, Miss S. C. : Grantham Mus.
Buddle, Rev. A. : Herb. Sloane.
Bunch, Rev. R. J. : Leicester University.
Cautley, Miss C. : in hand Miss E. J. Gibbons.
Carrington, Dr. B. : mostly lost.
Chandler, J. H. : in hand.
Cole, Rev. R. E. G. : Lincoln City and County Mus.
Dalton, Rev. J. : York Mus.
Dodsworth, Rev. J. : Lincoln City and County Mus.
Elwes, Miss S. : Elsham Hall.
Fisher, H. : Wollaton Hall, Nottingham; a few Grantham Mus.
Fowler, Rev. W. : Herb. Brit. Mus.
Gibbons, Miss E. J. : in hand.
Grantham, Dr. T. P. J. : lost.
Lees, F. A. : Bradford Mus.
Ley, Rev. A. : Lincoln City and County Mus. (Brambles).
Mason, J. L. : in hand (aliens).
Mason, Rev. W. W. : Oxford.
Mackinder, Miss and Newman, Mrs. M. : few in Lincoln City and County Mus.
Orchard, E. E. : in hand Miss E. J. Gibbons.
Owston, Rev. T. : a few in Lincoln City and County Mus.
Peacock, Rev. E. A. Woodruffe- : Lincoln City and County Mus.
Sandwith, Mrs. C. I. and Sandwith, N. Y. : Herb. Kew.
Salt, J. and Stovin, Mrs. and Miss : Sheffield Mus.
Seaward, Dr. M. R. D. : in hand.
Simpson, N. D. : Lincoln City and County Mus.
Simpson, C. : Lincoln City and County Mus.
Sneath, J. S. : Lincoln City and County Mus.
Stow, Miss S. C. : Lincoln City and County Mus.; at Grantham Mus.
Taylor, Miss R. E. : in hand Miss E. J. Gibbons.
Walker, Mrs. : few Lincoln City and County Mus.
Watkins, A. J. : Gainsborough Old Hall.
Watson, H. C. : Herb. Kew.
Wray, Rev. J. F.: in hand Miss E. J. Gibbons.
Lincs. Nat. Union : Lincoln City and County Mus.
Paintings — The first three are notable collections:
Rev. J. Mossop (Covenham, 1830-70) : Louth Public Libr.
Dr. J. Burgess (Spilsby, 1870-1900) : Lincoln City and County Mus.
Mrs. E. M. Cheales (née Lane-Claypon) (Wyberton, 1886-1900) : in hand Miss E. J. Gibbons.
Miss J. Cooke (Spalding, 1917-20) : Kew.

CHAPTER 7

COUNTY DIVISIONS

The eighteen Divisions arranged by Woodruffe-Peacock (see *Nat.*, 1895, p. 29) go by parish boundaries, in many cases by river beds and sometimes along roads. J. Burtt Davy produced a different geological map of fewer unequal shaped divisions, but this was not used. The geology of Lincolnshire is so masked by boulder clay and cover sands that no regular boundaries would be possible.

It would not have been possible to produce a first Flora on the 10 km grid system, as there are 85 squares in or partly in Lincolnshire, so it was decided to use the eighteen Divisions, with all the old records included. Parish boundaries may not always be ideal, but at any rate a hedge, road or watercourse is easier to follow than an imaginary straight line across country.

The relationship between County Divisions and 10 km grid squares can be seen from Figure 1. What follows here is a brief description of each Division.

Div. 1. This is the Isle of Axholme. A ridge of Keuper marl runs through the centre with several summits over 100 ft. (two are 133 ft.). Round this is a very low lying area adjoining Thorne Waste to the north-east. The R. Trent runs along the east where much of the land has been warped and is very fertile. Very extensive drainage was carried out after much opposition in the 17th century. At the time of Domesday there were woods too large to be described by acreage but which were measured by furlongs and leagues (probably 1½ miles). In 1377 Ralph Bassett was admonished for felling very many great trees called "Lindes" at Melwood so that 20 acres were wasted to the amount of £10. It was also recorded that "to this Island belong marshlands 10 leagues by 3." Two old turbaries still exist on the west and are nature reserves, and there are a few patches of sandy moorland. The disused railway, running north-south, provides steep-sided cuttings, and borrow pits on low ground.

Div. 2. North-west Lindsey is bounded by the Rivers Trent on the west, Humber on the north and Ancholme on the east. This is a very interesting Division botanically with two limestone ridges, much covered by blown sand. Formerly there were extensive rabbit warrens which provided pelts for a fur curing industry at Brigg. The large woods at Broughton had 300 acres of Lilies of the Valley and many of our best woodland plants. Very little limestone pasture exists, but its plants are good and so are those of the few remaining acid moorlands. Much opencast mining of ironstone has taken place and the disused quarries are becoming colonised by a few relic plants. The steep escarpment overlooking the outfall of the R. Trent is quite unlike anything else in Lincolnshire.

Div. 3. This Division is bounded by the R. Humber in the north and north-east and the R. Ancholme on the west, and is chiefly chalk Wolds covered by sand and boulder clay in parts. There are extensive brick clay pits along the estuary and now the industrial development is spreading rapidly. The woodlands are mainly man-made and their flora is chiefly chalk grassland plants in the rides. The blown sand moorlands used to have Bog Asphodel and Marsh Gentian up to the last 50 — 100 years.

Div. 4. This lies south of the last and is triangular in shape with silt coastline, and Wold mostly covered by boulder clay. Industrial building at Grimsby has restricted the possibility of botanical sites on the coast, but alien weeds of the docks have brought some enthusiasts to see them. The Wolds have some chalk grassland but valleys which have cut through into the Lower Cretaceous strata do not have an interesting flora. Not much remains of water plants in dykes and ponds. Freshney Bog was once a splendid habitat for calcareous bog plants.

Div. 5. Bounded by the R. Trent on the west, by Divs. 2 and 6 to north and south and by the R. Ancholme on the east. Sandy warrens at Scotter and Laughton have been planted up by the Forestry Commission since the 1920's but many of the rare plants are still on the nature reserve at Scotton Common. The limestone is chiefly covered by blown sand or is under cultivation. Waddingham Common, destroyed in 1963, was a last relic on the Ancholme peat fen growing Grass of Parnassus, Fragrant Orchid (var. *densiflora*), Butterwort, Black Bogrush (*Schoenus nigricans*) and Few flowered Spike-rush (*Eleocharis quinqueflora*).

Div. 6. Joining on to the last but having less uncultivated land. This Division is north of Lincoln with the R. Trent on the west and the Langworth river (Barlings Eau) on the east. A very little sandy moorland can be found at Torksey, chiefly on the golf course and a remnant of grassland beside the Trent. In Domesday there was much woodland on the rising ground stretching from Saxilby to Gainsborough, but only at the northern end is there any left. Some boulder clay woods on the east adjoin the Wragby woodlands with a similar flora. The little R. Till may have had marshes once but none remain. The limestone is mostly cultivated but the grass verges of Till Bridge Lane have had Purple Milk Vetch up to recent times. Burton Gravel pits have been lately declared a nature reserve. Marshes along the R. Witham had Fen Violet (*Viola stagnina*) and Marsh Orchid (*Dactylorchis praetermissa*) a few years ago.

Div. 7. East of the R. Ancholme and west of the Wolds, this Division runs north-south from the old Caistor canal to the R. Witham, Much woodland, now chiefly under the Forestry Commission, is round Wragby and Market Rasen. Blown sand warrens below the Wolds include Linwood Warren nature reserve. A little acid soil is also found near Panton and Holton Beckering. Marshes by the R. Witham would once have had interesting plants but nothing much now remains apart from railway ballast holes.

Div. 8. Chiefly chalk wolds much covered with glacial drift, with some undulating boulder clay to the east. Red Hill nature reserve includes chalk grassland. Some glacial gravels in the Bain Valley produce a very few acid soil plants.

Div. 9. Entirely coastal, with some very good sand dunes and dune slacks in the National Nature Reserve south of Saltfleet. To the north there are big areas of saltmarsh. Inland dykes had good aquatic plants.

Div. 10. This includes both Wold and fen, with the R. Witham as the south-west boundary. The Wolds are mainly Lower Cretaceous Spilsby Sandstone, but there is some chalk with a good flora. The Woodhall Spa sands and gravels give a fine area of heath and moorland, where there are two nature reserves on Kirkby Moor. The fenland peat is all under cultivation but there are extensive woods on the fen-edge gravels.

Div. 11. This joins Divisions 8 and 10 on the north and west and extends east to the coast. It includes a little of the chalk wolds with two quarries which are nature reserves. There is much boulder clay in the centre with woodlands by the Wold edge. On the coast is the internationally famous Gibraltar Point nature reserve with its extensive and growing system of sand dunes and salt marshes.

Div. 12. Fens with coastal saltings between Wainfleet and Boston. Before 1800 the East Fen was the last unreclaimed fenland, which included pools, decoys and a Cranberry moor. A nature reserve at Friskney Decoy Wood is almost the only natural habitat left. Climbing Fumitory seems to be a great feature of any remaining woodlands. Reclamation of saltmarsh is still proceeding and changing the coastal habitat.

Div. 13. South of Lincoln with the Nottinghamshire boundary to to west. The Lincoln Heath, a limestone plateau, was once all uncultivated calcareous grassland but was enclosed during the 18th century. From Lincoln to Newark is mainly sandy gravel and the only extensive acid soil in central and south Lincs. Some of the broad Delphs in the Witham Fen have a good aquatic flora. Old woodlands at Skellingthorpe, Potterhanworth and Nocton occur on gravel. The former woodlands at Hykeham, Branston and Metheringham mentioned in Domesday have been cleared and cultivated.

Div. 14. The Sleaford district, which includes limestone pasture, fen dykes and the interesting gravel-filled valley east from Wilsford where Rupturewort and the Elongated Thrift are special rarities. The former gravel pits at Rauceby Warren are now a nature reserve. A curious heavy clay rough pasture by Willoughby Gorse was of some interest in the 1940's.

Div. 15. The Grantham district with Divisions 13 and 14 to north and east and Division 16 to the south-east. The Leicestershire boundary is on the west along Sewstern Lane, which has the best kinds of limestone plants. Mostly undulating woodland country rising to 484 ft. South and south-west of Grantham there has been extensive opencast mining for ironstone. The old Grantham canal has good aquatic plants.

Div. 16. The Bourne district where very little remains of limestone grassland but there are many large woods and a good deal of cultivated fens. Baston Fen is now a nature reserve with grassland and borrow pits beside the R. Glen. Crowland and Deeping Fens would originally have been full of aquatic and fenland plants but now there are few relics to show what once grew there in profusion.

Div. 17. Extends from Boston to Spalding bounded by the Rivers Witham and Welland and the south Forty Foot Drain. Cultivation and drainage have swept away most of the original plants and little remains but plants of the saltmarsh, which also grow in a few places some distance inland which were formerly tidal, such as Bicker Haven and Surfleet Lows. The R. Glen at Surfleet had two miles near the estuary where the rare Ribbon-leaved Water-Plantain or Ribbon Weed (*Alisma gramineum*) grew. It is now hard to find after modern dredging of the river. Surfleet Lows and Reedbeds are now nature reserves.

Div. 18. East of the R. Welland, bounded by the Wash coast and the county boundary. Most of the reclaimed saltmarsh for about four miles inland has no indigenous plants besides coastal ones. Cowbit Wash had many aquatic plants but the lowering of the water table and cultivation of the water meadows has left very little of the former fen flora. On the Cambridgeshire border at Tydd Gote there is a gravelly bank with calcareous plants and a good patch of saltmarsh beside the R. Nene. A few scraps of woodland remained until 30 years ago.

CHAPTER 8

THE COUNTY FLORA

EXPLANATION AND KEY

The system and nomenclature is based on the *List of British Vascular Plants* by J. E. Dandy (London, 1958 and 1969). Plants are recorded on the basis of the natural history Divisions of the County as described in the previous chapter. The number in heavy type at the start of each line is the **Division number,** followed by the date of the record, the place and *name of finder*. A first **re**cord for the County is given in each case, but where recorded for more than nine Divisions only those with no records are given. If there are six or less recordings for the County, all are given. Ecological information is only given where there is a deviation from the normal pattern to be expected. Additional information is given for uncommon plants. A dagger (†) indicates the existence of a herbarium specimen (in the County Herbarium unless otherwise stated). A special key for the Roses is given at the beginning of that section. Aliens frequently recorded in the County have been included in the main list, but an appendix is also given which indicates specific collections of aliens from good sites as well as those of a more spasmodic nature. Entries in large square brackets have been included for interest.

PTERIDOPHYTA
LYCOPODIACEAE
1. Lycopodium L.

1. **L. selago** L. *Fir Clubmoss*
    1. 1815, Isle of Axholme, *Peck*.
    Unconfirmed; recorded for Notts.

2. **L. inundatum** L. *Marsh Clubmoss*
    2. 1858, Crosby, *Fowler*. Extinct.
    5. 1893, Scotton, *Peacock*.†
    7. 1878, Linwood Warren, *Lees*.
    10. 1954, Woodhall Spa, *M. E. Gibbons*.
    Native; rare; wet heaths.

4. **L. clavatum** L. *Common Clubmoss*
    1718, *Ray and Lister*. (Lees Outline Flora).
    2. 1875, Crosby, *Fowler*. Extinct.
    5. 1897, Scotton, *Fowler*. c. 1920, *A. J. Watkins*.
    7. 1905, Linwood, *A. Smith*.†
    13. 1967, Skellingthorpe, *B. Tear and I. Weston*.
    Native; rare.

5. **L. alpinum** L. *Alpine Clubmoss*
    2. 1857, Crosby, *Fowler*. (*Mr. E. Coates*).†
    (*J. Bot.*, **21**, 1883, p. 84). sp HBM.
    Native; extinct. A new plantation on this site, before the ironstone quarrying began, finished its survival.

SELAGINELLACEAE
2. Selaginella Beauv.

1. **S. selaginoides** (L.) Link *Lesser Clubmoss*
    2. 1879, Manton, *Fowler*. 1892, Twigmoor, *Peacock and Davy*†
    5. 1893, Scotton, *Peacock*.†
    Native; very rare. Found up to 1948 at Scotton.

EQUISETACEAE
4. Equisetum L.

1. **E. hyemale** L. *Dutch Rush*
    3. 1902, Nettleton, *Allett*.† Extinct.
    7. 1902, Moortown, *Allett*.†
    12. 1829, Wainfleet, *Oldfield*. Extinct.
    Native; very rare.

3. **E. ramosissimum** Desf.
    17. 1947, S. bank of R. Witham below Boston, *H. K. Airy Shaw*.
    Native? First British record. Artificial Channel 1884, Ballast?
    Ref: Watsonia, 1, p. 149, A. H. G. Alston, 1949.

5. **E. fluviatile** L. *Smooth Horsetail*
    12. 1829, Wainfleet, *Oldfield*.
    Recorded for all Divs.
    Native; shallow water and swampy places; frequent.

6. **E. palustre** L. *Marsh Horsetail*
    16. 1840, Bourne, *Dodsworth*.
    Recorded for all Divs.
    Native; frequent.

7. **E. sylvaticum** L. *Wood Horsetail*
    2. Scawby.
    3. 1905, N. Kelsey, *Allett and Peacock*.†
    6. c. 1840, Lea, *T. Owston*.†
    7. 1905, Moortown, *Allett*.† 1956, Legsby, *E. J. Gibbons*.
    12. 1829, Wainfleet, *Oldfield*. Extinct.
    13. 1890, Nocton, *Sneath*.† 1963, *E. J. Gibbons*.
    Native; rare; damp and shady places.

9. **E. arvense** L. *Common Horsetail,*
    Toad Pipes
    16. 1836, Morton, *Dodsworth*.
    Recorded for all Divs.
    Native; common; very persistent on acid soil.

10. **E. telmateia** Ehrh. *Great Horsetail*
    7. c. 1850, Claxby, *S. Skipworth*.
    Not recorded for Divs. 1, 4, 6, 9, 12, 17, 18.
    Native; boggy slopes near springs; locally common.

OSMUNDACEAE
5. Osmunda L.

1. **O. regalis** L. *Royal Fern*
    1. 1939, Haxey, *Allison*. Extinct?
    2. 1876, Santon, *Fowler*. Extinct.
    5. 1865, Laughton, *Charters*.† c. 1930, Blyton, *H. B. W. Smith*.
    6. 1957, Knaith, *L.N.U. Meeting*. Introduced? a very old stump.
    10. c. 1890, Woodhall Spa, *Alston*. Extinct.
    12. 1930, Friskney, *Cheales*. Extinct.
    13. 1807, Stapleford, *T. Ordoyno*. Extinct.
    1840, Bassingham, *Simpson Collection*. Extinct.
    1959, Skellingthorpe, *E. J. Gibbons*. Regenerating.
    Native; rare and practically extinct.

DENNSTAEDTIACEAE
8. Pteridium Scop.

1. **P. aquilinum** (L.) Kuhn *Bracken*
   16. 1836, Bourne, *Dodsworth.*
   Recorded for all Divs.
   Native; occurring on the chalk and limestone where blown sand is present; occasionally on walls.

ADIANTACEAE
11. Adiantum L.

1. **A. capillus-veneris** L. *Maidenhair Fern*
   14. 1952, Rauceby, *M. N. Read and Butcher.*
   ?Spores blown from garden, established on side of quarry.

13. Blechnum L.

1. **B. spicant** (L.) Roth *Hard Fern*
   13. 1807, Stapleford, *Ordoyno.*
   Recorded for Divs. 1, 2, 3, 5, 7, 8, 10, 11, 13.
   Native; occasional; mainly in woods W. of the Wolds on suitable acid soil.

ASPLENIACEAE
14. Phyllitis Hill

1. **P. scolopendrium** (L.) Newm. *Hart's-tongue Fern*
   13. 1855, Doddington, *Cole.*
   Not recorded for Div. 17.
   Native; colonist of walls and bridges; rarely in woods and quarries (4, 14, 15).

15. Asplenium L.

1. **A. adiantum-nigrum** L. *Black Spleenwort*
   10. 1820, Tattershall, *Ward.*
   Not recorded for Div. 13.
   Native; on walls and church stonework.

5. **A. trichomanes** L. *Maidenhair Spleenwort*
   1851, *H. C. Watson.*
   Not recorded for Div. 17.
   Native; on walls, churches, station platforms, etc.

6. **A. viride** Huds. *Green Spleenwort*
    11. 1860, Toynton St. Peter, church wall, *Dodsworth*. Extinct.

7. **A. ruta-muraria** L. *Wall-Rue*
    13. 1830, Doddington, *Allen*.
    Recorded for all Divs.
    Native; frequent on stonework, sides of quarries, station platforms, etc.

16. Ceterach DC.

1. **C. officinarum** DC. *Rusty-back Fern*
    6. 1962, Caenby, naturalised garden wall, *J. Gibbons*.
    10. 1958, Dogdyke Station, *M. N. Read*.
    14. 1889, Sleaford Church, *Wilkinson*. Extinct.
    15. 1927, Boothby Pagnall, *C. S. Carter*.
    16. 1930, Tallington, *Stow*. Extinct.
        1960, Edenham, *J. H. Chandler*.
    Doubtful native; colonist of stonework and brick walls.

ATHYRIACEAE

18. Athyrium Roth

1. **A. filix-femina** (L.) Roth *Lady Fern*
    1851, *H. C. Watson*. No locality.
    Not recorded for Divs. 9, 12, 17, 18.
    Native; frequent on acid soils; occasional in woods on calcareous soils in peat.

19. Cystopteris Bernh.

1. **C. fragilis** (L.) Bernh. *Brittle Bladder-fern*
    10. 1928, Salmonby church, *Mason*. Extinct.
    11. 1961, Spilsby cemetery, *J. Gibbons*.
    Colonist.

ASPIDIACEAE

21. Dryopteris Adans.

1. **D. filix-mas** (L.) Schott *Male Fern*
    1847, Saleby, *Dr. Grantham*.
    Recorded for all Divs.
    Native; frequent on both basic and acid soils.

2. **D. borreri** Newm. [**D. pseudo-mas** (Wollaston) Holdbz Ponzar] *Golden Scaled Male Fern*
   7. 1877, Middle Rasen, *Lees.*
   Recorded for Divs. 1, 2, 3, 4, 5, 7, 8, 10, 11, 13.
   Native; occasional on acid soils.

6. **D. carthusiana** (Villar) H. P. Fuchs *Narrow Buckler-fern*
   2. 1858, Winterton, *Fowler.*
   Not recorded for Divs. 9, 14, 17, 18.
   Native; infrequent, locally common in damp woods and on acid heaths.

7. **D. dilatata** (Hoffm.) A. Gray. *Broad Buckler-fern*
   1851, *H. C. Watson.* No locality.
   Not recorded for Divs. 9, 17, 18.
   Native; very common on peaty or acid soils.

22. **Polystichum** Roth.

1. **P. setiferum** (Forsk.) Woynar *Soft Shield-fern*
   4. 1955, Bradley, *E. J. Gibbons.*
   5. Check List.
   8. 1876, Burwell, *Fowler.* 1967, L.N.U. 1962, Muckton, L.N.U. 1965, Cadeby, *J. Gibbons.*
   11. 1887, Withern, *Allett.* 1890, Claxby (Alford), *Davy.*†
   1968, Hoplands Wood, L.N.U.
   Native; uncommon. On heavy clay E. of Wolds.

2. **P. aculeatum** (L.) Roth *Hard Shield-fern*
   1. Check List.
   3. 1952, Owmby, *E. J. Gibbons.* Extinct.
   4. 1933, Roxton, *E. J. Gibbons.* 1959, Bradley, *B. Woodliff.*
   7. 1953, Hainton, *L.N.U. Meeting.* 1960, S. Kelsey, *J. Gibbons.* 1973, Newball Wood, *G. F. Peterken.*
   8. 1856, Hallington, *Bogg.*† 1962, Muckton, L.N.U.
   11. 1887, Tothill, *Allett.*
   Native; scarce but more frequent than *P. setiferum.*

THELYPTERIDACEAE

24. **Thelypteris** Schmidel

1. **T. oreopteris** (Ehrh.) Slosson [**T. limbosperma** (All.) M. P. Fuchs] *Mountain Fern*
   13. 1851, Doddington, *H. C. Watson.*
   Recorded for Divs. 1, 3, 5, 7, 8, 10, 13.
   Native; uncommon; woods and acid heaths.

2. **T. palustris** Schott *Marsh Fern*
   1. 1974, Crowle, *A. Frankish & E. J. Gibbons*.
   2. 1895, Manton, *L.N.U.*
   3. 1878, Nettleton, *Lees*. Extinct.
   5. 1848 and 1919, Scotton, *Mason*.†
   7. Check List.
   10. Check List.
   12. 1797, East Fen, *A. Young*. Extinct.
   14. 1962, Sleaford, *J. Gibbons*.

   Native. Very rare and perhaps extinct.

4. **T. dryopteris** (L.) Slosson [**Gymnocarpium dryopteri** (L.) Newm.] *Oak Fern*
   1. *c.* 1890, Belshaw, *S. Hudson*.†
   5. 1838, Scotter, *A. Irvine* "Flora of London".
   10. 1957, Tattershall, *M. N. Read*. Extinct 1960.

   Native; very rare.

5. **T. robertiana** (Hoffm.) Slosson [**Gymnocarpium robertianum** (Hoffm.) Newm.] *Limestone Fern*
   3. 1892, Horkstow, *Firbank*.

   Colonist. Seeded into quarry from rockery?

POLYPODIACEAE

25. Polypodium L.

1. **P. vulgare** L. *Polypody*
   1789, *Gentleman's Magazine*.
   Not recorded for Divs. 4, 9, 17, 18.

   Native. Dry woods, churches and walls.

   ssp. *vulgare*. All specimens submitted in VC 54 and Div. 13. Most on vegetative habitats — a few on walls.

   ssp. *interjectum*. 15. 1961, Syston, *E. J. Gibbons*.
   1960, Denton, *M. Lowe*.
   16. 1961, Casewick, *J. H. Chandler*.

   All on stonework. (Det. C. Jermy).

MARSILEACEAE

26. Pilularia L.

1. **P. globulifera** L. *Pillwort*
   2. 1910, Risby, *W. D. Roebuck*.— Extinct.
   1972, Manton, *C. Walker*.

   Native; very rare; extinct. (*L.N.U. Trans.*, **2**, p. 233).

AZOLLACEAE
27. Azolla Lam.
1. **A. filiculoides** Lam.
   - 16. 1961, Billingborough, *D. Stollery*.
   - 17. 1957, Hammond Beck, Donnington, *M. Sayer*.
   - Introduced from N.W. America, colonizing in fen dykes.

OPHIOGLOSSACEAE
28. Botrychium Sw.
1. **B. lunaria** (L.) Sw. *Moonwort*
   - 13. 1746, Lincoln Heath, *Blackstone*.
   - Recorded from Divs. 1, 2, 3, 4, 5, 7, 8, 10, 13.
   - Native. Often overlooked. Old pastures; becoming scarce owing to cultivation.

29. Ophioglossum L.
1. **O. vulgatum** L. *Adder's Tongue*
   - 12. 1799, East Fen, *Young*.
   - Recorded from all Divs.
   - Native; becoming scarce.

SPERMATOPHYTA
GYMNOSPERMAE
PINACEAE
30. Pseudotsuga Carriere
1. **P. menziesii** (Mirb.) Franco *Douglas Fir*
   - Planted on private estates; distribution not known.

31. Picea A. Dietr.
1. **P. abies** (L.) Karst. *Norway Spruce*
   - Often planted, sometimes not flourishing on unsuitable soils.

Tsuga Carr
T. heterophylla (Raf.) Sarg. *Western Hemlock*
- Planted by Forestry Commission; distribution not known.

32. Larix Mill

1. **L. decidua** Mill — *European Larch*
   Extensively planted and regenerating.

2. **L. leptolepis** (Sieb Zucc) Gord. — *Japanese Larch*
   Planted on several private estates.

33. Pinus L.

1. **P. sylvestris** L. — *Scots Pine*
   Some evidence for presence of pines in Lincolnshire from prehistoric times:

   1797, *Phil. Trans.*, **89**, p. 145, Sir Joseph Banks and J. C. De Serra. (In submerged forest, Sutton-on-Sea, with oak and birch).

   1868, *History of the Fens of South Lincolnshire* (1st ed.), Wheeler, p. 16:
   "... near Bardney ... three and a half feet below the surface ... a great number of oak, yew and alder roots ..."
   "In Friskney, Wainfleet and Wrangle, and in the East Fen, great numbers of fir trees with their roots have been discovered in the moory soil" (1 — 6 ft. below the surface).

   1898, Wheeler (2nd ed.), p. 459:
   "In Thurlby Fen ... the timber found has been principally oak, yew and beech, lying 3 to 4 ft. below the surface".

   1893, *Rock Soil Flora*, Peacock:
   "Found in draining at Cadney; bushels of fir-cones and a fair quantity of trees, one over thirty feet long with the top broken off."
   "At Yaddlethorpe, 1898, with thin bark. The bark in earlier peat is of great thickness showing greater cold."
   "Ancient pinewoods, Bottesford, Caistor Moor, Linwood, Tattershall Thorpe."
   "Cadney church roof, put on in 1780, pine beams from Market Rasen."

CUPRESSACEAE

34. Juniperus L.

1. **J. communis** L. — *Juniper*
   No satisfactory record.
   J. sabina (Savin or Saffron) used in horse medicine; recorded Division 5, 1894, Lea Wood, *Miss Anderson*. (Peacock's M.S. flora and Victoria County History).

TAXACEAE
35. Taxus L.

1. **T. baccata** L. *Yew*
   Journ. of Science and Art, **2**, p. 244. 1816, Sir J. Banks, "Where the Oak, Yew and Fir (Pine) are found together, the Yew lies above the Oak, but below the Fir, as a rule." *Naturalist*, 1896, p. 245, Peacock.
   Formerly native. Dug up in carrs and fens and known as Wire-thorn. Not native now, planted in woods.

ANGIOSPERMAE

DICOTYLEDONES

RANUNCULACEAE
36. Caltha L.

1. **C. palustris** L. *Marsh Marigold, Kingcup*
   12. 1799, East Fen, *Young*.
   Recorded for all Divs.
   Native; ditches and wet fields.

   C. radicans. T. F. Forster.
   5. Scotter, *Mrs. C. C. Fowler*.
   Ref. *L.N.U. Trans.*, 1916, p. 14.

38. Helleborus L.

1. **H. foetidus** L. *Stinking Hellebore*
   11. 1840, Burgh, *Grantham*.
   Recorded for Divs. 3, 7, 8, 11, 13, 15, 16.
   Relic of cultivation; established at Blankney.

2. **H. viridis** L. *Green Hellebore*
   7. 1956, Claxby, *E. J. Gibbons*.
   11. 1890, Spilsby, *J. J. Rainey*.†
   16. 1836, Bourne, *Dodsworth*. Native? Extinct.‡
   Recorded for Divs. 1, 5, 7, 8, 10, 11, 13, 15, 16.
   Native and relic of cultivation. Established on the Wold-side, perhaps the remains of a herb garden.

39. Eranthis Salisb.

1. **E. hyemalis** (L.) Salisb. *Winter Aconite*
   5. 1838, Hemswell, *A. Irvine*.
   Recorded for Divs. 2, 3, 4, 5, 6, 8, 10, 13, 14, 15.
   Introduced before 1800; established away from gardens occasionally. Native of S. Europe.

40. Aconitum L.

1. **A. napellus** L. agg. *Monkshood*
   3. 1891, Bigby, *E. Bates.*† (*A. anglicum* Stapf).
   7. 1953, Panton, *E. J. Gibbons*. Established in plantation.
   8. 1890, Fanthorpe, *J. B. Davy*. Garden escape.
   10. 1909, Check List. 1961, Kirkby-on-Bain, *E. J. Gibbons*. Roadside.
   15. 1972, Ermine St., *J. H. Chandler*.
   Garden escape, except for Bigby where it appears to be native on the banks of a stream (150 yards).

41. Delphinium L.

1. **D. ambiguum** L. *Larkspur*
   13. 1862, Doddington, *Cole*.†
   Recorded for Divs. 3, 5, 7, 10, 11, 13, 16.
   Casual, formerly in foreign corn.

43. Anemone L.

1. **A. nemorosa** L. *Wood Anemone, Seam Cup*
   14. 1790, Newton, *Cragg*.
   Not recorded from Divs. 9, 12, 17, 18.
   Native. Locally abundant in old woods.

2. **A. ranunculoides** L. *Yellow Wood Anemone*
   13. 1840, Fulbeck, *Miss Stovin*.
   1853, Wellingore, *planted by Col. Noel*.

3. **A. apennina** L. *Blue Anemone*
   3. 1862, Brocklesby Park, *R. Bowstead*.
   1862, Hundon by Caistor, *J. Britten*.
   13. 1853, Fulbeck, *T. Bevan Rake*. (Planted by Col. Noel from Wellingore Wood).

   The last two species have been recorded as established in domestic woods in Div. 13 and elsewhere, 1840-1900. *Naturalist*, 1894, p. 87.

44. Pulsatilla Mill.

1. **P. vulgaris Mill.** *Pasque Flower*
    - **1.** 1895, Epworth, *Hudson*. Doubtful.
    - **2.** 1840, Broughton, *Elwes*.
    - **13.** 1746, Lincoln Heath, *Blackstone*.
    - **14.** 1855, Brauncewell, *Lowe*. Extinct.
    - **15.** 1795-1805, Colsterworth, *Crabbe*. Extinct.
      1891, Gt. Ponton, *H. Fisher*.
    - **16.** 1745, Stamford, *Stukeley*. Extinct.

    Many records exist for Divs. 13, 14, 15.

    Native. Formerly widespread on limestone grassland; now rare owing to cultivation.

45. Clematis L.

1. **C. vitalba L.** *Traveller's Joy*
    - **6.** 1836, Greetwell, *Simpson Collection*. 1963, E. J. Gibbons.

    Not recorded for Divs. 5, 18.

    Native in the south of the county and at Lincoln; perhaps native in the north in the Louth area; occasional as a colonist. On buckthorn bushes on the coast.

46. Ranunculus L.

1. **R. acris L.** *Meadow Buttercup*
    - **14.** 1790, Threckingham, *Cragg*.

    Recorded from all Divs.

    Native.

2. **R. repens L.** *Creeping Buttercup*
    - **16.** 1836, Carlby, *Dodsworth*.

    Recorded from all Divs.

    Native.

3. **R. bulbosus L.** *Bulbous Buttercup*
    - **16.** 1836, Bourne, *Dodsworth*.

    Recorded from all Divs.

    Native.

    R. acris, *R. repens* and *R. bulbosus* are disappearing under the plough and extensive spraying.

5. **R. arvensis L.** *Corn Crowfoot*
    - **11.** 1840, Burgh, *Dr. Grantham*.

    Recorded for all Divs.

    Native; not common; on heavy clay soils generally. Decreasing.

THE COUNTY FLORA

7. **R. sardous** Crantz   *Hairy Buttercup*
    6. 1851, Fiskerton, *H. C. Watson*.
    Not recorded from Divs. 10, 13, 14, 15, 16, 18.

Native. Uncommon, especially in the South; possibly overlooked. Patchy distribution on heavy soils chiefly near the coast.

9. **R. parviflorus** L.   *Small-flowered Buttercup*
    12. 1829, Wainfleet, *Oldfield*.
    Not recorded from Divs. 1, 2, 5, 6, 9, 10, 13.

Native; on chalky banks and casual elsewhere. Occasional.

10. **R. auricomus** L.   *Goldilocks*
    14. 1790, Threckingham, *Cragg*.
    Not recorded from Divs. 9, 12, 17, 18.

Native. Not very common; woods and roadsides; both apetalous and perfect forms occur.

11. **R. lingua** L.   *Great Spearwort*
    10. 1820, Kirkstead, *Ward*.
    Not recorded from Divs. 4, 12, 17, 18; probably once growing there before drainage.

Native. Uncommon; disappearing rapidly; planted by ornamental lakes.

12. **R. flammula** L.   *Lesser Spearwort*
    1. 1815, Isle of Axholme, *Peck*.
    Not recorded from Divs. 12, 17.

Native. Widespread on damp, peaty ground.

15. **R. sceleratus** L.   *Celery-leaved Crowfoot*
    7. 1829, Bardney, *J. F. Wray*.
    Recorded for all Divs.

Native; frequent in shallow ponds and ditches.

16. **R. hederaceus** L.   *Ivy-leaved Water Crowfoot*
    1851, *H. C. Watson*.
    Not recorded in Divs. 1, 4, 5, 9, 14, 16, 17, 18.

Native. Uncommon; few records in recent years.

19. **R. fluitans** Lam.   *Water Crowfoot*
    5. 1890, Morton, *Peacock*.†
    10. 1877, Tattershall, *W. Fowler*.
    12. 1885, Boston, *L. Gibbs*.
    15. 1920, Grantham, *S. C. Stow*. 1955, Claypole, *Howitt*.

Native; scarce; fast-moving water.

20. **R. circinatus** Sibth.
    4. 1851, Grimsby, *H. C. Watson.*
    Not recorded from Div. 7.
    Native. Fairly common; streams and ditches.

21. **R. trichophyllus** Chaix
    11. 1840, Burgh, *Dr. Grantham.*
    Not recorded for Div. 5.
    Native; widespread; ponds and slow-moving water generally.

22. **R. aquatilis** L. *Water Crowfoot*
    12. 1851, Boston, *H. C. Watson.*
    Not recorded for Divs. 9, 11, 17, 18.
    Native; frequent; ponds and rivers. (ssp. under recorded).

23. **R. baudotii** Godr.
    11. 1840, Ingoldmells, *Dr. Grantham.*
    Recorded for Divs. 2, 4, 9, 11, 12, 17.
    Native. Locally frequent; coastal.

24. **R. ficaria** L. *Lesser Celandine*
    14. 1790, Threckingham, *Cragg.*
    Recorded for all Divs.
    Native. Locally abundant; damp shady places. (ssp. under recorded).

47. Adonis L.

1. **A. annua** L. *Pheasant's Eye*
    2. 1874, Bottesford, *Peacock.*†
    6. 1948, Hackthorn, *M. Ruddock.*
    16. 1956, Aunby, *J. H. Chandler.*
    Not recorded for Divs. 1, 4, 5, 7, 9, 11, 17.
    Casual.

48. Myosurus L.

1. **M. minimus** L. *Mouse-tail*
    1. Check List.
    2. c. 1900, Scawby gull ponds (*Peacock's MS.*).
    12. 1856, Boston, *Thompson.*
    13. 1920, Harmston, *H. B. W. Smith.*
    15. 1901, Brandon, *S. C. Stow.*†
    17. 1894, Wyberton, *E. M. Lane-Claypon.*†
    Probably native. Rare; no recent record; weed of cultivation?

49. Aquilegia L.
1. **A. vulgaris** L. *Columbine, Granny Bonnets*
   - 2. 1842, Broughton, *Miller.* 1966, *J. Gibbons.*
   - 4. 1902, Irby, *C. B. Parker.†* 1954, *J. Gibbons.*
   - 1920 — 1940, Croxby, *W. J. Clayton.*
   - 14. 1893, Ancaster, *Stow.†*
   - 15. 1959, Stainby, *M. Lowe.*
   - 16. 1836, Bourne, *Dodsworth.*

   Also recorded from Divs. 7, 8, 9, 10, 13.

   Undoubtedly native in Divs. 2, 4, 14 and 15; garden escapes elsewhere. Rare; calcareous woodland.

50. Thalictrum L.
1. **T. flavum** L. *Common Meadow Rue*
   - 14. 1790, Threckingham, *Cragg.*
   - Recorded for all Divs.

   Native; widespread but becoming scarce.

3. **T. minus** L. *Lesser Meadow Rue*
   - *a* ssp. *minus.*
     - 1. 1897, Epworth, *S. Hudson.†*
     - 8. 1891, Tathwell, *J. B. Davy (BM). (E. Larder).*

   Doubtfully native.

   - *b* ssp. *arenarium* (Butcher) Clapham.
     - 4. 1892, Cleethorpes, *Lees.†* 1960, *J. Gibbons.*
     - 5. 1875, Saltfleetby, *W. Fowler.* 1965, *J. Gibbons.*
     - 11. 1882, Croft, *Dr. Burgess.*

   Native. Uncommon; sand dunes.

BERBERIDACEAE
53. Berberis L.
1. **B. vulgaris** L. *Barberry*
   - 14 1790, Threckingham, *Cragg.*
   - Not recorded for Divs. 9, 12, 17.

   Native? Occasional; hedges.

54. Mahonia Nutt.
1. **M. aquifolium** (Pursh) Nutt. *Oregon Grape*
   - 2 1874, Bottesford, *Peacock.*
   - Not recorded for Divs. 1, 9, 12, 17.

   Introduced; becoming established in many localities; pheasant cover.

NYMPHAEACEAE
55. Nymphaea L.
1. **N. alba** L. *White Water-lily*
    14. 1830, Haverholme, *Allen & Saunders*.
    Not recorded for Divs. 1, 4, 8, 9, 12, 17.
    Definitely native in Divs. 2, 5, 10, 13 and 14; probably introduced elsewhere.

56. Nuphar Sm.
1. **N. lutea** (L.) Sm. *Yellow Water-lily, Brandy-bottle*
    13. 1830, Skellingthorpe, *Allen & Saunders*.
    Not recorded for Div. 12.
    Native. Locally common; ponds, lakes and rivers.

CERATOPHYLLACEAE
57. Ceratophyllum L.
1. **C. demersum** L. *Horn-wort*
    13 1815, Bracebridge, *Miss Stovin*.
    Not recorded for Divs. 1, 4.
    Native. Frequent; ponds, ditches and dykes.

2. **C. submersum** L.
    2. Check List.
    7. Check List.
    8. 1906, Donington-on-Bain, *L.N.U.*
    9. Check List.
    10. 1913, Tattershall, *S. J. Hurst*.†
    11. 1888, Skegness, *Dr. Burgess*. (Ref: *Naturalist*, 1893, p. 332).
    Native. Rare; possibly confused with *C. demersum*.

PAPAVERACEAE
58. Papaver L.
1. **P. rhoeas** L. *Field Poppy*
    14. 1790, Threckingham, *Cragg*.
    Recorded for all Divs.
    Native or introduced. Weed of cultivation; common, but greatly reduced by spraying.

THE COUNTY FLORA

2. **P. dubium** L. *Long-head Poppy*
    16. 1836, Stamford, *Dodsworth*.
    Recorded for all Divs.
    Native or introduced; occasional although widespread.

3. **P. lecoqii** Lamotte *Babington's Poppy*
    7. 1877, Tealby, *Lees*.
    Recorded for Divs. 1, 3, 7, 8, 10, 11, 12, 15, 16.
    Native or introduced; occasional on chalk, etc.; possibly overlooked.

4. **P. hybridum** L. *Round Prickly-headed Poppy*
    2. 1889, Sawcliffe, *W. Fowler*.
    Recorded for Divs. 1, 2, 3, 4, 7, 10, 13, 15, 17.
    Casual or native? Uncommon; weed of cultivation.

5. **P. argemone** L. *Long Prickly-headed Poppy*
    16. 1836, Carlby, *Dodsworth*.
    Not recorded for Div. 17.
    Native or introduced. Occasional; weed of cultivation.

6. **P. somniferum** L. *Opium Poppy*
    12. 1856, Boston, *Thompson*.
    Not recorded for Divs. 13, 17, 18.
    Introduced. Few recent records; cultivated in the Isle of Axholme and Whitton areas in the early 1800's.

61. **Glaucium** Mill.

1. **G. flavum** Crantz *Yellow Horned-poppy*
    4. 1885, Cleethorpes, *E. M. Browne*.†
    9. 1884, Mablethorpe, *E. M. Browne*.
    11. 1877, Skegness, *Lees* (Still there?). 1955, *A. E. Smith*.
    Impermanent native; rare; colonizer from the Norfolk coast?

62. **Chelidonium** L.

1. **C. majus** L. *Greater Celandine*
    14. 1790, Threckingham, *Cragg*.
    Recorded for all Divs.
    Locally frequent; established near habitations, on walls and banks.

FUMARIACEAE

65. Corydalis Medic.

3. **C. claviculata** (L.) DC. *White Climbing Fumitory*
   - **7.** 1877, Tealby, *Lees*. 1966, Linwood, *J. Gibbons*.
     1965, Holton Beckering, *J. Gibbons*.
   - **10.** 1876, Tattershall, *W. Fowler*; 1966, *J. Gibbons*.
     1880, Keal, *Burgess*.
   - **12.** 1881, Wainfleet, *W. Fowler*; 1965, *J. Gibbons*.
   - **13.** 1909, Nocton, *W. W. Mason*. 1961, Bloxholme, *J. Gibbons*.
   - **14.** 1894, Billinghay, *Walker*.†
   - **15.** 1885, Harrowby, *Fisher*; 1904, *S. C. Stow*.†
   - **17.** 1898, Wyberton, *E. M. Lane-Claypon*.

   Native; uncommon; peat-loving.

4. **C. lutea** (L.) DC. *Yellow Fumitory*
   - **2.** 1864, Bottesford, *Peacock*.

   Not recorded for Divs. 1, 4, 9, 11.

   Established near buildings.

66. Fumaria L.

2. **F. capreolata** L.
4. **F. bastardii** Bor.
6. **F. muralis** Sond. ex Koch.
9. **F. vaillantii** Lois.

   Single records exist for these four species, but are unconfirmed.

8. **F. officinalis** L. *Common Fumitory*
   - **14.** 1790, Threckingham *Cragg*.

   Recorded for all Divs.

   Native. Common; light and calcareous soils.

CRUCIFERAE

67. Brassica L.

2. **B. napus** L. *Rape, Cole*
   - **2.** 1822, Appleby, *Strickland*.

   Not recorded for Divs. 6, 7, 8, 9, 11, 14, 17, 18.

   Introduced, uncommon; Wash area — used in land reclamation?

3. **B. rapa** L. *Turnip*
   - **8.** 1856, Yarburgh, *Bogg*.

   Not recorded for Divs. 4, 12, 14, 15, 17.

   Introduced.

4. **B. nigra** (L.) Koch *Black Mustard*
    1851, *Watson*.
    Not recorded for Divs. 1, 6, 13.
    Probably native; scattered; chiefly in fen districts.

70. Sinapis L.
1. **S. arvensis** L. *Charlock, Wild Mustard*
    16. 1836, Bourne, *Dodsworth*.
    Recorded for all Divs.
    Probably native; widespread; absent from acid sand.
2. **S. alba** L. *White Mustard*
    12. 1856, Boston, *Thompson*.
    Not recorded for Divs. 1, 5, 8.
    Introduced; relic of fen reclamation.

72. Diplotaxis DC.
1. **D. muralis** (L.) DC. *Sand Rocket*
    3. 1835, nr. Barton-on-Humber, *E. J. Nicholson*.
    Not recorded for Divs. 1, 9, 14, 18.
    Introduced. Common near railway lines, occasional elsewhere.
2. **D. tenuifolia** (L.) DC. *Perennial Wall Rocket*
    14. 1893, Billinghay, *Walker*.†
    Recorded for Divs. 3, 4, 5, 10, 13, 14, 15, 16.
    Native? Scarce; waste places; established at Barton-on-Humber.

74. Raphanus L.
1. **R. raphanistrum** L. *Wild Radish, White Charlock*
    13. 1851, Boultham, *H. C. Watson*.
    Not recorded for Divs. 17, 18.
    Introduced? Weed of cultivation; locally abundant on poor soil.
2. **R. maritimus** Sm. *Sea Radish*
    9. 1909, Tetney Haven, *G. A. Grierson*.
    Casual.
3. **R. sativus** L. *Radish*
    3. 1903, Cadney, *Peacock*.†
    4. 1963, Cleethorpes, *E. J. Gibbons*.
    7. 1953, Great Sturton, *E. J. Gibbons*.
    Casual. Scattered; occuring on rubbish dumps.

75. Crambe L.

1. **C. maritima** L. *Seakale*
    1805, Botanists Guide, *Banks*.
    "Among the sand hills on the coast in abundance."
    Extinct — Sir Joseph Banks may have encouraged local enterprise to market it or graze sheep on it.

77. Cakile Mill.

1. **C. maritima** Scop. *Sea Rocket*
    9 or 11. 1834, *Bayley*. (No locality).
    4. Grimsby.
    5. 1914, Gainsborough, *L. Gilliam*. (*L.N.U. Trans.*, 1914, p. 176).
    9. 1855, Saltfleet-Mablethorpe, *Bogg*.†
    11. 1867, Skegness-Huttoft, *Mason*.
    Native; coastal sandhills. Abundant after 1953 floods.

78. Conringia Adans.

1. **C. orientalis** (L.) Dumort. *Hare's-ear Cabbage*
    13. 1896, Lincoln, *Sneath*.
    Recorded for Divs. 3, 4, 7, 8, 11, 12, 13, 15.
    Casual. Occasional

79. Lepidium L.

1. **L. sativum** L. *Garden Cress*
    14. 1895, Sleaford, *Sneath*.
    Recorded for Divs. 3, 4, 12, 13, 14, 16.
    Casual.

2. **L. campestre** (L.) R. Br. *Pepperwort*
    2. 1875, Winterton, *Fowler*.
    Not recorded for Divs. 5, 8.
    Native; widespread, but not common. Drybanks and stubble fields.

3. **L. heterophyllum** Benth. *Smith's Cress*
    16. 1840, Witham on the Hill, *Dodsworth*.
    Recorded for Divs. 2, 7, 9, 10, 12, 13, 15, 16.
    Native; uncommon; no recent records.

4. **L. ruderale** L. *Narrow-leaved Pepperwort*
    5. 1895, Kirton Lindsey, *Peacock*.
    Recorded for Divs. 3, 4, 5, 10, 11, 12, 15.
    Casual; dumps and waste places.

80. Coronopus Zinn

1. **C. squamatus** (Forsk.) Aschers. *Swine-cress, Wart-cress*
   16. 1836, Bourne, *Dodsworth.*
   Recorded for all Divs.
   Native. Frequent on hard soil in stackyards and gateways.

2. **C. didymus** (L.) Sm. *Lesser Swine-cress*
   4. 1866, Scartho, *W. H. Daubney.*
   Recorded for Divs. 3, 4, 9, 12, 13.
   Introduced. Docks and waste places. Scarce.

81. Cardaria Desv.

1. **C. draba** (L.) Desv. *Hoary Cress, Hoary Pepperwort*
   10. 1893, Horncastle, *Fowler.*
   Recorded for all Divs.
   Introduced; scattered; established but not spreading rapidly.

2. **C. chalepensis** (L.) Hand.-Mazz.
   3. 1953, New Holland, *E. J. Gibbons.*
   4. 1956, Grimsby Docks, *E. J. Gibbons.*
   7. 1930, Holton le Moor, *E. J. Gibbons.*
   Introduced. Persistent casual of railway ballast at Holton and New Holland. Flowering two weeks later than *C. draba.*

82. Isatis L.

1. **I. tinctoria** L. *Woad*
   Recorded for Div. 4, 12, 14, 17. (Check List).
   Cultivated from prehistoric times until 1931 (See *Lincs. Mag.*, Vol. II, p. 73, 1935). Chiefly around Boston and Great Coates (*B.E.C. Report*, 1919, p. 641). 'Woad Lanes' at Great Coates, Skirbeck, Algarkirk and Long Sutton.

83. Iberis L.

1. **I. amara** L. *Wild Candytuft*
   12. 1908, Boston Docks, *Hurst.* Introduced.
   14. 1920, Wilsford/Ancaster, *K. Brown.* 1927, *G. Bacon* (*Foggitt*) B.E.C. Rep. 1927. 184.
   Native, very rare. No recent record. To be looked for.

84. Thlaspi L.

1. **T. arvense** L. *Field Penny-cress, Muzzlejimp*
   12. 1856, Boston, *Thompson.* (As T. perfoliatum L.) Nat. 1894, p. 135.
   Recorded for all Divs.
   Introduced?

85. Teesdalia R. Br.

1. **T. nudicaulis** (L.) R. Br. *Shepherd's Cress*
   3. 1856, Caistor, *J. Daubney*.
   Recorded for Divs. 1, 2, 3, 5, 6, 7, 10, 13.
   Native. Occasional on acid sand in N. Lincs.; rare in S. Lincs.

86. Capsella Medic.

1. **C. bursa-pastoris** (L.) Medic. *Shepherd's Purse*
   14. 1790, Threckingham, *Cragg*.
   Recorded for all Divs.
   Native; widespread in waysides and waste places.

2. **C. rubella** Reut.
   11. 1945, Huttoft, *N. D. Simpson*.
   Casual; possibly overlooked.

88. Cochlearia L.

1. **C. officinalis** L. *Scurvy-grass*
   12. 1957, Boston District, *Gerarde*.
   Recorded for Divs. 1, 3, 4, 9, 11, 12, 17, 18.
   Native; locally common; salt marshes and mud sea-shore. Records may have included *C. anglica*.

5. **C. danica** L. *Danish Scurvy-grass*
   7. 1961, Holton le Moor, *E. J. Gibbons*. Casual on railway.
   11. 1908, Skegness, *B. Reynolds*. Native.
   13. 1961, Hykeham, *Gibbons*.
   1965, Branston, *N. Read*. Casual on railway.
   Native on coast. Rare.

6. **C. anglica** L. *Long-leaved Scurvy-grass*
   12. 1806, Boston (Salt Herbarium).
   Recorded for Divs. 4, 9, 11, 12, 17, 18.
   Native. Coastal mud; locally common; commoner than *C. officinalis* L. around the Wash.

90. Bunias L.

2. **B. orientalis** L.
   5. 1893, Kirton Lindsey, *Peacock*. Flour mills.
   11. 1956, Welton-le-Marsh, *Read*. Quarry.
   15. 1957, Woolsthorpe Colsterworth, *J. H. Chandler*. Quarry.
   Casual.

91. Alyssum L.

1. **A. alyssoides** (L.) L. *Small Alison*
    14. 1855, Brauncewell, *G. Lowe*.
    Not recorded for Divs. 1, 5, 6, 12, 16, 17, 18.
    Introduced. Cornfields; no recent records.

92. Lobularia Desv.

1. **L. maritima** (L.) Desv. *Sweet Alison*
    9. 1958, Humberston, *E. J. Gibbons*.
    13. 1912, Lincoln, *Sneath*.
    On rubbish dumps; hardly established.

95. Erophila DC.

1. **E. verna** (L.) Chevall *Whitlow Grass*
    14. 1790, Threckingham, *Cragg*.
    Recorded for all Divs.
    Native. Frequent in dry places; walls and paths.

3. **E. praecox** (Stev.) DC.
    2. 1904, Hibaldstow, *Peacock*.
    4. 1958, Cleethorpes, *E. J. Gibbons*.
    6. 1950, Fiskerton, *E. J. Gibbons*.
    7. 1877, Walesby, *Lees*. 1953, Holton le Moor, *E. J. Gibbons*.
    9. 1958, Mablethorpe, *E. J. Gibbons*.
    11. 1956, Gibraltar Point, *E. J. Gibbons*.
    15. 1900, Castle Bytham, *Fisher*.
    Native. Occasional in dry places, particularly on the coast. Rosettes noticeably smaller than in *E. verna*.

96. Armoracia Gilib.

1. **A. rusticana** Gaertn., Mcy & Scherb. *Horse-Radish*
    16. 1936, Bourne, *Dodsworth*.
    Recorded for all Divs.
    Introduced. Waysides; roadside dumping of soils has increased spread in recent years.

97. Cardamine L.

1. **C. pratensis** L. *Cuckoo Flower, Lady's Smock*
    12. 1820, Boston, *Thompson*.
    Recorded for all Divs.
    Native. Frequent; damp meadows.

2. **C. amara** L. *Large Bitter-cress*
    1. 1948, Crowle, *E. J. Gibbons*. (var. erubescens. Peterm.) det. *A. J. Wilmott*.
    4. 1910, Freshney Bog, *A. Roebuck*.
    6. 1896, Kettlethorpe, *Burton*. 1965, Saxilby, *B. Howitt*.
    7. 1953, Hainton, *E. J. Gibbons*. Naturalised?
    9. 1968, Saltfleetby, *L.N.U.*
    11. 1968, Partney, *G. Phillips*.
    16. 1831, Market Deeping, *R. J. Bunch*. 1838, Stamford, *Browning*. 1960, *J. H. Chandler*.
    Native; Rare; River banks and alder swamps.

[3. **C. impatiens** L. *Narrow-leaved Bitter-cress*
    Mistaken identification; 1885 record for Mablethorpe is *C. hirsuta* L.]

4. **C. flexuosa** With. *Wavy Bitter-cress*
    15. 1879, Grantham, *Browne*.
    Not recorded for Divs. 1, 12, 17, 18.
    Native. Occasional; damp woodrides and bogs.

5. **C. hirsuta** L. *Hairy Bitter-cress*
    8. 1820, Hemingby, *Ward*.
    Recorded for all Divs.
    Native. Common; gardens and waste places.

8. **C. bulbifera** (L.) Crantz. *Coral-wort*
    13. 1962, Wellingore, *E. J. Gibbons*.
    Introduced.

98. Barbarea R. Br.

1. **B. vulgaris** R. Br. *Winter Cress, Yellow Rocket*
    14. 1797, Threckingham, *Cragg*.
    Recorded for all Divs.
    Native; common on ditch banks.

2. **B. stricta** Andrz. *Small-flowered Yellow Rocket*
    2. 1877, Brigg, *Lees*.
    Recorded for Divs. 1, 3, 5, 6, 7, 10, 13, 15, 16.
    Native? Scattered; Canal and river banks. Brought by barges.

3. **B. intermedia** Bor. *Intermediate Yellow Rocket*
    3. Check List.
    7. 1903, S. Kelsey, *Peacock*.† 1967, Horsington, *J. Gibbons*.
    8. 1896, Louth, *Lees*.†
    10. 1958, Tumby, *E. J. Gibbons*. 1961, Langrick, *E. J. Gibbons*.
    11. 1963, Claxby, *E. J. Gibbons*.
    17. 1896, Wyberton, *E. M. Lane Claypon*.
    Introduced? Rare; cultivated fields.

4. **B. verna** (Mill.) Aschers. *Early-flowering Yellow Rocket*
Land Cress
    5. 1893, Kirton Lindsey, *Peacock*.†
    7. 1890, Hatton, *Jarvis*.
    8. 1904, Donington-on-Bain, *Mason*.
    14. 1904, Dorrington, *Mason*.
    Casual; no recent records. Formerly cultivated.

100. Arabis L.

1. **A. turrita** L. *Tower Rock-cress, Tower-cress*
    13. 1896, S. Lincoln, *Sneath*.†
    Grain alien.

4. **A. hirsuta** (L.) Scop. *Hairy Rock-cress*
    13. 1829, Canwick Pits, *Bunch*.
    Not recorded for Divs. 1, 7, 9, 10, 17.
    Native. Infrequent; chalk and limestone.

101. Turritis L.

1. **T. glabra** L. *Tower Mustard*
    10. 1897, Sibsey, *Howard*. (Unconfirmed).

102. Rorippa Scop.

1. **R. nasturtium-aquaticum** (L.) Hayek *Watercress*
    14. 1790, Threckingham, *Cragg*.
    Recorded for all Divs.
    Native. Common in ditches, streams, ponds; particularly calcareous waters.

2. **R. microphylla** (Boenn.) Hyland. *One-rowed Watercress*
    5. 1947, Waddingham, *J. Gibbons*.
    6. 1961, Welton, *J. Gibbons*.
    10. 1945, Reedham, *N. D. Simpson* and *A. H. Alston*.
    11. 1945, Huttoft, *N. D. Simpson* and *A. H. Alston*.
    14. 1947, Aswarby, *Butcher* and *J. Gibbons*.
    15. 1958, Corby, *Gibbons*. 1959, Gunby, *Gibbons*.
    16. 1958, Thurlby, *Gibbons*.
    Native. Distribution unknown; in similar places to *R. nasturtium-aquaticum*.

3. **R. sylvestris** (L.) Bess *Creeping Yellow-cress*
    5. 1840, Gainsborough, *Miller*.
    Not recorded for Divs. 3, 4, 8, 17.
    Native. Occasional; mainly along the Rivers Trent and Witham. Becoming rare after spraying.

4. **R. islandica** (Oeder) Borbas *Marsh Yellow-cress*
   1. 1840, Wroot, *Miller*.
   Not recorded for Divs. 8, 9, 11, 15.
   Native. Frequent.

5. **R. amphibia** (L.) Bess. *Great Yellow-cress*
   12. 1799, East Fen, *Young*.
   Not recorded for Divs. 3, 4, 8, 15, 17.
   Native. Uncommon; ponds and low marshy places by rivers.

105. Erysimum L.

1. **E. cheiranthoides** L. *Treacle Mustard*
   6 or 13. 1849, Lincoln, *Carrington*.
   Not recorded for Divs. 4, 8, 18.
   Introduced? Scattered; in peaty, arable fields.

106. Cheiranthus L.

1. **C. cheiri** L. *Wallflower*
   3. 1835, Thornton Abbey, *E. J. Nicholson*.
   Not recorded for Divs. 1, 4, 9, 13, 17.
   Introduced. Infrequent; well established on old quarries and stonework.

107. Alliaria Scop.

1. **A. petiolata** (Bieb.) Cavara & Grande *Garlic Mustard, Jack-by-the-Hedge*
   10. 1820, Poolham, *Ward*.
   Recorded for all Divs.
   Native. Frequent in hedgerows and shady places, but absent on acid soils.

108. Sisymbrium L.

1. **S. officinale** (L.) Scop. *Hedge Mustard*
   14. 1790, Threckingham *Cragg*.
   Recorded for all Divs.
   Native. Frequent hedgerows, roadsides, waste places; but rare in some districts.

2. **S. irio** L. *London Rocket*
    4. 1900, Grimsby, *Arthur Smith*.†
    6. 1908, Langworth Station, *Sneath*.
    Recorded for Divs. 4, 5, 6, 8, 13.
    Introduced; casual.

4. **S. orientale** L. *Eastern Rocket*
    6. 1944, Lincoln, *J. Gibbons*.
    8. 1927, Louth, *D. Marsden*.
    9. 1968, Mablethorpe, *J. Gibbons*.
    10. 1936, Horsington, *D. Marsden*.
    11. 1935, Willoughby, *D. Marsden*.
    12. 1912, Boston, *B. Reynolds*.
    15. 1963, Corby, *J. Gibbons*.
    16. 1958, Uffington, Stamford, *J. Gibbons*.
    Introduced. Weed of railways and dumps; on the increase.

5. **S. altissimum** L. *Tall Rocket*
    5. 1895, Kirton Lindsey, *Peacock*.†
    Recorded for Divs. 2, 4, 5, 6, 8, 9, 10, 11, 13, 16.
    Introduced. Waste places.

109. Arabidopsis (DC.) Heynh.

1. **A. thaliana** (L.) Heynh. *Common Wall Cress, Thale Cress*
    10. 1820, Horncastle, *Ward*.
    Not recorded for Divs. 4, 9, 12, 17, 18.
    Native. Infrequent; sandy banks and gardens.

110. Camelina Crantz.

1. **C. sativa** (L.) Crantz *Gold of Pleasure*
    10. 1785, Spalding, *Banks*.
    Recorded for Divs. 2, 4, 5, 7, 9, 13, 16, 18.
    Introduced. Usually a casual of dumps.

2. **C. microcarpa** Andrz. ex DC.
    7. 1944, N. Owersby, *J. Gibbons*.
    8. 1918, Calcethorpe, *Mason*.
    10. 1917, Scrivelsby, *F. S. Alston*.
    13. 1900, Lincoln, *Mason*.†
    Introduced. Rare; growing amongst Flax.

111. Descurainia Webb & Berth.

1. **D. sophia** (L.) Webb ex Prantl *Flixweed*
    16. 1831, Crowland Abbey, *R. J. Bunch*.
    Not recorded for Divs. 8, 11, 18.
    Native? Uncommon; dry, sandy places.

RESEDACEAE
112. Reseda L.

1. **R. luteola** L. *Weld, Dyer's Rocket*
    14. 1790, Threckingham, *Cragg*.
    Not recorded for Divs. 17, 18.
    Native. Usually on calcareous soils, occasionally as a casual elsewhere.

2. **R. lutea** L. *Wild Mignonette*
    15. 1726, Grantham, *V. Bacon*.
    Not recorded for Div. 18.
    Native. Usually on chalk but found in calcareous fen at Waddingham and elsewhere.

3. **R. alba** L. *White Mignonette*
    8. 1923, Hallington, *D. Marsden*.
    11. 1922, Halton Holgate, *Mason*.
    Introduced. Casual of waste places.

VIOLACEAE
113. Viola L.

1. **V. odorata** L. *Sweet Violet*
    10. 1820, Horncastle, *Ward*.
    Recorded for all Divs.
    Native. Absent on blown sand, frequent on chalk and clay; white-flowered form occurring more on the clay.

2. **V. hirta** L. *Hairy Violet*
    10. 1820, Tetford, *Ward*.
    Not recorded for Divs. 1, 9, 12, 17, 18.
    Native. Calcareous soils and clay. Uncommon. Where *V. hirta* and *V. odorata* grow together, hybrids often occur.
    ssp. *calcarea* (Bab.) E. F. Warb.
    15. 1949, Wyville, *F. Rose*. 1963, Holywell, *J. H. Chandler*.

4. **V. riviniana** Reichb. *Common Dog Violet*
    10. 1820, Horncastle, *Ward*.
    Not recorded for Divs. 12, 18.
    Native. Common; woods and hedgebanks.

5. **V. reichenbachiana** Jord. ex Bor. *Wood Dog Violet*
    7. 1877, Claxby, *Lees*.
    Not recorded for Divs. 9, 12, 18.
    Native. Not very common; woods on basic soils.

6. **V. canina** L. *Heath Dog Violet*
    2. 1895, Broughton, *Peacock*.
    Not recorded for Divs. 6, 17.
    Records for Divs. 4, 8, 9, 12, 14, 15, 18 (not vouched for) may be the aggregate.
    Native. Acid soils.

    ssp. *montana* (L.) Hartm.
    2. 1890, Manton, *R. Hutchinson*. 1952, *J. Gibbons*
    6. 1954, Torksey, *E. J. Gibbons*.
    7. 1950, Linwood Warren, *E. J. Gibbons*.
    13. 1893, Skellingthorpe Ferry, *R. J. Owston*.
    Native. Very rare plant of wet heaths.

7. **V. lactea** Sm. *Pale Heath Violet*
    1839, mistake (*Naturalist*, 1894, p. 136).

8. **V. stagnina** Kit. [**V. persicifolia** Schoeb.] *Fen Violet*
    6. 1936, Fiskerton, *Steele*.
    10. 1926, Nr. Woodhall Spa, *Stewart* (Rep. B.E.C., 1926-7).
    13. 1833, Boultham, *Cautley*.† 1836, *Dr. John Nicholson*, M.D. Simpson collection. (First published British record. *Annals of Nat. Hist.* 1839, p. 383, *Sir W. J. Hooker*.)
    13. 1852-3, Potterhanworth, *J. Lowe*. 1864, Branston Fen, *Burtt*.†
    Native. Rare; damp peat; perhaps extinct.
    Naturalist, 1894, p. 136 and 1897, p. 136.

9. **V. palustris** L. *Marsh Violet*
    12. 1829, Wainfleet, *Oldfield*.
    Recorded for Divs. 1, 2, 5, 6, 7, 10, 12, 13.
    Native. Uncommon, except in Divs. 2 and 5 where it is locally frequent.

12. **V. tricolor** L. *Wild Pansy*
    ssp. *tricolor*.
    15. 1929, West Willoughby, *S. C. Stow* conf. *Drabble*.
    16. 1934, Carlby, *G. C. Druce*.
    Native. Probably elsewhere.

    ssp. *saxatilis* (Schmidt), *E. F. Ward*.
    6. 1950, Lea, *R. W. Butcher*.
    8. 1953, Farforth, *J. Hope-Simpson*.
    10. 1948, Tetford, *E. J. Gibbons* det. *R. Meikle*.
    Native. Rare; dry calcareous banks.

13. **V. arvensis** Murr. *Field Pansy*
    1851, *H. C. Watson*.
    Recorded for all Divs.
    Native. Light soils; common on cultivated and waste ground.

POLYGALACEAE
114. Polygala L.

1. **P. vulgaris** L. *Common Milkwort*
   - 14. 1790, Threckingham, *Cragg*.
   - Recorded for all Divs.
   - Native. Old pasture land; dune slacks at Mablethorpe, 1958; decreasing; possibly confused with *P. serpyllifolia*.

2. **P. serpyllifolia** Hose *Heath Milkwort*
   - 2. 1877, Santon, *Fowler*.
   - 3. 1894, Howsham, *C. Skipworth*.†
   - 7. 1895, Linwood, *Lees*.†
   - 10. 1949, Woodhall Spa, *E. J. Gibbons*.
   - 13. 1856, Doddington, *Cole*.†
   - Native. Acid heaths, very local.

3. **P. calcarea** F. W. Schultz
   - 15. 1949, Hungerton, *F. Rose*.
   - Native. Very rare; records for Div. 2, Broughton and Div. 8, Louth districts (W. W. Mason) not confirmed.

GUTTIFERAE
115. Hypericum L.

1. **H. androsaemum** L. *Tutsan*
   - 10. 1820, Tattershall, *Ward*.
   - 11. 1830-70, Well Vale, *Mossop* (painting).
   - 14. 1952, Rauceby, *L.N.U. meeting* (*Trans.*, 1952, 38).
   - Needs confirmation as a native; no herbarium specimen.

2. **H. elatum** Ait. [**H. inodorum** Mill.]
   - 10. 1884, Woodhall Spa, *Mackinder*.
   - Introduced; destroyed.

4. **H. calycinum** L. *Rose of Sharon, Aaron's Beard*
   - 3. 1892, S. Ferriby, *Firbank*. Chalk-pit.†
   - 11. No date, Langton, *Mason's M.S. 1926*.
   - 13. 1865, Leadenham, *Burtt*.
   - 15. 1960, Grantham Canal, *J. Gibbons*.
   - Introduced. Escape from cultivation?

5. **H. perforatum** L. *Common St. John's Wort*
   - 14. 1790, Threckingham, *Cragg*.
   - Not recorded for Div. 18.
   - Native. On basic soil; scattered distribution; introduced on road sides with stone.

THE COUNTY FLORA

6. **H. maculatum** Crantz *Imperforate St. John's Wort*
   7. 1878, Middle Rasen, *Lees*.
   Recorded for Divs. 2, 3, 4, 6, 7, 11, 13, 15, 16.
   Native. Rare. Acid soil; damp places by dykes but not in bogs.

8. **H. tetrapterum** Fr. *Square-stemmed St. John's Wort*
   12. 1799, East Fen, *Young*.
   Recorded for all Divs.
   Native. Wet places; frequent.

9. **H. humifusum** L. *Trailing St. John's Wort*
   5. 1840, Morton, *Miller*.
   Not recorded for Divs. 9, 12, 17, 18.
   Native. Uncommon; rides, open woods and pastures.

11. **H. pulchrum** L. *Slender St. John's Wort*
    10. 1820, Tattershall, *Ward*.
    Not recorded for Divs. 1, 9, 17, 18.
    Native. Uncommon; warm banks and dry woods.

12. **H. hirsutum** L. *Hairy St. John's Wort*
    16. 1836, Bourne, *Dodsworth*.
    Not recorded for Divs. 9, 12, 17, 18.
    Native. Frequent in wooded districts on basic soils.

13. **H. montanum** L. *Mountain St. John's Wort*
    2. 1878, Broughton, *Fowler*.
    4. 1970, Hatcliffe, *E. J. Gibbons*.
    Native. Very rare; in more than one part of Broughton wood but sparingly.

14. **H. elodes** L. *Marsh St. John's Wort*
    2. 1875, Sawcliff, *Fowler* (B.M.). 1959, Twigmoor, *S. Monk*.
    1964, Manton, *E. J. Gibbons*.
    5. *c.* 1840, Laughton, *Owston*. 1896, Scotton, *Mason*.
    7. 1877, Linwood, *Lees*. Extinct.
    1908-35, Holton le Moor, *M. E. Gibbons*. Killed by drought.
    10. 1820, Hemingby Lane, *Ward*. Extinct.
    Native. Very rare; dying out; perhaps only survives in two localities of Div. 2.

CISTACEAE

118. Helianthemum Mill.

1. **H. chamaecistus** Mill. *Common Rock-rose*
   15. 1726, Grantham, *Bacon.*
   Not recorded for Divs. 1, 9, 12, 17, 18.
   Native. Not very common in the North, more frequent in the South.

FRANKENIACEAE

121. Frankenia L.

1. **F. laevis** L. *Sea Heath*
   11. 1973, Gibraltar Point, *K. R. Payne.*
   18. 1763, Tydd Gowt, *Martyn.* 1852, Bray, sp. *Lees* MS.
   Native.

CARYOPHYLLACEAE

123. Silene L.

1. **S. vulgaris** (Moench) Garcke *Bladder Campion*
   2. 1822, Appleby, *Strickland.*
   Recorded for all Divs.
   Native. Occasional, not common, on basic soil.

2. **S. maritima** With. *Sea Campion*
   4. 1836, Cleethorpes (Simpson collection) and 1873, *Lees* (B.M.). Extinct due to building development.
   9. 1861, Theddlethorpe, *Wallis.* 1929, Saltfleetby, *D. Marsden.* 1963, Humberstone, *G. Newton.*
   11. 1910, Ingoldmells, *B. Reynolds.* (*J. of Bot.*, 57, 1910)†. 1891, Skegness, *Jarvis Rainey.* 1949, Gibraltar Point, *A. E. Smith.*
   Native. As shingle is scarce along the coast, this is a rare plant. Two habitats in Div. 11 have been on the inland side of the dunes.

3. **S. conica** L. *Striated Catchfly*
   4. 1902, Grimsby Docks, *A. Smith.*†
   13. 1897, Lincoln, *Grierson.*† Flour mill. (*Naturalist*, 1897, p. 226).
   Introduced. Casual of waste places.

5. **S. dichotoma** Ehrh. *Forked Catchfly*
   1. 1895, Epworth, *S. Hudson.*†
   5. 1853, Gainsborough, *J. Lowe. Science Gossip*, I, p. 258.
   8. 1900, Louth, *Mason.*
   Introduced. Native of E. and S.E. Europe.

6. **S. gallica** L. *Small-flowered Catchfly*
   13. 1955, Boultham, *E. J. Gibbons.*
   var. anglica L.
   10. 1785, Moorby, *Banks.* H.B.M.
   Recorded for Divs. 3, 4, 5, 7, 10, 13, 14, 15.
   Native. Occasional, in stubble on light soils.

   var. quinquevulnera (L.). *Mert.* and *Koch.*
   2. 1951, Brumby, *E. J. Gibbons.*
   10. 1842, Woodhall Spa, *Walter.*† *Nat.*, 1893, p. 311.

10. **S. nutans** L. *Nottingham Catchfly*
    4. 1891, Grimsby, *A. Smith.*
    10. 1915, Stickney, *Stewart* and *Hamond.*
    Introduced. Grain alien.

11. **S. italica** (L.) Pers. *Italian Catchfly*
    2. 1905, Broughton, *Peacock.*†   1920, *Havelock.* Pheasant food.
    4. 1905, Grimsby, *Smith and Peacock.*†
    Introduced. Grain alien.

12. **S. noctiflora** L. *Night-flowering Catchfly*
    16. 1836, Bowthorpe, *Dodsworth.*
    Not recorded for Divs. 9, 17.
    Native. Frequent, in arable fields and stubble, especially on lighter soils.

13. **S. dioica** (L.) Clairv. *Red Campion, Red Candlestick*
    16. 1836, Bourne, *Dodsworth.*
    Recorded for all Divs.
    Native. Locally abundant, but noticeably absent from some districts; woods and hedgebanks.

14. **S. alba** (Mill.) E. H. L. Krause *White Campion, White Candlestick*
    2. 1822, Appleby, *Strickland.*
    Recorded for all Divs.
    Native. Frequent on both cultivated and waste ground.

124. Lychnis L.

3. **L. flos-cuculi** L. *Ragged Robin*
   12. 1799, East Fen, *Young.*
   Recorded for all Divs.
   Native. Decreasing.

125. Agrostemma L.

1. **A. githago** L.   *Corn Cockle*
    16. 1837, Bourne, *Dodsworth*.
    Recorded for all Divs. except 18.
    Introduced. Seldom seen in recent years; native of the Mediterranean region.

127. Dianthus L.

1. **D. armeria** L.   *Deptford Pink*
    10. 1894, Coningsby, *Sinclair*. 1965, Woodhall Spa, *Z. Porter*.
    Native. Very rare.

8. **D. deltoides** L.   *Maiden Pink*
    7. 1910, Stixwould, *Hawley*.†
    10. 1918, Woodhall Spa, *F. S. Alston*.†
    13. 1763, Lincoln Heath, *Martyn*. 1959, N. Hykeham, *W. M. Peet*.
    Native. Rare; dry grassy localities.

128. Vaccaria Medic.

1. **V. pyramidata** Medic.   *Cowherb*
    3. 1893, Brigg, *Peacock*.†
    Recorded for Divs. 2, 3, 4, 5, 6, 8, 12, 13, 14, 15.
    Introduced; in foreign seed, and cultivated by florists.

129. Saponaria L.

1. **S. officinalis** L.   *Soapwort*
    7. 1820, Baumber, *Ward*.
    Not recorded for Divs. 13, 14, 17, 18.
    Introduced. Established on railway banks and roadsides.

131. Cerastium L.

2. **C. arvense** L.   *Field Mouse-ear Chickweed*
    15. 1726, Manthorpe, *V. Bacon*.
    Recorded for all Divs.
    Native. Common in dry places, blown sand and red chalk.

[4. **C. alpinum** L.   *Alpine Mouse-ear Chickweed*]
    Record of 1726, Manthorpe, *V. Bacon*, is a mistake for *C. arvense* L.
    (Quoted in Gough's Camden, 1789).

7. **C. holosteoides** Fr.   *Common Mouse-ear Chickweed*
    14. 1790, Threckingham, *Cragg*.
    Recorded for all Divs.
    Native. Very common throughout the county.

8. **C. glomeratum** Thuill.   *Sticky Mouse-ear Chickweed*
    16. 1837, Bourne, *Dodsworth*.
    Recorded for all Divs.
    Native. Frequent; waste dry places.

10. **C. atrovirens** Bab. [**C. diffusum** Pers.]
                                 Dark-green Mouse-ear Chickweed
    4. 1865, Cleethorpes, *Britten*.
    Recorded for Divs. 4, 7, 9, 10, 11, 12, 13, 14.
    Native. Uncommon; sandy and stony places; railway lines.

12. **C. semidecandrum** L.   *Little Mouse-ear Chickweed*
    1851, *H. C. Watson*. (No locality).
    Not recorded for Divs. 12, 17.
    Native. Frequent; open dry places, sandy banks, etc.

132. Myosoton Moench.

1. **M. aquaticum** (L.) Moench.   *Water Chickweed*
    16. 1836, Bourne, *Dodsworth*.
    Not recorded for Divs. 4, 8, 9.
    Native. Uncommon; local plant of peaty fens.

133. Stellaria L.

1. **S. nemorum** L.   *Wood Stitchwort*
    2. 1856, Broughton, *Fowler*. B.M.
    7. 1877-9, Tealby, *Lees*.
    10. Check List. (*Banks?*).
    13. 1890, Fulbeck, *Miss Venables*. (Doubtful and unconfirmed).
    Native. Damp woods; no recent records; north-western species.

2. **S. media** (L.) Vill.   *Chickweed*
    14. 1790, Threckingham, *Cragg*.
    Recorded for all Divs.
    Native. Abundant; weed of cultivated ground.

3. **S. pallida** (Dumort.) Pire   *Lesser Chickweed*
    2. 1877, Frodingham, *Fowler*.
    Not recorded for Divs. 1, 6, 8, 10, 14, 17.
    Native. Distribution underworked.

4. **S. neglecta** Weihe   *Greater Chickweed*
    2, 3, 7, 8. Check List.
    10. 1785, Revesby, *Banks*. 1902, Woodhall Spa, *S. C. Stow*.
    15. 1960, Claypole, *E. J. Gibbons*.
    Native. Uncommon; hedgerows, shady places, etc.

5. **S. holostea** L.   *Greater Stitchwort*
    16. 1790, Sempringham, *Cragg*.
    Not recorded for Divs. 12, 17, 18.
    Native. Locally common but absent from many likely places; woods and hedgerows.

6. **S. palustris** Retz.   *Marsh Stitchwort*
    2. 1822, Appleby Millfield, *Strickland*.
    Not recorded for Divs. 2, 4, 7, 8, 12, 15, 17.
    Native. Scarce and local; base-rich peat.

7. **S. graminea** L.   *Lesser Stitchwort*
    16. 1836, Thurlby, *Dodsworth*.
    Recorded for all Divs.
    Native. Frequent; becoming less common through cultivation; grassy places on acid soils.

8. **S. alsine** Grimm   *Bog Stitchwort*
    5. 1840, Morton, *J. K. Miller*.
    Not recorded for Divs. 9, 12, 14, 17, 18.
    Native. Frequent; wet places on acid soils.

134. Holosteum L.

1. **H. umbellatum** L.   *Jagged Chickweed*
    "I have notes of it as a 'seed' field species on sands twice but have no specimens". *Peacock*, Rock Soil Flora, Cambridge MS. (*c.* 1920).
    Introduced.

135. Moenchia Ehrh.

1. **M. erecta** (L.) Gaertn., Mey & Scherb.   *Upright Chickweed*
    Not recorded for Lincolnshire, but there is a record for Belvoir on the Leicestershire/Lincolnshire boundary, and at Langham, Rutland, 1956, *B. M. Howitt*.
    Possibly overlooked. Damp gravelly places in pastures.

136. Sagina L.

1. **S. apetala** Ard. *Annual Pearlwort*
   12. 1851, Boston, *H. C. Watson.*
   Not recorded for Divs. 9, 18.
   Native. Occasional; distribution underworked.

2. **S. ciliata** Fr. *Ciliate Pearlwort*
   3. 1877, Wrawby, *Lees.*†
   Not recorded for 6, 11, 12, 13, 14, 15, 17, 18.
   Native. Occasional; dry grassland and bare patches; often overlooked.

3. **S. maritima** Don. *Sea Pearlwort*
   4. 1865, Cleethorpes, *Britten.*
   Recorded for Divs. 3, 4, 9, 11, 12, 17.
   Native. Occasional around the coast.

4. **S. procumbens** L. *Procumbent Pearlwort*
   16. 1836, Bourne, *Dodsworth.*
   Recorded for all Divs.
   Native. Common; pathways, woods, lawns, etc.

10. **S. nodosa** (L.) Fenzl. *Knotted Pearlwort*
    1695, *Ray.*
    Not recorded for Div. 18.
    Native. Damp sandy places; rare or occasional.

137. Minuartia L.

1. **M. verna** (L.) Hiern. *Vernal Sandwort*
   3. 1892, Barton-on-Humber, *Firbank.* (*Naturalist,* 1908, 288). Water-carried from Yorkshire valley?
   11. 1847, Ingoldmells, *Grantham* and 1910-13, *Nash.*
   These records are puzzling.

4. **M. hybrida** (Vill.) Schischk. *Fine-leaved Sandwort*
   14. 1855, Leasingham, *Bloxam.* B.M.
   Recorded for Divs. 3, 4, 8, 10, 13, 14, 15, 16.
   Native on dry limestone; also railway line casual.

139. Honkenya Ehrh.

1. **H. peploides** (L.) Ehrh. *Sea Sandwort*
   12. 1826, Freiston, *Howitt.*
   Recorded for Divs. 3, 4, 5, 9, 11, 12, 17, 18.
   Native. Fairly frequent on sand along coast.

140. Moehringia L.
1. **M. trinervia** (L.) Clairv. *Three-nerved Sandwort*
   16. 1836, Bourne, *Dodsworth.*
   Not recorded for Divs. 17, 18.
   Native. Frequent; woods and hedge-bottoms.

141. Arenaria L.
1. **A. serpyllifolia** L. *Thyme-leaved Sandwort*
   16. 1836, Bourne, *Dodsworth.*
   Recorded for all Divs.
   Native. Frequent; on walls, bare and arable lands.

2. **A. leptoclados** (Reichb.) Guss. *Lesser Thyme-leaved Sandwort*
   2. 1876, Broughton, *Fowler.*
   Not recorded for Divs. 1, 4, 6, 8, 9, 11, 17.
   Native. Overlooked.

142. Spergula L.
1. **S. arvensis** L. *Corn Spurrey, Pickpurse, Dother*
   16. 1836, Bourne, *Dodsworth.*
   Not recorded for Div. 18.
   Native. Abundant; troublesome weed in sandy fields.

143. Spergularia (Pers.) J. & C. Presl.
1. **S. rubra** (L.) J. & C. Presl. *Sand-spurrey*
   16. 1836, Bourne, *Dodsworth.*
   Not recorded for Divs. 3, 4, 12, 17.
   Native. Uncommon; very dry and open sand.

4. **S. media** (L.) C. Presl.
   12. 1826, Freiston, *Howitt.*
   Recorded for Divs. 2, 3, 4, 9, 11, 12, 17, 18.
   Native. Frequent along the entire coastline.

5. **S. marina** (L.) Griseb.
   12. 1826, Freiston, *Howitt.*
   Recorded for Divs. 2, 3, 4, 9, 11, 12, 17, 18.
   Native. Similar to *S. media*, but growing further away from the sea.

144. Polycarpon L.

1. **P. tetraphyllum** (L.) L. *Four-leaved All-seed*
    4. 1899, Grimsby, *C. Parker.*†
    Introduced. Grain alien.

ILLECEBRACEAE
146. Herniaria L.

1. **H. glabra** L. *Glabrous Rupture-wort*
    4. 1930, Grimsby Docks, *S. A. Cox.* (Casual).
    14. 1805, Quarrington, *Crabbe.* 1836, Wilsford, *Dr. Latham* (Sp. Cambridge). 1895, Rauceby, *Stow.*† 1970, *J. Gibbons.*
    15. 1900, Ancaster, *Goulding.*† 1894, West Willoughby, *Stow.* 1969, *J. H. Chandler.*
    Native. Very rare; restricted distribution in Divs. 14 and 15.

3. **H. hirsuta** L. *Hairy Rupture-wort*
    4. 1902, Grimsby, *A. Smith.*†
    9. 1893, Mablethorpe, *Mackinder.*
    Introduced. Grain alien.

148. Scleranthus L.

1. **S. annuus** L. *sensu lato* *Annual Knawel*
    6 or 13. 1848, Brayford, Lincoln, *Forster.*
    Not recorded for Divs. 9, 12, 15, 18.
    Native. Dry sandy places.

PORTULACACEAE
149. Montia L.

1. **M. fontana** L. *Blinks*
    13. 1829, Canwick Common, *R. J. Bunch.*
    Recorded for Divs. 2, 3, 5, 6, 7, 8, 10, 13, 15.
    Native. Occasional; moist, acid soil and dry ant hills according to Peacock's *Rock Soil Flora*, Cambridge M.S.

2. **M. perfoliata** (Willd.) Howell *Perfoliate Claytonia*
    13. 1886, Skellingthorpe, *Newman.*
    Not recorded for Divs. 1, 8, 15, 17.
    Introduced. Sand-dunes; weed of gardens and waste places; native of N. America. Naturalised and locally abundant.

3. **M. sibirica** (L.) Howell
    3. 1927, Limber, *R. May.*
    7. *c.* 1894, Toft, *H. C. Brewster.*
    Recorded for Divs. 2, 4, 5, 9, 12, 14, 18.
    Introduced, native of N. America; garden weed.

AMARANTHACEAE
153. Amaranthus L.

1. **A. retroflexus** L.
    7. 1954, Tealby, *H. Proctor.*
    9. 1956, Humberstone, *J. Gibbons and D. McClintock.*
    10. 1919, Woodhall Spa, *F. S. Alston.* L.N.U., 55, 1919.
    Introduced, native of N. America; possibly overlooked.

CHENOPODIACEAE
154. Chenopodium L.

1. **C. bonus-henricus** L.   *Lincolnshire Spinach, Mercury*
    14. 1790, Threckingham *Cragg.*
    Recorded for all Divs.
    Introduced. Relic of cultivation near buildings.

2. **C. polyspermum** L.   *Many-seeded Goosefoot*
    3. 1829, Wootton, *E. J. Nicholson.*
    Not recorded for Divs. 1, 2, 4, 9.
    Native. Waste places, gardens, arable fields.

3. **C. vulvaria** L.   *Stinking Goosefoot*
    14. 1790, Threckingham, *Cragg.* 1872, Watson's Supplement, *Bogg.* (*Naturalist*, 1896, 183).
    4. 1902, Grimsby Docks, *A. Smith.*
    6. and **13.** 1898, Fossdyke Bank Shipyard, *Sneath and Peacock.*
    Introduced. Casual of waste places.

4. **C. album** L.   *Fat Hen*
    16. 1836, Bourne, *Dodsworth.*
    Recorded for all Divs.
    Native. Very common; waste and cultivated lands.

8. **C. hircinum** Schrad.
    9. 1960, Humberstone, *J. A. Lowe;* det *Kew.*
    Alien.

THE COUNTY FLORA

9. **C. ficifolium** Sm. *Fig-leaved Goosefoot*
   9. 1830-70, Tetney, *Mossop*. (Paintings).
   Not recorded for Divs. 1, 3, 4, 5, 10.
   Sandy or peaty soils. Overlooked until recently.

10. **C. pratericola** Rydb.
    10. 1919, Woodhall Spa, *F. S. Alston*. See *L.N.U. Trans*. 55, 1919.
    Alien.

11. **C. murale** L. *Nettle-leaved Goosefoot*
    12. 1885, Boston, *Gibbs*.†
    Recorded for Divs. 4, 5, 6, 9, 10, 12, 15, 18.
    Introduced? Casual of waste places, ruins, etc.

12. **C. urbicum** L. *Upright Goosefoot*
    16. 1839, Bourne, *Dodsworth*.
    Recorded for Divs. 3, 9, 10, 11, 12, 15, 16, 18.
    Introduced. Casual of waste places.

13. **C. hybridum** L. *Sowbane*
    7. 1890, Great Sturton, *A. Jarvis*.
    Recorded for Divs. 6, 7, 10, 12, 15, 16.
    Casual.

14. **C. rubrum** L. *Red Goosefoot*
    16. 1836, Bourne, *Dodsworth*.
    Recorded for all Divs.
    Native. Dry margins of ponds; waste places.

15. **C. botryodes** Sm.
    11. 1861, Ingoldmells, *Dodsworth*.
    17. 1959, Fosdyke, *Fitter*.
    Native. Maritime.

16. **C. glaucum** L. *Glaucous Goosefoot*
    1. 1943, Wroot, *John Brown* (B.E.C. Report, 1944).

17. **C. capitatum** (L.) Aschers. *Strawberry Blite*
    3. 1942, Limber, *R. May*.
    10. 1917, Scrivelsby, *F. S. Alston*.†
    15. 1965, Syston, *L.N.U.*
    Introduced. Rare.

155. Beta L.

1. **B. vulgaris** L. *Beet*
   12. or 17. 1666, Near Boston, *C. Merrett.* ("Pinax Rerum Naturalium Britannicarum", 1667, Londini.).
   Recorded for Divs. 3, 4, 9, 11, 17, 18.
   Usually as an impermanent casual. Rare in N. Lincs. (V.C. 54).

156. Atriplex L.

1. **A. littoralis** L. *Shore Orache*
   12 or 17. 1728, *Stukeley*.
   Recorded for Divs. 3, 4, 9, 11, 12, 17, 18.
   Native; locally abundant.

2. **A. patula** L. *Common Orache*
   16. 1836, Bourne, *Dodsworth*.
   Recorded for all Divs.
   Native; common on cultivated land.

3. **A. hastata** L. *Hastate Orache*
   6 or 13. 1851, Lincoln, *H. C. Watson*.
   Not recorded for Divs. 5, 8.
   Native.

4. **A. glabriuscula** Edmondst. *Babington's Orache*
   4. 1851, Grimsby, *H. C. Watson*.
   6. 1907, Newton Cliff, *L.N.U. Meeting*.
   Native. Occasional; seashore near the high tide mark; prefers mud rather than sand.

5. **A. laciniata** L. *Frosted Orache*
   12. 1826, Freiston, *Howitt*.
   Recorded for Divs. 3, 4, 9, 11, 12.
   Native; shows a preference for sand.

157. Halimione Aellen

1. **H. portulacoides** (L.) Aell *Sea Purslane*
   12. 1826, Freiston, *Howitt*.
   Recorded for Divs. 3, 4, 9, 11, 12, 17, 18.
   Native. Locally abundant, becoming rare in the Humber area; mud, salt-marshes flooded at high tide.

2. **H. pedunculata** (L.) Aell. *Boston Purslane*
    11. 1886, Croft, *Burgess*. (Paintings).
    12. 1691, Skirbeck, *Plukenet*. "Almagestum Botanicum".
        1696, London (*L. Plukenet, M.D.*). 1727, *Blair*.
    17. 1805, Fosdyke, *Turner and Dilwyn*.
    18. 1805, Cross Keys Wash, *Turner and Dilwyn*.

Native. Very rare; probably extinct. 13 specimens in B.M. up to 1871.

158. Suaeda Forsk. ex. Scop.

1. **S. maritima** (L.) Dum. *Herbaceous Seablite*
    12 or 17. 1840, Boston, *Dodsworth*.
    Recorded for Divs. 3, 4, 9, 11, 12, 17, 18.

Native. Common; seashore and salt-marsh.

2. **S. fruticosa** Forsk. [**S. bera** J. S. Gmel.] *Shrubby Seablite*
    11. 1928, Gibraltar Point, *H. B. Willoughby Smith*. Increasing.
    12. 1836, Boston, *Dodsworth*.
    17. 1951, South of Boston, *Butcher*. L.N.U. meeting.
    18. 1911, Dawsmere, *H. A. Healey;* destroyed by fire in 1919.

Native. Rare; colonizer, perhaps from the Norfolk coast.

159. Salsola L.

1. **S. kali** L. *Saltwort*
    1666, *Lister;* L.N.U. Trans., 1927, p. 5: "In abundance within the sea banks."
    4. 1891, Cleethorpes.
    9. 1892, Mablethorpe, *Mackinder*.
    11. 1890, Skegness, *Burgess*. (Paintings).
    12. 1686, Boston, *Ray*.
    Not recorded for Divs. 17, 18.

Native. Locally frequent; on sand along coastal tide-line.

160. Salicornia L.

The agg. is recorded by *Lister*, 1666, and others later. Divs. 2, 3, 4, 9, 11, 12, 17, 18.

1. **S. perennis** Mill *Glasswort, Marsh Samphire*
    12. 1919, N. of Boston, *Newman and Walworth. J. Ecol.*, vol. 7, 208.
        1960, Freiston, *E. Seppings*.
    17. 1947, Kirton Marsh, *M. Haslam*.

Native. Rare; *Puccinellia maritima* zone of salt-marshes.

2. **S. dolichostachya** Moss
   11. 1948, Gibraltar Point, *E. J. Gibbons.*
   Recorded for Divs. 3, 9, 11, 12, 17, 18.
   Native. Not usually above mean high water mark. Locally abundant.

   S. fragilis, P. W. Ball and Tutin.
   9. 1957 and 1964, Horseshoe Pt., *E. J. Gibbons.* (Det. P. W. Ball).
   11. 1956, Gibraltar Pt., *P. W. Ball.*
   17. 1956, S. of Boston, *P. W. Ball.*
   Native.

3. **S. europaea** L.
   11. 1951, Gibraltar Pt., *E. J. Gibbons.*
   Recorded for Divs. 2, 3, 4, 9, 11, 12, 17, 18.
   Native. Common.

4. **S. ramosissima** Woods
   4. 1949, Cleethorpes, *E. J. Gibbons.* (Det. Wilmott).
   Recorded for Divs. 3, 4, 9, 12, 17, 18.
   Native. Common on higher saltmarsh, variable.

5. **S. pusilla** Woods
   9. 1957, N. Somercotes Haven, *E. J. Gibbons.* (Erect and bushy).
   11. 1951, Gibraltar Pt., *E. J. Gibbons.* 1956, *P. W. Ball* (Prostrate).
   Native. Rather scarce; usually at or above mean high water mark, sometimes even above extreme high water springs.

TILIACEAE

162. Tilia L.

2. **T. cordata** Mill *Small-leaved Lime*
   7. 1670, Wragby, *Ray.* See Critical Catalogue, 1894, *Nat.*, p. 214.
   Not recorded for Divs. 8, 9, 12, 17, 18.
   Native. In many large woods; hedgerow trees, occasionally composing part of the hedge. "1377. Many great trees called Lindes felled in Melwood" Div. 1. (See Fig. 8, p. 35).

   1. T. platyphyllos Scop. Frequently planted.

   2a. T.x. vulgaris Hayne. Frequently planted.

MALVACEAE
163. Malva L.

1. **M. moschata** L. *Musk Mallow*
   6. 1829, Saxilby, *Rev. J. F. Wray*.
   Not recorded for Divs. 17, 18.
   Native. Occasional in dry places; often with white flowers.

2. **M. sylvestris** L. *Common Mallow*
   14. 1790, Threckingham, *Cragg*.
   Recorded for all Divs.
   Native. Common on waste places, roadsides, etc.

3. **M. nicaeansis** All.
   13. 1896, Lincoln, *Lees*.

4. **M. neglecta** Wallr. *Dwarf Mallow*
   16. 1836, Bourne, *Dodsworth*.
   Recorded for all Divs.
   Native. Less common than *M. sylvestris;* present near houses.

5. **M. pusilla** Sm.

6. **M. parviflora** L.
   3. 1898, Brigg, *Claye*.

7. **M. verticillata** L.
   The latter form have occurred as casuals at Humberston and elsewhere.

164. Lavatera L.

1. **L. arborea** L. *Tree Mallow*
   11. 1943, Chapel St. Leonards, *McClintock*.
   16. 1962, Uffington, *J. H. Chandler*. (Established in gravel pit).
   Not native. Grows on shingle.

165. Althaea L.

1. **A. officinalis** L. *Marsh Mallow*
   2. *Check List*.
   4. 1924, Cleethorpes. *Miss Linley*. (*R. Martyn and E. B. Miller*). Most northern record — extinct.
   7 and 10. *Check List*. Garden escape?
   11. Before 1950, Gibraltar Point, *E. Rudkin*.
   1962, Ingoldmells, *E. J. Gibbons*. (Extinct).
   12. 1666, Wainfleet, *Lister*.
   1937, Frieston, *B.E.C.*
   17. 1960, Kirton, *E. J. Gibbons*.
   18. 1958, Whaplode, *E. J. Gibbons and F. Perring*. Extinct, 1970.
   Native. Many records for Divs. 12, 17, 18.

2. **A. hirsuta** L. *Hispid Mallow*
   - **2.** 1935, Broughton Wood, *L.N.U. Meeting* (pheasant food).
   - **9.** 1957, Humberston, *E. V. Wray*. (Sea bank).
   - **10.** 1905, Coningsby, *F. S. Alston*.

   Casual. Rare and local.

LINACEAE
166. Linum L.

1. **L. bienne** Mill *Pale Flax*
   - **9.** 1961, Tetney, *J. A. Lowe*. (First record).
   - **16.** 1961, Stamford, *J. H. Chandler*. (Grass seed).
   - **18.** 1961, Wingland, *E. J. Gibbons*. (Modern sea bank).

   Doubtfully native here, on man-made banks.

2. **L. usitatissimum** L. *Cultivated Flax*
   - **1.** 1661, Axholme, *Childrey*.

   Not recorded for Divs. 6, 7, 8, 9, 10, 12, 14, 17.

   Impermanent casual from cultivation.

3. **L. anglicum** Mill *Perennial Flax*
   - **2.** 1900, Hibaldstow, *Peacock*†
   - **10.** 1820, High Toynton, *Ward*.
   - **12.** No date, Freiston, *Mason's* M.S. (Unconfirmed perhaps *L. bienne*).
   - **15.** 1904, Somerby, *Stow*.† 1897, Ropsley, *Woolward*.†
   - **16.** 1666, Stamford, *Merrett*. 1948, *E. J. Gibbons*.

   Native. Local; on limestone roadsides; often associated with Roman roads.

4. **L. catharticum** L. *Purging Flax*
   - **16.** 1836, Bourne, *Dodsworth*.

   Recorded for all divisions.

   Native. Common; dry banks, meadows and turfy bogs.

167. Radiola Hill

1. **R. linoides** Roth. *All-seed*
   - **1.** Axholme, *Check-List*.
   - **2.** 1875, Crosby Warren, *Fowler*. B.R.C. 1900, Scunthorpe, *Mason*. 1915, Manton, *Claye*.
   - **5.** 1805, Nr. Gainsbrough, Salt Herbarium, Sheffield. 1894, Scotton, *Sneath*.† 1905, *Peacock*.†
   - **7.** 1877, Osgodby Lane, *Lees*.† 1930 and 1954, Linwood, *E. J. Gibbons*.
   - **10.** Before 1909, Woodhall, again in 1957, *E. J. Gibbons*.
   - **13.** 1807, Stapleford, *Ordoyno*. 1862, Doddington, *Cole*.†

   Native. Rare, and only found in two localities in divisions 7 and 10 recently.

GERANIACEAE
168. Geranium L.

1. **G. pratense** L.   *Meadow Cranesbill*
   6 or 13. 1828, Lincoln, *Rev. J. F. Wray.*
   Not recorded for Div. 18.
   Native; local on roadsides, generally calcareous; more common to the west of the county, especially along the Trent bank.

3. **G. endressii** Gay
   10. 1961, Woodhall Spa, *J. Bell.*
   Alien; garden escape.

4. **G. versicolor** L.
   2. 1875, Bottesford, *Peacock.*
   Recorded for Divs. 2, 3, 8, 10, 11 and 12.
   Introduced. Garden escape, becoming naturalised along hedgebanks, etc. Native of Europe.

6. **G. phaeum** L.   *Dusky Cranesbill*
   13. 1843, Fulbeck, *Rake.*
   Recorded for Divs. 2, 3, 10, 11, 13, 15, 18.
   Introduced. Garden escape.

7. **G. sanguineum** L.   *Bloody Cranesbill*
   2. 1789, Broughton, *Gough's Camden's Britannia.*
   3. *Check-list.*
   6. 1965, Welton, *W. Heath.*
   10. 1893, Woodhall Spa, *E. M. Lane-Claypon* (Paintings).
   11. 1909, Skegness, *H. C. Brewster.* (Escape?)†
   13. 1746, Lincoln Heath, *Blackstone.*
   15. 1930, Belton, *E. E. Orchard.*
   Native on the limestone; surviving in small quantity, but not recently observed in S. Lincs.

9. **G. pyrenaicum** Burm. f.   *Mountain Cranesbill*
   14. 1855, Ruskington, *J. Lowe.*
   Not recorded for Divs. 8, 9, 11, 18.
   Native? Not uncommon, usually found on roadsides.

10. **G. columbinum** L.   *Long-stalked Cranesbill*
    12 or 17. 1806, Near Boston, *Hailstone.*
    Not recorded for Div. 18.
    Native. Uncommon, mainly on chalk and limestone.

11. **G. dissectum** L. *Cut-leaved Cranesbill*
    2. 1822, Appleby, *Strickland*.
    Recorded for all Divs.
    Native. Common, particularly on basic soils.

12. **G. rotundifolium** L. *Round-leaved Cranesbill*
    3. 1942, Caistor, *A. M. Smith*.
    11. 1934, Welton Wood, *L.N.U. Meeting*.†
    13. 1903, Coleby, *Miss Stow*. 1961, *E. J. Gibbons*.
    15. 1930, Ancaster, *E. Orchard*.
    Native. Uncommon; found on wall-tops and hedgebanks; possibly overlooked.

13. **G. molle** L. *Dove's-foot Cranesbill*
    16. 1790, Threckingham, *Cragg*.
    Recorded for all Divs.
    Native. Common; generally in cultivated and waste places.

14. **G. pusillum** L. *Small-flowered Cranesbill*
    16. 1790, Threckingham, *Cragg*.
    Recorded for all Divs.
    Native. Common in stubble on light soils.

15. **G. lucidum** L. *Shining Cranesbill*
    1836, Simpson Collection.
    Not recorded for Divs. 1, 9, 12, 13, 14, 17, 18.
    Introduced into churchyards from Derbyshire; established and spreading in some places, mainly in north Lincs.; probably native in Div. 16.

16. **G. robertianum** L. *Herb Robert*
    16. 1790, Threckingham, *Cragg*.
    Recorded for all Divs.
    Native. Generally common, but scarce on more acid soils.

169. **Erodium** L'Hérit.

2. **E. moschatum** (L.) L'Hérit. *Musk Storksbill*
    1. 1900, Axholme, *Hudson*.
    9. c. 1860, Saltfleetby, *Mossop* (painting). 1956, Humberstone, *E. J. Gibbons*.
    Casual; in clover seed.

3. **E. cicutarium** (L.) L'Hérit. *Common Storksbill*
    2. 1822, Appleby, *Strickland*.
    Recorded for all divs.
    Native. Frequent; in both cultivated and waste places; near the sea and on light soils.
    ssp. *dunense* Andreas.
    9. 1962, Mablethorpe, *J. Gibbons*.
    11. 1962, Skegness, *J. Gibbons*.
    Native, growing with *E. cicutarium*, needs confirmation.

OXALIDACEAE
170. Oxalis L.

1. **O. acetosella** L. *Wood Sorrel*
    10. 1820, Tetford, *Ward*.
    Not recorded for Divs. 9, 17, 18.
    Native. Uncommon; in sandy deciduous woodland.

2. **O. corniculata** L. *Procumbent Yellow Sorrel*
    5. 1893, Kirton Lindsey, *Peacock*.†
    Recorded for Divs. 3, 5, 7, 8, 10, 15, 16.
    Introduced. Distribution unknown; garden weed.

3. **O. europaea** Jord. *Upright Yellow Sorrel*
    16. 1836, Bourne, *Dodsworth*.†
    Recorded for Divs. 3, 7, 8, 10, 13, 16.
    Introduced. Distribution unknown; garden weed.

BALSAMINACEAE
171. Impatiens L.

1. **I. noli-tangere** L. *Touch-me-not*
    The only herbarium specimens available have, in each instance, proved to be *I. parviflora* DC.; thus the record for division 13 in Peacock's List is incorrect. This may be true for the following records:
    7. 1885-6, South Kelsey, *Brewster*.
    15. 1884, Harlaxton, *Browne*.
    17. Pinchbeck, *Naturalist*, 1902.
    Casual.

2. **I. capensis** Meerb *Orange Balsam*
    16. 1940, Market Deeping, *Burchnall*.
    Established in Div. 16 only.
    Denizen. Native of N. America.

3. **I. parviflora** DC. *Small Balsam*
    10. 1889, East Keal, *Burgess.*
    Recorded for Divs. 5, 6, 10, 13, 15.
    Introduced into shrubberies where it survives. Confused in the past with *I. noli-tangere* L. Native of Siberia and Turkistan.

4. **I. glandulifera** Royle *Policeman's Helmet*
    12. 1945, Boston, *A. H. G. Alston* and *N. D. Simpson.*
    1. 1967, Owston Ferry, *B. Howitt.*
    Recorded for Divs. 1, 2, 4, 6, 10, 12, 13, 15 and 17.
    Introduced? Established on waste ground at Boston and Lincoln; causuals recorded in other localities, but not fully naturalized yet.

ACERACEAE

173. Acer L.

1. **A. pseudoplatanus** L. *Sycamore*
    12 or 17. 1820, Boston, *Thompson.*
    Recorded for all Divs.
    Introduced. Widely planted and seeding freely.

2. **A. platanoides** L. *Norway Maple*
    Recorded for Divs. 2, 3, 4, 7, 8, 13, 16.
    Introduced. Distribution unknown; ornamental woodlands mainly.

3. **A. campestre** L. *Common Maple*
    16. 1836, Bourne, *Dodsworth.*
    Recorded for all Divs.
    Native. Occasional; but more frequent in hedges on basic clay.

STAPHYLEACEAE

174. Staphylea L.

1. **S. pinnata** L. *Bladder-nut*
    16. 1633, Gerarde's Herbal. "Nux Vesicaria. The Bladder Nut. In the Frier Yarde without St. Paule's gate in Stamford, and about Spalding Abbey. It groweth also in my garden."
    Introduced. Planted in shrubberies.

HIPPOCASTANACEAE
175. Aesculus L.
1. **A. hippocastanum** L. *Horse-chestnut*
   Introduced. Planted and seeding in all Divs.
   Native of Greece and Albania.

AQUIFOLIACEAE
176. Ilex L.
1. **I. aquifolium** L. *Holly*
   2. Dragonby. Prehistoric excavations 1968, in vegetative remains identified by Dr. A. J. Hayes, Edinburgh.
   11. 1796, Sutton-on-Sea, *Banks*, "Submerged Forest".

   J. Correia da Serra, "Account of the Submarine Forest on the Coast of Lincoln," *Phil. Trans.* Vol. 89.
   Not recorded for Divs. 5, 9, 12, 14.
   Native in preglacial times but doubtfully native now. Frequently bird-sown from gardens.

CELASTRACEAE
177. Euonymus L.
1. **E. europaeus** L. *Spindle-tree*
   16. 1790, Bridge End, *Cragg*.
   Not recorded for Divs. 5, 9, 12, 17, 18.
   Native. Calcareous soils in woods and hedges.

BUXACEAE
178. Buxus L.
1. **B. sempervirens** L. *Box*
   2. 1874, Bottesford, *Peacock*.
   Recorded in Divs. 2, 3, 11, 14.
   Doubtfully wild in the Ancaster valley; naturalised and seeding in Brocklesby Woods near Caen Hill; much planted in shrubberies.

RHAMNACEAE
179. Rhamnus L.
1. **R. catharticus** L. *Buckthorn*
   14. 1790, Threckingham *Cragg*.
   Not recorded for Divs. 4, 8, 9, 17.
   Native. Frequent on limestone and calcareous clay; hedges.

180. Frangula Mill

1. **F. alnus** Mill *Alder Buckthorn/Black Dogwood*
   13. 1836, Doddington Lane, *Simpson Collection.*
   Not recorded for Divs. 3, 8, 9, 11, 12, 17, 18.
   Native. Occasional; on acid heaths, also open woods.

LEGUMINOSAE
184. Laburnum Medic.

1. **L. anagyroides** Medic. *Golden Rain, Laburnum*
   2. 1879, Bottesford, *Peacock.*
   Introduced. Established in Barton Chalk-pit, 1958; parks and woodlands elsewhere.

185. Genista L.

1. **G. tinctoria** L. *Dyer's Greenweed*
   15. 1780, Easton, *Sibthorp.*
   Not recorded for Divs. 1, 10, 12, 17, 18.
   Native. Frequent in Div. 14, rare elsewhere; calcareous clay.

2. **G. anglica** L. *Needle Furze, Petty Whin*
   15. 1780, Easton, *Sibthorp.*
   Recorded for Divs. 2, 3, 5, 6, 7, 10, 13, 15.
   Native. Scarce; decreasing through drainage, etc.

187. Ulex L.

1. **U. europaeus** L. *Furze, Gorse (Ling*)*
   14. 1790, Threckingham, *Cragg.*
   Not recorded for Divs. 17, 18.
   Native. Formerly abundant on wolds. Used for fuel.
   *Ling is a localised Lincolnshire name.

2. **U. gallii** Planch *Dwarf Furze*
   6. 1879, Marton, *F. A. Lees (J. Bot. 3).*
   7. 1877 and 1895, Linwood, *F. A. Lees.* 1877, Osgodby, *Lees.* "Much more abundant than *minor.*"
   13. 1928, Doddington Road, Hykeham, *E. J. Gibbons.*
   1931, Fossway, Swinderby, *E. J. Gibbons.*
   None of these records are confirmed.
   13. 1959, Thurlby Moor, *G. Posnett* (Conf. M. C. F. Proctor).

3. **U. minor** Roth. [**U.** *nanus* Forst.] *Dwarf Furze*
    6. 1878, between Gainsborough and Marton, *Lees*.†
    7. 1858, *M. E. Dixon* (this refers to *U. nanus*).
    1877, Holton le Moor, *Lees*. 1877, Linwood, *Lees*.
    13. 1807, Thurlby Moor (var 2 Withering), *Ordoyno*. (This record may be *U. gallii*).
    1855, Doddington, *Cole*.† 1855, Skellingthorpe, *Lowe*.
    1905, Hykeham Station, *Peacock*.† (Conf. M. C. F. Proctor).

No recent record; apparently extinct.

188. Sarothamnus Wimm.

1. **S. scoparius** (L.) Wimm. ex Koch   *Broom*
    10. 1820, Harrington, *Ward*.
    Not recorded for Divs. 9, 12, 17, 18.

Native. Uncommon except in division 13; gravelly soils.

189. Ononis L.

1. **O. repens** L.   *Restharrow*
    10. 1820, Horncastle, *Ward*.
    Not recorded for Divs. 1, 17.

Native. Calcareous soil; also on sand dunes.

2. **O. spinosa** L.   *Spiny Restharrow*
    15. 1726, Ropsley, *Bacon*.
    Recorded for all Divs.

Native. Stiff clay soil; decreasing.

190. Medicago L.

1. **M. falcata** L.   *Sickle Medick*
    4. 1897, Grimsby Docks, *Woods*.†
    8. 1957, Scamblesby, *M. N. Read* (Native?).
    10. 1917, Woodhall, *Alston*. 1967, *Z. Porter*.
    13. 1880, Boultham, *Sneath*.†
    15. Check list.

Introduced. Grain alien.

2. **M. sativa** L.   *Lucerne, Alfalfa*
    13. 1858, Boultham, *Cole*.

Introduced. An escape, or relic, of cultivation.

3. **M. lupulina** L.   *Black Medick*
    16. 1836, Bourne, *Dodsworth*.
    Recorded for all Divs.

Native. Common, especially on the chalk; cultivated with clover.

4. **M. minima** (L.) Bartal   *Small Medick*
Incorrect. See *Naturalist*, 1894, 217.

5. **M. polymorpha** L. [**M. hispida** Gaertn.]   *Toothed Medick*
    7. 1879, Market Rasen, *W. Allen.*†
    Recorded for Divs. 3, 4, 7, 9 as a grain alien.

6. **M. arabica** (L.) Huds   *Spotted Medick*
    2. 1875, Scabcroft, *Fowler* (BM).
    5. 1855, Gainsborough, *J. Lowe.*
    6. 1894, Kettlethorpe, *Fowler.*† 1901, Newton-on-Trent, *Mason.* 1951, Marton, *E. J. Gibbons.*
    8. 1885, Louth, *J. W. Chandler.*†
    9. 1956, Humberstone Dump, *E. J. Gibbons.*
    11. 1961, Gibraltar Point, *E. J. Gibbons and A. Lodge.* (Casual).
    12. 1909, Skirbeck, *L.N.U. Meeting.*
    15. 1953, Little Bytham, *J. H. Chandler.*
    17. 1894, Wyberton, *Lane-Claypon.*

Native. Rare; dry banks usually near the R. Trent.

191. **Melilotus** Mill.

1. **M. altissima** Thuill.   *Tall Melilot*
    11. 1847, Burgh, *Dr. Grantham.*
    Recorded for all Divs.
    Native? Occasional; in waste places.

2. **M. officinalis** (L.) Pall   *Common Melilot*
    7. 1890, Hatton, *Mrs. Jarvis.*
    Recorded for all Divs.
    Denizen.

3. **M. alba** Medic   *White Melilot*
    16. 1836, Bourne, *Dodsworth.*
    Not recorded for Divs. 9, 17.
    Introduced. Occasional; on roadsides and in clover.

4. **M. indica** (L.) All   *Small-flowered Melilot*
    5. 1893, Kirton Lindsey, *Miss M. Peacock.*†
    Not recorded for Divs. 1, 10, 14, 15, 17, 18.
    Introduced. Occasional; gardens, dumps and waste places.

192. **Trifolium** L.

1. **T. ornithopodiodes** L. [**Trigonella ornithopodioides** L. DC.]   *Birdsfoot Fenugreek*
    13. 1897, Fulbeck, *Goodall.*
    Native. Possibly overlooked elsewhere.

2. **T. pratense** L. *Red Clover*
   12. 1820, Boston, *Thompson*.
   Recorded for all Divs.
   Native. Abundant; frequently cultivated.

3. **T. ochroleucon** Huds. *Sulphur Clover*
   Ref: W. *Fowler, Nat.*, Dec. 1889.
   4. *c.* 1900, Cleethorpes, *A. Smith*. Dock alien.
   11. 1889, Rigsby, *Davy*.
   12. 1820, Boston, *Thompson*. 1933, Freiston, *Kime*.†
   15. 1879, Grantham, *Browne*.† 1904, Holywell, *Mason*.
   16. 1789, Stamford, *Gough*. 1954, *Howitt*.
   Native in Divs. 15 and 16; fringe area of the British distribution. In foreign seed in other divisions.

4. **T. medium** L. *Zigzag Clover*
   12. 1823, Wainfleet, *Sinclair*.
   Recorded for all Divs.
   Native. Not common; calcareous clay.

5. **T. squamosum** L. *Sea Clover*
   3. 1957, Killingholme, *E. J. Gibbons*.
   4. 1881, Grimsby, *Searle*.
   9. 1957, Mablethorpe, *E. J. Gibbons*.
   11. 1890, Skegness, *Lane-Claypon*.
   12. 1893, Benington, *Disbrowe*.
   Native. Rare; dry banks on coast.

7. **T. incarnatum** L. *Crimson Clover*
   13. 1864, Leadenham, *Burtt*.
   Recorded for Divs. 2, 3, 7, 10, 11, 13, 14.
   Introduced. Casual; relic of cultivation.

9. **T. arvense** L. *Hare's-foot Trefoil*
   16. 1836, Witham-on-the-Hill, *Dodsworth*.
   Recorded for all Divs.
   Native. Uncommon; sandy fields and dunes.

10. **T. striatum** L. *Knotted Trefoil*
    5. 1868, Gainsborough, *Charters*.
    Not recorded for Divs. 1, 8, 16, 17, 18.
    Native. Occasional; dry places.

11. **T. scabrum** L. *Rough Trefoil*
    14. 1838, Wilsford, *Latham*.
    Recorded for Divs. 2, 4, 9, 10, 11, 14, 16.
    Native. Less common than *T. striatum* L.

13. **T. subterraneum** L. *Subterranean Trefoil*
    6. 1840, Lea, *J. K. Miller*. 1908, Torksey, *F. H. Mills*.†
       1925, Knaith, *Sneath*. 1957, *L.N.U. Meeting*.
    13. 1865, Doddington, *Cole*.† 1894, North Hykeham, *Burton*.†

Native. Rare; sandy turfy places west of Lincoln; northern limit for the British Isles.

17. **T. hybridum** L. *Alsike Clover*
    3. 1865, Caistor, *Britten*.
Recorded for all Divs.
Much cultivated.

18. **T. repens** L. *White or Dutch Clover*
    7. 1824, Acre House, Claxby, *Sinclair*.
Recorded for all Divs.
Native. Common; grassy places.

19. **T. fragiferum** L. *Strawberry Clover*
    10. 1724, Tattershall, *Stukeley*.
Not recorded for Div. 5.
Native. Widespread but not common; heavy clay soil, both inland and by the sea.

20. **T. resupinatum** L. *Reversed Clover*
    4. 1906, Grimsby, *A. Smith*.
    9. 1956, Humberstone, *M. and E. J. Gibbons*.
    13. 1896, Boultham, *Sneath*.†
Grain alien.

21. **T. campestre** Schreb. *Hop Trefoil*
    12. 1820, Boston, *Thompson*.
Recorded for all Divs.
Native. Not really common, and absent from blown sand.

23. **T. dubium** Sibth *Lesser Yellow Trefoil*
    16. 1836, Bourne, *Dodsworth*.
Recorded for all Divs.
Native. Very common.

24. **T. micranthum** Viv *Slender Trefoil*
    12 or 17. 1856, Boston, *Thompson*.
Not recorded for Divs. 1, 3, 6, 11, 13, 14, 18.
Native. Scarce, perhaps overlooked; dry places.

193. Anthyllis L.

1. **A. vulneraria** L. *Kidney Vetch, Ladies' Fingers*
   15. 1726, Grantham, *Bacon.*
   Not recorded for Divs. 1, 9, 17.
   Native. Uncommon on the chalk, more frequent on the southern limestone; occasionally introduced.

195. Lotus L.

1. **L. corniculatus** L. *Birdsfoot-trefoil*
   14. 1790, Threckingham, *Cragg.*
   Recorded for all Divs.
   Native. Abundant.

2. **L. tenuis** Waldst. & Kit. ex Willd. *Slender Birdsfoot-trefoil*
   13. 1863, Doddington, *Cole.*†
   Not recorded for Divs. 1, 12, 15, 18.
   Native and colonist; sometimes on roadsides; occasional but perhaps overlooked.

3. **L. uliginosus** Schkuhr. *Greater Birdsfoot-trefoil*
   7. 1829, Bardney, *J. F. Wray.*
   Recorded for all Divs.
   Native. Frequent; chiefly in wet places; acid bogs.

200. Astragalus L.

1. **A. danicus** Retz. *Purple Milk Vetch*
   13. 1780, Lincoln Heath, *J. Sibthorp.*
   Recorded for Divs. 2, 3, 4, 5, 6, 13, 14, 15, 16.
   Native. Very rare on the chalk, more frequent on limestone, absent from coastal sands.

3. **A. glycyphyllos** L. *Milk Vetch, Wild Liquorice*
   1. *c.* 1900, Haxey, *Val Palmer.* (Lees' MS). 1948, *F. Rose.*
   2. 1895-6, Broughton, *Peacock.*† 1950, *J. Gibbons.*
   5. 1840, Gainsborough, *J. K. Miller.* 1955, *J. Gibbons.*
   6. 1902, Newton-on-Trent, *Mason.* 1947, *J. Gibbons.*
   15. 1805, Grantham, *Crabbe.* 1904, Saltersford, *Stow.*†
       1960, *E. J. Gibbons.*
   16. 1883, Uffington, *Fowler* (Bot. Rec. Club). 1963, *J. Gibbons.*
   Native. On limestone and lias chiefly; distribution in the west of the county.

202. Ornithopus L.

1. **O. perpusillus** L. *Birdsfoot Vetch*
    13. 1829, Skellingthorpe, *R. J. Bunch.*
    Not recorded for Divs. 9, 18.
    Native. Locally frequent on acid soils.

203. Coronilla L.

1. **C. varia** L. *Crown Vetch*
    3. 1960, Ferriby, *E. J. Gibbons.*
    4. 1901, Grimsby, *A. Smith.*† 1938, Immingham, *L.N.U. Meeting.* 1957, *J. Gibbons.*
    5. 1960, Gainsborough, *E. J. Gibbons.*
    7. 1933, Hainton, *S. J. Hurst.*
    10. 1918, Woodhall Spa, *F. S. Alston.* 1964, *W. Heath.*
    13. 1909, Lincoln, *Sneath.*†
    15. 1910, Saltersford, Grantham, *Peacock and Preston.*†
    Introduced. Persisting; well established on Humber bank at Immingham and Woodhall Spa.

204. Hippocrepis L.

1. **H. comosa** L. *Horse-shoe Vetch*
    15. 1726, Grantham, *Bacon.*
    Recorded for divs. 2, 5, 8, 10, 13, 14, 15, 16, 17.
    Native. Rather scarce on limestone; records for Div. 8 on the chalk unconfirmed.

205. Onobrychis Mill.

1. **O. viciifolia** Scop. *Sainfoin*
    15. 1726, Grantham, *Bacon.*
    Not recorded for Div. 12.
    Native? Probably native in the S.W. on the limestone; elsewhere, a relic of cultivation.

206. Vicia L.

1. **V. hirsuta** (L.) Gray *Hairy Tare*
    13. 1829, Boultham, *R. J. Bunch.*
    Recorded for all Divs.
    Native. Widespread; fairly common on acid soils.

2. **V. tetrasperma** (L.) Schreb. *Smooth Tare*
    16. 1836, Bourne, *Dodsworth.*
    Not recorded for Div. 5.
    Native. Frequent on clay soils; scarce on sand.

THE COUNTY FLORA

3. **V. tenuissima** (Bieb.) Schinz & Thell. *Slender Tare*
    1. 1950, Isle of Axholme, Haxey, *Dr. J. Dony and Mrs. B. Welch.* (B.S.B.I., 1951, 71) Introduced?

4. **V. cracca** L. *Tufted Vetch*
    12. 1820, Boston, *Thompson.*
    Recorded for all Divs.
    Native. Common.

10. **V. sylvatica** L. *Wood Vetch*
    8. 1834, Louth, *Bayley.*
    Recorded for Divs. 5, 7, 8, 10, 11, 13, 15, 16.
    Native. Scarce, but in several of the larger woods.

11. **V. sepium** L. *Bush Vetch*
    16. 1836, Bourne, *Dodsworth.*
    Not recorded for Div. 17.
    Native. Not very common; bushy places on calcareous soils.

12. **V. lutea** L. *Yellow Vetch*
    2. 1908, Broughton Wood, *A. N. Claye.*†
    4. 1902, Grimsby, *A. Smith.*†
    12. 1856, Boston, *Thompson.* 1909, *S. J. Hurst.*†
    13. 1896, Lincoln, *Lees, Sneath and Peacock.*†
    14. 1966, Wilsford, *Z. Porter.*
    Casual. Rare; perhaps native in the Boston area.

13. **V. hybrida** L.
    13. 1842, Swanpool, Lincoln, *John Nicholson.*
    5th edition of Hooker and Arnott's *British Flora.* 1842, 88.
    1896, Lincoln, *Lees, Sneath and Peacock*†.
    Casual.

14. **V. sativa** L. *Common Vetch*
    4. 1851, Grimsby, *H. C. Watson.*
    13. 1851, Boultham.
    Recorded for all Divs.
    Introduced? Common; naturalised in waste places; much cultivated.

15. **V. angustifolia** L. *Narrow-leaved Vetch*
    5. 1840, Laughton, *Miller.*
    Recorded for all Divs.
    Native. Often confused with *V. sativa* L. Common on green sand.
    var. bobartii (Forst.) Koch.
    10. 1883, Woodhall Spa, *W. Fowler* (Bot. Rec. Club).
    Recorded for Divs. 2, 3, 7, 10, 13.
    Native. Common on acid soil.

16. **V. lathyroides** L. *Spring Vetch*
    11. 1872, Gibraltar Point, *Streatfeild.*†
    Recorded for divs. 2, 3, 6, 7, 9, 10, 11, 12, 13.
    Native. Occasional; in dry places, particularly along coast.

17. **V. bithynica** (L.) L. *Bithynian Vetch*
    2. 1915, Broughton Woods, *Claye.*
    3. 1906, Melton Ross, *Claye.*†
    11. 1912, Strubby, *S. Allett.*†
    Introduced.

207. Lathyrus L.

1. **L. aphaca** L. *Yellow Vetchling*
    6 or 13. 1849, Lincoln, *B. Carrington.*
    Recorded for Divs. 4, 5, 7, 8, 9, 13, 15.
    Casual. Flour mills; also as pheasant food.

2. **L. nissolia** L. *Grass Vetchling*
    6 or 13. 1831, Lincoln, *Drury.*
    6. *c.* 1950, Scothern, *R. Hull.* Not confirmed.
    7. 1963, Bucknall, *Read.* 1967, Langton, *J. K. Shaw.*
    13. 1944, Caythorpe, *L.N.U. Meeting.*
    15. 1950, Hougham, *W. A. Burton.*† 1957, Gunby, *J. H. Chandler.*
    17. 1935, Wyberton, *Hurst* (L.N.U. Trans., 1935).
    18. 1892, Lutton, *Welbourn.*†
    Native. Rare; chiefly in the south; grassy places and sea banks; casual for Gainsborough. *c.* 1930.

4. **L. pratensis** L. *Meadow Vetchling*
    14. 1790, Threckingham, *Cragg.*
    Recorded for all Divs.
    Native. Common; grassy places.

THE COUNTY FLORA

5. L. tuberosus L. *Earth-nut Pea*
    6. 1965, Caenby, *E. Hargrave.*
    13. 1708, Blankney, *Sedgewick*; in Buddle Herbarium, B.M. (Ref: *Phytol.*, 1861, p. 188; and *Nat.*, 1891, 190).
    14. 1964, Rauceby, *S. Lyon. Com. C. J. Allerton.*
Introduced. Relic of cultivation?

6. L. sylvestris L. *Narrow-leaved Everlasting Pea*
    4. c. 1870, Cleethorpes, *Lees.* Extinct.
    5. 1862, Gainsborough, *Charters.* Casual?
    11. 1861, Skegness, *M. Walcott.*
    13. 1896, Lincoln, *Sneath*† Casual?
    15. 1892, Woolsthorpe, Belvoir, *D. M. Craster.*† 1961, *M. Lowe.*
    16. 1956, Bourne, *J. H. Chandler.*
Native in the S.W.; possibly casual elsewhere. Rare.

8. L. latifolius L.
    3. 1903, Cadney, *Peacock.*†
Recorded for Divs. 3, 5, 6, 9, 10, 11, 13, 14.
Casual. Naturalised on railway banks, etc.

9. L. palustris L. *Marsh Pea*
    6. 1891—1896, Bishop Bridge, Saxilby Road, *Sneath.*‡
    9. 1857, Granthorpe, *T. W. Bogg.**
        1972, Theddlethorpe, *C. Walker.*
    12. 1789, East Fen, *Gough.*
    13. 1896, Lincoln, *Peacock.*† 1945, Saxilby, *E. Redfern.*
    15. 1780, Easton, Grantham, *J. Sibthorp.*
Native.

10. L. japonicus Willd. *Sea Pea*
    9. 1849, South of Saltfleetby, *Carrington.*† (*Bot. Gaz.*, 1849, 323-4).
    11. 1640, Ingoldmells, *Parkinson.* (Theat. Bot., 1640, 1060; H. C. Watson, *Cybele* 3, 415).
Extinct native. Lost by erosion; shingle scarce.

11. L. montanus Bernh. *Tuberous Bitter Vetch*
    10. 1820, Horncastle, *Ward.*
Recorded for Divs. 1, 2, 3, 5, 6, 7, 8, 10, 13.
Native. Scarce, except for division 13 where it is frequent; woods and hedgebanks.

ROSACEAE
209. Spiraea L.

1. **S. salicifolia** L. *Willow Spiraea*
    - 2. 1927, Manton, *E. J. Gibbons*. Planted.
    - 3. 1865, Brocklesby Woods, *Britten*. Planted.
    - 5. 1882, Laughton, *W. Fowler*. Established. (*S. tomentosa* L. in Peacock's Check List).
    - 15. 1963, Twyford Forest, *E. J. Gibbons*.
    - 16. 1959, Tallington, *E. J. Gibbons*. (Gravel Pit).

    Introduced. Planted for game cover, etc.

210. Filipendula Mill.

1. **F. vulgaris** Moench. *Dropwort*
    - 15. 1726, Grantham, *Bacon*.

    Not recorded for Divs. 1, 11, 12, 17, 18.

    Native. Occasional; most often on limestone and calcareous clay; rare on chalk.

2. **F. ulmaria** (L.) Maxim. *Meadow-sweet*
    - 12 or 17. 1820, Boston, *Thompson*.

    Recorded for all Divs.

    Native. Abundant; in damp places.

211. Rubus L.

2. **R. saxatilis** L. *Stone Bramble*
    - 2. 1789, Broughton, *Gough*. 1958, *Dunn*.
    - 6. 1870, Gate Burton, *F. A. Lees*.
    - 15. 1865, Little Ponton, *Miss S. C. Brooks*.

    Native. On limestone only. Rare.

6. **R. idaeus** L. *Raspberry*
    - 10. 1820, Somersby, *Ward*.

    Recorded for all Divs.

9. **R. caesius** L. *Dewberry*
    - 16. 1836, Bourne, *Dodsworth*.

    Recorded for all Divs.

11. **R. fruticosus** L. sensu lato.
    All checked by E. S. Edees.
(Sect. **Suberecti** P. J. Muell.)

1. **R. nessensis** W. Hall *Bramble*
    6. 1907, Burton, *Ley*.
    10. 1965, Mareham-le-Fen, *Mrs. Z. Porter*.
    13. 1856, Skellingthorpe, *Cole*.
    Uncommon.

2. **R. scissus** W. C. R. Wats.
    6. 1907, Burton, *Ley*.
    7. 1965, Linwood Warren, *Edees*.
    10. 1932, Tumby, *Fisher*.
    13. 1913, Skellingthorpe, *Fisher*.
    Local.

5. **R. plicatus** Weihe and Nees.
    7. 1907, College Wood, Apley, *Ley*.
    13. 1965, Stapleford Moor, *Edees*.
    Uncommon.

(Sect. **Triviales** P. J. Muell).
14. **R. conjungens** (Bab.) W. C. R. Wats.
    16. 1964, Aunby; 1972, Thurlby, *J. H. Chandler*.

16. **R. eboracensis** W. C. R. Wats.
    2. 1936, Winterton, *Edees*.
    6. 1965, Burton, *Edees*.
    8. 1965, Muckton, *Edees*.
    10. 1965, Tattershall, *Edees*.
    14. 1965, Ancaster, *Edees*.
    16. 1961, Stamford, *J. H. Chandler*.
    Common.

17. **R. sublustris** Lees.
    3. 1907, Near Elsham, *Ley*.
    6. 1965, Burton, *Edees*.
    8. 1892, Stenigot, *Larder*.
    13. 1907, Skellingthorpe, *Ley*.
    Not common.

30. **R. scabrosus** P. J. Muell.
    8. 1965, Muckton, *Edees*.
    Rare.

(Sect. **Sylvatici** P. J. Muell).

34. **R. gratus** Focke.
    7. 1907, Market Rasen district, *Ley*.
    10. 1926 Woodhall Spa, *Fisher*.
    13. 1930, North Scarle, *Fisher*.
    Locally common.

42. **R. calvatus** Lees ex. Bloxam
    7. 1907, Stainton Wood, *Ley*.
    10. 1965, Tattershall, *Edees*.
    13. 1931, Blankney, *Fisher*.
    Uncommon.

47. **R. carpinifolius** Weihe and Nees.
    7. 1907, Market Rasen, *Ley*.
    10. 1965, Tattershall, *Edees*.
    13. 1907, Skellingthorpe, *Ley*.
    Uncommon.

52. **R. nemoralis** P. J. Muell.
    5. 1907, Scotton Common, *Ley*.
    Recorded for Divs. 1, 2, 3, 5, 7, 10, 13, 15.
    Acid moors. Pink notched petals; large juicy fruit.

59. **R. lindleianus** Lees
    7. 1907, Stainton Wood, *Ley*.
    Recorded for Divs. 6, 7, 10, 11, 13, 15, 16.
    Widely distributed.

64. **R. robii** (W. Wats.) A. Newton.
    10. 1965, Woodhall Spa, *Edees*.
    13. 1965, Stapleford Moor, *Edees*.
    Rare.

66. **R. macrophyllus** Weihe & Nees.
    11. 1965, Mother Wood, *Edees*.
    Rare.

74. **R. silvaticus** Weihe & Nees.
    14. 1931, Ancaster, *Fisher*.
    Very rare.

77. **R. amplificatus** Lees.
    10. 1965, Kirkby-on-Bain, *Edees*.
    13. 1965, Eagle, *Edees*.
    14. 1931, Heydour, *Fisher*.
    Uncommon.

79. **R. belophorus** Muell. & Lefèv.
   7. 1907, Stainton Wood, *Ley.*   1964, *B. A. Miles.*
   Very Rare.

80. **R. pyramidalis** Kalt.
   7. 1965, College Wood, Apley, *Edees.*
   13. 1907, Skellingthorpe, *Ley.*
   15. 1963, Lincs. Gate, *J. H. Chandler.*
   Uncommon.

88. **R. macrophylloides** Genev. Sensu Watson.
   7. 1907, Apley, *Ley.*
   Local.

113. **R. polyanthemus** Lindeb.
   3. 1965, Wrawby Moor, *Edees.*
   7. 1965, Holton le Moor, *Edees.*
   10. 1965, Tattershall, *Edees.*
   11. 1965, Mother Wood, *Edees.*
   13. 1892, Fossway; 5 miles south of Lincoln, *Fisher.*
   Widely distributed and probably common.

123. **R. cardiophyllus** Muell & Lefèv.
   7. 1907, Moortown, *Ley.*   1965, Holton-le-Moor, *Edees.*
   Rare.

125. **R. lindebergii** P. J. Muell.
   7. 1907, College Wood, Apley, *Ley.*   1965, *Edees.*
   13. 1907, Skellingthorpe, *Ley.*
   15. 1912, Harlaxton, *Fisher.*
   Uncommon.

(Sect. **Discolores** P. J. Muell).

129. **R. ulmifolius** Schott.
   8. 1894, Louth, *Rogers.*
   Not recorded for Divs. 1, 4, 9, 12, 17, 18.
   On basic soils, widely distributed. Small leaves with white backs, neat bushes, pink flowers with felted sepals, small fruit very sweet.

139. **R. procerus** P. J. Muell.
   10. 1965, Tattershall, *Edees.*
   Rare.

141. **R. falcatus** Kalt.
    3. 1907, Wrawby Moor, *Ley*.
    13. 1892, Fossway; 5½ miles south of Lincoln, *Fisher*.
    14. 1965, Wilsford, *Edees*.
    15. 1970, Ancaster, *J. H. Chandler*.
    Uncommon.

(Sect. **Appendiculati** (Genev.) Sudre).

165. **R. vestitus** Weihe & Nees.
    7. 1907, Market Rasen, *Ley*.
    Recorded for Divs. 7, 8, 10, 11, 13, 14.
    Locally common.

173. **R. boraeanus** Genev.
    16. 1931, Edenham, *Fisher*.
    Rare.

183. **R. drejeri** Jensen.
    15. 1912, Harrowby Gorse, *Fisher*.
    Rare.

191. **R. leyanus** Rogers
    16. 1965, Bourne Wood, *Edees*.
    Rare.

194. **R. mucronatoides** A. Ley.
    7. 1965, College Wood and Stainton Wood, *Edees*.

204. **R. radula** Weihe ex. Boenn.
    6. 1907, Burton, *Ley*. 1965, *Edees*.
    7. 1907, Stainton Wood and College Wood, *Ley*. 1965, *Edees*.
    13. 1907, Skellingthorpe, *Ley*. 1965, *Edees*.
    Local.

212. **R. echinatus** Lindl.
    10. 1965, Woodhall Spa, *Edees*.
    13. 1965, Skellingthorpe, *Edees*.
    15. 1913, Harlaxton, *Fisher*.
    16. 1926, Bourne Wood, *Fisher*.
    Local.

213. **R. echinatoides** (Rogers.) Sudre.
    2. 1907, Scunthorpe, *Ley*.
    13. 1965, Skellingthorpe, *Edees*.
    Rare.

216. **R. rudis** Weihe & Nees.
    10. 1926, Woodhall Spa, *Fisher.*
    11. 1965, Claxby, *Edees.*
    15. 1932, Ropsley, *Fisher.* 1963, Lincs. Gate, *J. H. Chandler.*
    16. 1926, Bourne Wood, *Fisher.*
    Rare in north; locally plentiful in south.

223. **R. flexuosus** Muell. & Lefèv.
    13. 1931, Skellingthorpe, *Fisher.*
    16. 1931, Bourne, *Fisher.*
    Rare.

243. **R. pallidus** Weihe & Nees.
    16. 1926, Bourne, *Fisher.*
    Rare.

246. **R. newbouldii** Bab.
    3. 1907, Grasby, *Ley.*
    13. 1926, North Scarle, *Fisher.*
    Local.

284. **R. rufescens** Muell. & Lefèv.
    7. 1907, College Wood, *Ley.*
    13. 1907, Skellingthorpe, *Ley.*
    15. 1965, Ponton, *Edees.*
    Locally plentiful.

(Sect. **Glandulosi** P. J. Muell).

348. **R. hylocharis** W. C. R. Wats.
    7. 1965, College Wood, *Edees.*
    13. 1907, Skellingthorpe, *Ley.*
    Local.

356. **R. dasyphyllus** (Rogers) Rogers.
    2. 1907, Twigmoor, *Ley.*
    Recorded for Divs. 2, 3, 7, 13, 14, 15, 16.
    Thinly distributed.

375. **R. bellardii** Weihe & Nees. [**R. glandulosus** Bellardi]
    15. 1912, Boothby, *Fisher.*
    16. 1930, Castle Bytham, *Fisher.* 1965, Bourne Wood, *Edees.*
    Local.

212. Potentilla L.

2. **P. palustris** (L.) Scop. *Marsh Cinquefoil*
   12. 1799, East Fen, *Young*.
   Not recorded for Divs. 4, 9, 14, 15, 16, 17, 18.
   Native. Fairly common in the N.W.; uncommon elsewhere.

3. **P. sterilis** (L.) Garcke *Barren Strawberry*
   15. 1790, Ropsley, *Cragg*.
   Not recorded for Divs. 9, 12, 17, 18.
   Native. Uncommon; basic soil and open woodland.

5. **P. anserina** L. *Silverweed*
   14. 1790, Threckingham, *Cragg*.
   Recorded for all Divs.
   Native. Common; damp roadsides and arable land.

6. **P. argentea** L. *Hoary Cinquefoil*
   2. 1915, Broughton, *A. N. Claye*.†
   5. 1931, Scotton, *L.N.U. Meeting*.
   6. 1950, Laughterton, *E. J. Gibbons*.
   10. 1893, Salmonby, *Larder*. 1954, Roughton, *N. Read*.
   11. 1870, Halton Holgate, *Streatfeild*.†
   14. 1938, Rauceby, *E. J. Gibbons*.
   Native. Rare; dry places; found to be more widespread in recent years.

7. **P. recta** L.
   3. 1957, Elsham, *Dunn*.
   4. 1956, Swallow, *Cox*. 1956, Waltham, *Hopkins*.
   6. 195-, Fiskerton, *E. E. Steel*.
   16. 1958, Uffington, *J. H. Chandler*.
   Casual. Airfields, etc.

8. **P. norvegica** L.
   10. 1917, Woodhall Spa, *F. S. Alston*. 1965, Dogdyke, *Z. Porter*.
   Casual.

11. **P. tabernaemontani** Aschers. *Spring Cinquefoil*
    14. 1953, Wilsford, *B. Howitt and E. J. Gibbons*.
    Native. Very rare.

13. **P. erecta** (L.) Rausch. *Common Tormentil*
    1. 1815, Isle of Axholme, *Peck*.
    Not recorded for Divs. 9, 18.
    Native. Common, although drainage, forestation and ploughing have reduced its frequency.

14. **P. anglica** Laichard. *Trailing Tormentil*
   1. 1906, Belton turbary, *L.N.U. Meeting*.†
   2. 1912, Manton, *Peacock*.†
   3. 1877, Wrawby, *Lees*.†
   8. 1967, Burwell, *J. Gibbons*.
   10. 1900, Woodhall Spa, *L.N.U. Meeting*.
   11. 1890, Belleau, *Mackinder*. 1961, Welton Wood, *S. M. Walters*.
   13. 1866, Doddington, *Cole*.† 1964, Stapleford, *L.N.U.*
   15. 1961, Holywell, *E. J. Gibbons*.

   Native. Uncommon; peaty soil mostly.

15. **P. reptans** L. *Creeping Cinquefoil*
   10. 1820, Scrivelsby, *Ward*.
   Recorded for all Divs.
   Native. Locally common, but absent from acid sand.

215. Fragaria L.

1. **F. vesca** L. *Wild Strawberry*
   10. 1820, Somersby, *Ward*.
   Not recorded for Divs. 1, 9, 12, 18.
   Native. Frequent, but more common on calcareous soils.

216. Geum L.

1. **G. urbanum** L. *Herb Bennet, Wood Avens*
   10. 1686, Tetford, *Lister*.
   Recorded for all Divs.
   Native. Common; shady places, woods and hedgebanks on all soils.

3. **G. rivale** L. *Water Avens*
   1670, *Ray*.
   Not recorded for Divs. 1, 5, 9, 12, 17, 18.
   Native. Abundant in some districts on the Wolds on basic clay and limestone. Less common in the N.W.

   G. x intermedium occurs in all Divs. where both parents are present, in small quantity.

217. Dryas L.

[
1. **D. octopetala** L. *Mountain Avens*
   c. 1890, Great Coates — Freshney Bog, *Clement Reid* (Cordeaux). Peacock's *Rock-Soil Flora* MS. (Cambridge).
   Native in the Inter-glacial period. In peat.
]

218. Agrimonia L.

1. **A. eupatoria** L. *Common Agrimony*
    1820, Hemingby, *Ward*.
    Recorded for all Divs.
    Native. Common; absent from acid soil.

2. **A. odorata** (Gouan) Mill. [**A. procera** Wallr.]
    Fragrant Agrimony
    10. 1900, Woodhall Spa, *Stow*.† (*Nat.*, 1900, p. 241).
    Native. Very rare; possibly overlooked.

220. Alchemilla L.

3. **A. vulgaris** L. sensu lato *Lady's Mantle*
    15. 1666, Ropsley, *Merrett*.

2. **A. vestita** (Buser) Raunk
    1836, Witham-on-the-Hill, *Dodsworth*.†
    Not recorded for Divs. 6, 9, 12, 14, 17, 18.
    Native. Fairly frequent; grassland and woods.

8. **A. xanthochlora** Rothm.
    2. 1958, Broughton, *E. J. Gibbons*. Conf. S. M. Walters.
    3. 1957, Great Limber, *E. J. Gibbons*. Conf. S. M. Walters.
    7. 1958, South Willingham, *E. J. Gibbons*. Conf. S. M. Walters.
    Native. Rare; calcareous soils; wolds.

10. **A. glabra** Neygenf.
    2. 1903, Broughton, *Peacock*.† Conf. S. M. Walters.
    10. 1957, Woodhall Spa, *M. Stephenson and E. J. Gibbons*.
    Native. Rare; still present in both localities; open woods and sandy soils.

221. Aphanes L.

1. **A. arvensis** L. *Parsley Piert*
    16. 1836, Bourne, *Dodsworth*.
    Not recorded for Divs. 17, 18.
    Native. Light soil, often basic.

2. **A. microcarpa** (Boiss. & Reut.) Rothm.
    7. 1948, Holton le Moor, *E. J. Gibbons*. Conf. S. M. Walters
    Recorded for Divs. 1, 2, 3, 5, 6, 7, 10, 11, 13.
    Native. Acid sand. Locally abundant.

222. Sanguisorba L.

1. **S. officinalis** L. *Great Burnet*
   15. 1726, Grantham, *Bacon*.
   Not recorded for Divs. 17, 18.
   Native. Formerly quite common, but decreasing with the ploughing up of damp grassland.

223. Poterium L.

1. **P. sanguisorba** L. *Salad Burnet*
   15. 1726, Grantham, *Bacon*.
   Not recorded for Divs. 1, 17, 18.
   Native. Frequent; Calcareous soils.

2. **P. polygonum** Waldst. & Kit.
   8. 1896, Louth, *Lees*.
   Casual.

224. Acaena Mutis ex L.

1. **A. anserinifolia** (J. R. & G. Forst.) Druce *Pirri-pirri-bur*
   10. 1963, Kirkby Moor, *S. Monk*.
   Introduced on R.A.F. camp and established. On gravel.

225. Rosa L.

(This account follows Wolley-Dod's "Revision of British Roses, 1930-31").
C=J. H. Chandler. G=Revd. G. G. Graham. F=H. Fisher.
P=Peacock's "Check List", 1909.
*=Specimens checked or determined by Dr. R. Melville of Kew.
Names in [ ] are those used in "Flora Europaea", Vol. 2, 1968.
These are given as a general guide only. They are not always strictly synonymous.

1. **R. arvensis** Hudson *Field Rose*
   16. 1836, Thurlby, *Dodsworth*. [*R. arvensis* Hudson]
   Not recorded for Div. 9.
   var. *vulgaris* Ser., f. major Coste **16.** 1963, Careby Wood, C.
   var. *ovata* (Lej.) Desv. **16.** 1963, Careby Wood, C.*
   var. *biserrata* Crép. **15.** 1964, Heydour Quarry, C.*
   "This is probably *R. arvensis* x *canina*" — R.M.
   var. *gallicoides* (Bak.) Crép. **15.** 1965, Morkeny Wood, C.*
   Native. Fairly common on the clay; rare in the fens; not on acid sand. Hedges and woods on clay.
   N.B.—This var. is now known to be the hybrid *R. arvensis* x *rubiginosa*.*

2. **R. pimpinellifolia** L. [**R. spinosissima** L.] *Burnet Rose*
   var. *pimpinellifolia*.
   6. 1842, Newton Cliff, *Hawkins*. [*R. pimpinellifolia* L.]
   1855, *Cole*.*
   8. 1893, Raithby, *E. Larder**. (Garden escape?).
   Doubtfully native in Lincs. Rare. Now extinct.

3. **R. stylosa** Desv.
   9. 1855, Saltfleet, *Bogg*.* [*R. stylosa* Desv.]
   var. *desvauxiana* Ser. **9.** 1855, Saltfleet, *Bogg*.*
   The only record in the County for this Southern rose. See under *R. canina* for a hybrid.

4. **R. canina** L. agg. *Dog Rose*
   16. 1836, Bourne, *Dodsworth*.
   The aggregate species recorded for all divs.
   Native. Generally common, but scarce on the chalk.

   GROUP I **Lutetianae** [*R. canina* L. s.s. pro parte]

   var. *lutetiana* (Lem.) Baker
   Common.
   7. 1877, Bleasby, *F. A. Lees*.
   15. —, Caythorpe, *S. C. Stow*.*
   16. 1964, Aunby, C.
   var. *sphaerica* (Gren.) Dum.
   3. 1890-1900, Barrow-on-Humber, *E. M. Uppleby*.*
   8. 1893, Legbourne, *J. Larder*.
   var. *flexibilis* (Déségl.) Rouy.
   8. 1893, Kenwick, *J. Larder*.*
   16. 1964, Uffington, C. and G.*
   var. *senticosa* (Ach.) Baker.
   8. 1893, Legbourne, *J. Larder*.*

   GROUP II **Transitoriae**. [*R. canina* L. s.s. pro parte]

   var. *spuria* (Pug.) W. Dod. **16.** 1967, Carlby, C.
   var. *spuria f. syntrichostyla* (Rip.)
   Rouy. **16.** 1964, Uffington, C. & G.*
   var. *globularis* (Franch.) Dum. **12.** 1909, Boston, *L.N.U.**
   15. 1964, Holywell, C.

   GROUP III **Dumales** [*R. squarrosa* (Rau.) Boreau]

   var. *dumalis* (Bechst.) Dum.
   Common.
   13. 1868, Doddington, *Cole*.*
   15. { ? ? P.
   { 1964, Denton, C.
   16. { ? ? P.
   { 1964, Uffington, C. & G.
   var. *dumalis f. vividicata* (Pug.)
   Rouy.
   15. 1900, Caythorpe, *C. S. Stow*.†
   16. 1964, Stamford, C.
   var. *dumalis f. cladoleia* (Rip.)
   W. Dod.
   16. 1964, Uffington, C. & G.

THE COUNTY FLORA

var. *stenocarpa* (Déségl.) Rouy.   **16.** 1964, Aunby, C.
var. *medioxima* (Déségl.) Rouy.   **15.** 1964, Holywell, C.
var. *biserrata* (Mér.) Baker.   **7.** 1877, Linwood, *F. A. Lees.*
   15. 1964, Holywell, C.
var. *carioti* (Chab.) Rouy.   **16.** 1964, Uffington, C.
var. *fraxinoides* H. Br.   **16.** 1964, Uffington, C.
var. *fraxinoides f. recognita* Rouy.   **8.** 1893, Louth, *J. Larder.**
   15. { 1904, Holywell, *L.N.U.**
      { 1964, Hollywell, C.
   16. 1964, Uffington, C. & G.*
var. *schlimperti* Hofm.   **15.** 1964, Holywell, C.
   16. 1964, Uffington, C. & G.
var. *sylvularum* (Rip.) Rouy.   **8.** 1894, Tathwell, *J. Larder.**
   15. { 1900, Caythorpe, *C. S. Stow.**
      { 1964, Little Bytham, C.*
var. *sylvularum f. adscita* (Déségl.) Rouy.   **15.** 1964, Holywell, C.
   16. 1967, Carlby, C.

GROUP IV Andegavenses.

[*R. andegavensis* Bast.]

var. *andegavensis* (Bast.) Desp.   **13.** { 1891, Swinderby, F.
      { 1932, Bassingham, F.
   15. { 1934, Belton Park, F.
      { 1934, Boothby Pagnell, F.
var. *andegavensis f. surculosa* (Woods) Hook.   **8.** 1893, Cawthorpe, *J. Larder.*
var. *verticillacantha* (Mév.) Baker.   **13.** { 1891, Swinderby, F.
      { Norton Disney, F.

GROUP V Scabratae

[*R. nitidula* Besser]

var. *blondaeana* (Rip.) Rouy.   **7.** P.
   13. { 1868,   Cole.
      { 1934, Norton Disney, F.
   14. 1934, Dembleby, Wilsford, N. & S. Rauceby, F.
   15. { 1934, Ancaster, Barkston, Castle Bytham,
      {   Gt. Ponton, Pickworth, F.
      { 1964, Heydour Quarry, C.
      { 1965, Leadenham Quarry, C.
   18. 1945, Gedney, *N. D. Simpson.*
var. *blondaeana f. vinacea* (Bak.) Rouy.   **7.** P.
   13. 1934, Norton Disney, F.
   14. { 1934, Dembleby, F.
      { ,, Haydor, F.
      { ,, Wilsford, F.
   15. { 1933, Ancaster, F.
      { ,, Pickworth, F.

HYBRID

R. canina x stylosa.   **16.** 1964, Uffington, C. & G.*

5. **R. dumetorum** Thuill. *Dog Rose*
13. 1855, Doddington, *Cole*.
Probably generally distributed, but not so common as *R. canina*.
Native. Hedges and scrub.

GROUP I **Pubescentes**   [*R. corymbifera* Borkh.]
var. *typica* W. Dod.   15. 1964, Holywell, C.
var. *typica f. urbica* (Lem.)
W. Dod.   7.   P.
    8. 1893, Hubbard's Valley,
         J. Larder.
    9, 13.   P.
    15. 1964, Heydour Quarry, C.
    16. 1964, Uffington, C. & G.*
var. *typica f. semiglabra* (Rip.)
W. Dod.   15. 1964, Holywell, C.*
    16. 1964, Uffington, C. & G.
var. *ramealis* (Pug.) W. Dod.   16. 1964, Uffington, C. & G.*
var. *calophylla* Rouy.   15. 1964, S. Witham, C.
var. *hemitricha* (Rip.) W. Dod.   13. 1868, Doddington, *Cole*.*
    16. 1964, Uffington, C.

GROUP II **Déséglisei**   [*R. déséglisei* Boreau]
var. *fanasensis* R. Kell.   15. 1965, Leadenham, C.*
    16. 1964, Uffington, C. & G.*
(N.B.—This is probably the hybrid R. afzeliana x rubiginosa—Dr. R. Melville).

6. **R. afzeliana** Fries [**R. glauca** Vill. non Pourret; **R. dumalis** (Bechst.) sensu C.T.W. pro parte]
(N.B.—*R. dumalis* (Bechst.) sensu C.T.W. includes both *R. afzeliana* Fr. and *R. coriifolia* Fr., and must not be confused with *R. canina* var. *dumales* (Bechst.) Dum as used by W. Dod. See (4) Group III).
1868, Doddington, *Cole* (P.).
Distribution uncertain; stated by F. A. Lees (1877) to be common in the North; now uncommon in the South of the County.
The aggregate is recorded by Peacock from Divs. 7, 13, 15.
Native. Hedges and scrub.

GROUP I **Reuterianae**.   [*R. vosaglaco* Desportes]
var. *reuteri* (Godet) Cott.   13. 1891, Swinderby, F.
var. *glauco* phylla (Winch)   6. 1907, Newton Cliff, *Peacock*.*
W. Dod.   13. 1891, Swinderby, F.
    (untypical).
         P.
15. { 1968, Cabbage Hill, C.
    16. 1969, Stamford, C.

GROUP II **Subcaninae**.
         [*R. subcanina* (Christ.) Dalla Torre and Sarnth]
var. *glandulifera*. R. Kell.   15. 1970, Holywell, C.

7. **R. coriifolia** Fries [R. **dumalis** (Bechst.) sensu C.T.W. pro parte.]
    7. 1877, F. A. Lees.
    Distribution unknown. Uncommon-rare, Native.

    GROUP I **Typicae**.   [R. caesia Sm.]
    var. *typica* (Christ.) f. *implexa* (Gren.). W. Dod.
    13. { 1891, Swinderby, F.
          „ Thurlby, F.
    15. 1969, Ancaster, C.

    GROUP II **Subcollinae**.
    var. *caesia* (Sm.). W. Dod.   7.   , Lees (P.).

8. **R. obtusifolia** Desv.
    [R. *obtusifolia* Desv.]
    11. 1890, Skegness, H. Fisher.
    Distribution not worked out; probably widespread but not common. Native. Hedges and scrub.

var. *tomentella* (Lem.) Baker.	8.	1894, Raithby, *Baker*.*
(Burrt, Davy and Lees, 1891, but	10.	1900, Caythorpe, *Stow*.*
no locality; Outline Flerg 1892)	15.	1934, Castle Bytham, Ropsley Rise, Sedgebrooke, F.
var. *decipiens* (Dum.) W. Dod.		
f. *glandulosa* (Crép.) W. Dod.	15.	1934, Ropsley Rise, F.
var. *borreri* (Woods) W. Dod.	8.	P.
	11.	1890, Skegness, F.
	12.	P.
	13.	1891, Thurlby Moor, F.
	14.	1912, Wilsford, Peacock.*
	15.	1934, Ropsley Rise, F.
var. *rothschildii* (Druce) W. Dod.	16.	1933, Nr. Stamford, Mrs. C. L. Wilde
var. (?)	15. {	1964, Holywell, G. VC 54, 1891, ?*Burtt-Davy and Lees*.

9. **R. villosa** L.   *Downy Rose*
    7. 1877, Market Rasen, F. A. Lees.   [R. *mollis* Sm.]
    Acid woodland.
    Probably native. Rare. A Northern rose. Probably confined to VC 54.

var. *mollis* (Sm.) W. Dod.	2.	1894, Broughton Wood, *Peacock*.*
	7.	1877, Market Rasen, *Lees*.
var. *mollis* f. *glandulosa*. W. Dod.	8.	1893, Little Welton, *J.Larder**
(var. ?)	7.	1967, Middle Rasen, E. J. Gibbons.
(See note after R. *tomentosa*).		
f. *annesiensis* (Déségl.) R. Kell.	7.	Walesby.*

10. **R. sherardii** Davies

[*R. sherardii* Davies]

13. 1868, Doddington, *Cole*.

Another Northern rose near its southern limits and of rare occurrence.

Native. Gravel.

var. *typica*. W. Dod.	6.	1895, Newton-on-Trent, *Peacock*.*
f. *uncinata* (Lees.) W. Dod.	10.	Woodhall Spa.*
var. *omissa* (Déségl.) W. Dod,		
f. *resinosoides* (Crép.) W. Dod.	10.	1945, Woodhall Spa, *N. D. Simpson*
var. *suberecta* (Ley) W. Dod.	13.	1868, Doddington, *Cole*.*
f. *glabrata* Ley (untypical)*		

11. **R. tomentosa** Sm.

13. 1868, Eagle, *Rev. R. E. G. Cole*.

Distribution unknown. Rather uncommon.

Native. Limestone.

GROUP I **Typicae**.

[*R. tomentosa* Sm.]

var. *typica*. W. Dod.	15.	⎧ 1904, Holywell, *L.N.U.** ⎨ 1933, Pickworth, Ropsley Rise, Witham, F. ⎩ 1964, Holywell, C.*
var. *pseudocuspidata* (Crep.) Rouy.	15.	1900, Caythorpe, *C. S. Stow*.*
var. *dimorpha* (Bess.) Déséglise	15.	1963, Twyford Forest, C.*

GROUP II **Scabriusculae**

[*R. scabriuscula* Sm.]

var. *scabriuscula* Sm.   14. 1882, Sleaford, *G. Webster*.

Note: Peacock ("Check List of Lincolnshire Plants", 1909) combines *R. mollis*. Sm. with *R. tomentosa* Sm. to give an aggregate species *R. mollissima* Willd. for which he gives a divisional distribution of 2, 6, 7, 9, 12-14 and 16, but these records cannot be allocated to the species in the present account.

12. **R. rubiginosa** L.   *Sweet Briar*

16. 1837, Thurlby, *Dodsworth*.

Native. Locally common on the limestone about Ancaster and Holywell; rare elsewhere. Peacock recorded (1909) the aggregate from Divisions 1, 2, 3, 8, 9, 10, 15, 16.

var. *typica*. W. Dod.	9.	1890, Mablethorpe, *F. A. Lees*.*
	15.	⎧ 1964, Holywell, C.* ⎩ 1964, Heydour Quarry, C.

13. **R. micrantha** Sm. [*R. micrantha* Borrer ex Sm.]
  10. 1882, Woodhall, *Melvill* (P.).
  Status and distribution unknown. No recent record.

14. **R. agrestis** Savi.
  13. 1851, Thorpe-on-the-Hill, *Watson*.
  Native. Local and rare.
  The only other records are by H. Fisher *viz*. 1892, Swinderby to Thorpe-on-the-Hill and 1893, N. Scarle. No recent record.

226. Prunus L.

1. **P. spinosa** L. *Blackthorn, Sloe*
  12. 1820, Boston, *Thompson*.
  Recorded for all Divs.
  Native. Frequent on heavy clay soil but rare on sand. Used on Lincolnshire coast for making "mattresses" for the maintenance of sea banks.

2. **P. domestica** L.
  2. 1866, Bottesford, *Peacock*.
  Distribution unknown.
  Introduced. Much planted in hedges.

4. **P. avium** (L.) L. *Gean, Wild Cherry*
  6. 1836, Lea. Simpson Collection.
  Not recorded for Divs. 9, 12, 14, 18.
  Native. Extensively planted in many areas; bird sown; mainly basic soils.

6. **P. padus** L. *Bird Cherry*
  2. 1877, Frodingham, *Parsons*.†
  Recorded for Divs. 2, 4, 7, 8, 11, 13.
  Planted. Possibly native in some localities.

227. Cotoneaster Medic.

1. **C. microphyllus** Wall. ex Lindl.
  Introduced. Colonist; established in quarry at Hibaldstow (1958, E. J. Gibbons). Site destroyed 1962.

229. Crataegus L.

1. **C. oxyacanthoides** Thuill   *Midland Hawthorn*
   7. 1855, Hainton, *E. B. Bogg.*†
   Not recorded for Divs. 1, 2, 9, 17.
   Native. Abundant in the Wragby/Horncastle district; occasional elsewhere; in old woods.

2. **C. monogyna** Jacq.   *Hawthorn*
   12. 1820, Boston, *Thompson.*
   Recorded for all Divs.
   Native. Colonizing as scrub occasionally; extensively planted for hedges.

232. Sorbus L.

1. **S. aucuparia** L.   *Rowan, Mountain Ash*
   16. 1836, Bourne, *Dodsworth.*
   Not recorded for Divs. 9, 18.
   Native. Woods and hedges, chiefly on acid soils.

5. **S. aria** (L.) Crantz sensu lato   *White Beam*
   7. 1856, Benniworth, *T. W. Bogg.*
   Introduced. Occasionally planted and bird sown.

7. **S. torminalis** (L.) Crantz   *Wild Service Tree*
   2. 1894, Scawby, *Mason.*
   6. 1965, Lea, *J. Gibbons.*
   7. 1830-70, Bardney, *Mossop.* 1890, Hatton, *Jarvis.*
   10. 1958, Woodhall Spa, *E. J. Gibbons.*
   15. 1904, Holywell, *F. Woolward.*
   16. 1836, Thurlby, *Dodsworth.* 1895, Dunsby, *Mason and Peacock.*†
       1904, Careby Wood, *S. C. Stow.*†
   Native. Locally common in many woods around Wragby and Bourne. Planted in Divs. 4 and 14.

233. Pyrus L.

1. **P. communis** L.   *Pear*
   7. 1878, Walesby, *Lees.*
   Recorded for divs. 1, 2, 7, 10, 11.
   Casual. Bird-sown; isolated trees mostly.

234. Malus Mill.

1. **M. sylvestris** Mill.   *Crab Apple*
   18. 1200, *William of Malmesbury.*
   Recorded for all divs.
   Native. Used in hedges; also as isolated trees.

CRASSULACEAE
235. Sedum L.

2. **S. telephium** L. *Orpine, Livelong*
   1. *c.* 1948, Haxey, *E. J. Gibbons.* Casual?
   2. 1875, Broughton, *Fowler.*
   7. 1950, Middle Rasen, *M. Gibbons.*
   8. 1878, Louth, *W. W. Mason.*
   13. 1941, Nocton, *L.N.U. Meeting.*
   15. 1962, Irnham, *L.N.U. Meeting.* 1965, Morkery Wood, *L.N.U.*
   16. 1838, Bourne, *Dodsworth.*

   Native. Rare; occasionally as a casual.

   ssp. *purpurascens* (Koch) Syme.
   2. 1895, Broughton, *Lees, Fowler and Peacock.*‡

5. **S. anglicum** Huds. *English Stonecrop*
   3. 1966, Killingholme, *G. S. Phillips.*
   9. 1883, Mablethorpe, *Mackinder.*
   15. 1897 Great Ponton, *Mason.*
   16. Check List.

   Garden escape. Probably.

6. **S. album** L. *White Stonecrop*
   3. 1865, Hundon, *Britten.*
   10. 1927, Salmonby, *Mason.* Naturalized garden escape.
   15. 1894, Hough, *S. C. Stow.*†
   16. 1906, Rippingale, *Peacock.*†

   Garden escape.

8. **S. acre** L. *Wall-pepper*
   13. 1829, Lincoln, *Rev. R. J. Bunch.*
   Recorded for all Divs.

   Native. Dry places, especially railway stations.

9. **S. sexangulare** L. *Stonecrop*
   15. 1956, Sewstern, *N. Saunders.*
   16. 1959, Gedney, *J. H. Chandler.*

   Garden escape.

10. **S. forsteranum** Sm. *Rock Stonecrop*
    8. *c.* 1920, Louth, *D. Marsden.*
    9. 1891, Mablethorpe, *Davy.*

    Garden escape.

11. **S. reflexum** L.
    17. 1838, Swineshead, *Dodsworth.*
    Recorded for Divs. 2, 8, 10, 11, 12, 15, 17.

    Garden escape. Naturalized in many localities.

236. Sempervivum L.

1. **S. tectorum** L. *Houseleek*
    18. 1661, Spalding, *Ray*.
    Not recorded for Divs. 1, 5, 6, 13.
    Introduced; well established. "Regarded as a protection against lightning" — hence found on the roofs of thatched houses in the past.

237. Crassula L.

1. **C. tillaea** L.-Garland
    13. 1949, Stapleford, *B. Chalk, Butcher and E. J. Gibbons*.
    Native. A continuation of the Nottinghamshire locality.

238. Umbilicus DC.

1. **U. rupestris** (Salisb.) Dandy *Pennywort, Navelwort*
    7. 1877, Tealby, *Lees*.
    10. 1727, Tattershall Castle, *Blair*.
    Introduced. According to W. W. Mason (1904), "Two hot summers rendered it extinct in these habitats." (*British Plant Life*, W. B. Turrill, p. 257 is misleading).

SAXIFRAGACEAE

239. Saxifraga L.

8. **S. tridactylites** L. *Rue-leaved Saxifrage*
    18. 1728, Spalding, *Stukeley*. 1959, *J. Gibbons*.
    Not recorded for Divs. 4, 8, 17.
    Native. Occasional; sandy ground and walls.

9. **S. granulata** L. *Meadow Saxifrage*
    14. 1790, Threckingham, *Cragg*.
    Not recorded for Divs. 9, 17, 18.
    Native. Locally abundant, but decreasing.

242. Chrysosplenium L.

1. **C. oppositifolium** L. *Opposite-leaved Golden Saxifrage*
    12. 1597, Boston, *Gerarde*.
    Recorded for Divs. 3, 5, 7, 10, 12, 14, 15.
    Native. Locally frequent in divisions 7, 10, 15; rare elsewhere; wet shady places by streams.

THE COUNTY FLORA

2. **C. alternifolium** L. *Alternate-leaved Golden Saxifrage*
   - **7.** 1867, Claxby, *Brewster.*
   - **10.** 1892, Keal, *Dr. Burgess.* 1958, *M. N. Read.* 1965, Tattershall, *E. J. Gibbons.*
   - **15.** *c.* 1880, Near Grantham, *Browne.* Doubtful.

   Native. Rare; similar habitats to *C. oppositifolium* L.

PARNASSIACEAE
243. Parnassia L.

1. **P. palustris** L. *Grass of Parnassus*
   - **1.** 1847, Craiselound, *T. V. Wollaston.* Extinct.
   - **2.** 1875, Hemp Dyke, *Crosby, Fowler.* Extinct.
   - **4.** 1896, Freshney Bog, *Dr. Burgess.* Extinct?
   - **5.** 1893, Scotter, *Peacock;*† last seen 1932. 1937, Waddingham, *L.N.U. Meeting.* Destroyed 1963. Extinct.
   - **6.** 1876 and 1895, Hackthorn, *W. W. Mason.*† Extinct.
   - **8.** 1820, Hemingby, *Ward.* Extinct.
   - **10.** 1920, High Toynton, *H. Carlton.*
   - **11.** 1888, Tothill, *S. Allett.* Skendleby, *Mossop* (painting).
   - **13.** 1829, Canwick, *J. F. Wray.* Extinct. *c.* 1836, Branston, *Simpson Collection.* Extinct.
   - **15.** 1906, Stroxton Bog, *S. C. Stow.*
   - **16.** 1839, Dunsby Springs, *Dodsworth.* Extinct.

   Native. Becoming extinct through drainage; wet, boggy places.

GROSSULARIACEAE
246. Ribes L.

1. **R. sylvestre** (Lam.) Mert. and Koch [**R. rubrum** L.]
   Red Currant
   - **16.** 1836, Thurlby, *Dodsworth.*
   - Not recorded for Divs. 5, 12, 15, 17.

   Native? Scattered; commoner than *R. nigrum* L.; woods and waste places.

3. **R. nigrum** L. *Black Currant*
   - **16.** 1878, Bourne, *Fowler.*
   - Recorded for Divs. 2, 3, 4, 7, 8, 10, 11, 12, 13.

   Native? Scattered; wet places and willow holts.

5. **R. alpinum** L. *Mountain Currant*
   - **3.** 1893, Barrow-upon-Humber, Uppleby.† *c.* 1930, Brocklesby, *Noel. B.S.B.I. Report.*
   - **6.** 1949, Lincoln, *R. E. Taylor.* Male.
   - **7.** 1954, Girsby, *L.N.U. Meeting.* Male.

   Not native.

6. **R. uva-crispa** L. *Gooseberry*
    16. 1836, Thurlby, *Dodsworth*.
    Not recorded for Divs. 15, 17.

Native? Often as escape; woods and hedges near villages.

DROSERACEAE
247. Drosera L.

1. **D. rotundifolia** L. *Sundew*
   1. 1898, Epworth, *Sam Hudson*. 1950, B.S.B.I. 1959, *E. J. Gibbons*.
   2. 1900, Scunthorpe, *Mason*. Extinct. 1880's, Manton, *Fowler*. Extinct. 1964, Messingham, *A. E. Smith*.
   3. 1840, Wrawby, *Elwes*. Extinct.
   5. 1847, Scotter, *T. V. Wollaston*.
   7. 1877, Linwood, *Lees*.
   10. 1724, Tattershall, *Stukeley*. 1893, Woodhall, *Fowler*.†
   12. 1829, Wainfleet, *Oldfield*. Extinct.
   13. 1807, Stapleford, *Ordoyno*. Extinct. 1851, Thorpe, *Watson*. Extinct. 1855, Doddington, *Cole*. Extinct.†
   15. *Check list* (details unknown).

Native. Uncommon; decreasing.

2. **D. anglica** Huds. *Great Sundew*
   1. 1890, Epworth, *Hudson*. (Peacock M.S. Flora).
   2. 1880, Manton, and at Bagmoor, *Fowler*.
   5. 1840, Laughton, *Owston*.† 1843, Scotter, *Wollaston*. 1887, *Fowler*.

Native. Extinct. (Not seen since drought — April 4th to October 21st, 1893).

3. **D. intermedia** Hayne *Long-leaved Sundew*
   1. Haxey and Epworth, see *The Naturalist*, 1898, p. 336. Extinct.
   2. 1880, Manton, *Fowler*. 1847, Twigmoor, *Anderson*. Extinct.
   3. 1855, Nettleton, *J. Daubney*. Extinct.
   5. 1847, Scotter and Laughton, *T. V. Wollaston*.
   7. 1879, Linwood, *Lees*.
   10. 1820, Tattershall, *Ward*. Extinct. 1891, Woodhall, *F. Alston*.† (See *L.N.U. Trans.*, 1919, p. 55). Extinct.
   12. 1820, Wainfleet, *Oldfield*. Extinct.
   13. 1839, Stapleford, *G. Howitt*. Extinct.

Native. Rare; decreasing.

LYTHRACEAE
249. Lythrum L.
1. **L. salicaria** L. *Purple Loosestrife*
   12. 1799, East Fen, *Young*.
   Recorded for all Divs.
   Native. Decreasing due to spraying by River Authority and mechanised drainage.
   L. hyssopifolia L. and **L. junceum** (Banks) Sol. occur on dumps.

250. Peplis L.
1. **P. portula** L. *Water Purslane*
   13. 1851, Thorpe, *Watson*.
   Recorded for Divs. 2, 5, 6, 7, 12, 13, 14, 16.
   Native. Scarce; damp edges of pools.

THYMELAEACEAE
251. Daphne L.
1. **D. mezereum** L. *Mezereon*
   10. 1879, Hundleby, *Burgess*. 1960, Tattershall, *Porter*.
   15. 1856, Humby, *Browne*.
   Introduced. Regenerating.

2. **D. laureola** L. *Spurge Laurel*
   12. 1829, Wainfleet, *Oldfield*. 1957, Fishtoft, *Gibbons*.
   Not recorded for Divs. 1, 5, 10, 17, 18.
   Native. Rather scarce; in woods and hedges on strong clay or chalk soils.

ELAEAGNACEAE
252. Hippophae L.
1. **H. rhamnoides** L. *Sea Buckthorn*
   1669, J. Ray to M. Lister, *Phil.* Letters of J. Ray, Derham Edition, ed. 1748. "Dr. Mapletoft was informed it grew wild on the sea coast of Lincs. which you now confirm."
   4. 1963, Cleethorpes, *G. Newton*.
   9. 1893, Mablethorpe, *E. Uppleby*.†
   11. 1862, Skegness, *Mason*.
   1970, S. Thoresby Gravel Pit, inland. *E. J. Gibbons*.
   12. 1957, Wainfleet, *B.S.B.I.*
   Native. Locally abundant on sand dunes.
   (Pearson & Rogers, *J. Ecol.* **50**, p. 501-513; Groves, *Proc. B.S.B.I.*, **3.**1.1958).

ONAGRACEAE
254. Epilobium L.

1. **E. hirsutum** L. *Great Hairy Willow-herb*
    12. 1799, East Fen, *Young*.
    Recorded for all Divs.
    Native. Locally abundant.

2. **E. parviflorum** Schreb. *Hoary Willow-herb*
    16. 1836, Bourne, *Dodsworth*.
    Recorded for all Divs.
    Native. Less frequent than the last, and less gay.

3. **E. montanum** L. *Broad-leaved Willow-herb*
    3. 1835, Wootton, *E. J. Nicholson*.
    Recorded for all Divs.
    Native. A frequent weed in gardens; also in woods.

5. **E. roseum** Schreb. *Pale Willow-herb*
    5. 1951, Gainsborough, *L.N.U.*
    6. 1945, Lincoln Waterside, *N. D. Simpson and A. H. Alston*.
    16. 1896, Bourne, *Stow*.
    Native. Rare and sometimes confused with other spp.

6. **E. adenocaulon** Hausskn. *American Willow-herb*
    15. 1956, Skillington, *Chandler*.
    Recorded for Divs. 3, 7, 11, 15 (and maybe more?)

    Spreading in forestry plantations and quarries; native of North America.

7. **E. adnatum** Griseb [**E. tetragonum** L.]
    Square-stemmed Willow-herb
    13. 1851, Boultham, *Watson*.
    Not recorded for Divs. 12, 18.
    Native, uncommon, decreasing.

9. **E. obscurum** Schreb. *Short-fruited Willow-herb*
    6. or 13. 1837, Lincoln, *Deakin*.
    Not recorded for Divs. 3, 4, 5, 6, 9, 14.
    Native. Should be looked for.

10. **E. palustre** L. *Marsh Willow-herb*
    16. *c.* 1822, Braceborough, *Twopeny*.
    Not recorded for Divs. 17, 18.
    Native; wet places on acid soil.

THE COUNTY FLORA

255. Chamaenerion Adams. [Epilobium L.]

1. **C. angustifolium** (L.) Scop.   *Rose-bay Willowherb, Fireweed*
   1686, *Ray*.
   Recorded for all Divs.
   Native. Dominant in felled woodland.

256. Oenothera L.

1. **O. biennis** agg.   *Evening Primrose*
   11. 1886, Spilsby, *Burgess*.
   Not recorded for Divs. 1, 7, 8, 12, 15, 17.
   Introduced.

258. Circaea L.

1. **C. lutetiana** L.   *Common Enchanter's Nightshade*
   8. 1834, Louth, *Bayley*.
   Not recorded for Div. 1.
   Native. In calcareous woods; ineradicable as a garden weed.

HALORAGACEAE

259. Myriophyllum L.

1. **M. verticillatum** L.   *Whorled Water-milfoil*
   12. 1799, East Fen, *Young*.
   Not recorded for Divs. 2, 3, 4, 5, 7, 8, 10, 17.
   Native. Decreasing and becoming rare; peat fen, in still water.

2. **M. spicatum** L.   *Spiked Water-milfoil*
   12 or **17.** 1670, Near Boston, *Ray*.
   Not recorded for Div. 8.
   Native. Frequent; still water ponds.

4. **M. alterniflorum** DC.   *Alternate-flowered Water-milfoil*
   13. 1851, Lincoln, *Watson*.
   16. 1883, Deeping Fen, *Beeby*. 1911, Crowland, *Druce*.
   Native. Possibly overlooked; not fully worked.

HIPPURIDACEAE

261. Hippuris L.

1. **H. vulgaris** L.   *Mare's-tail*
   16. *c.* 1822, Braceborough, *Twopeny*.
   Recorded for all Divs.
   Native. Frequent; ponds and dykes.

CALLITRICHACEAE
262. Callitriche L.

1. **C. stagnalis** Scop. *Starwort*
   1851, *Watson*.
   Recorded for all Divs.

   Native. Common; wet mud, ponds and dykes.

2. **C. platycarpa** Kutz. [**C. polymorpha** auct. **C. verna** auct]
   12. 1856, Boston, *Thompson*.
   Not recorded for Divs. 1, 2, 4, 6, 7, 8, 17.

   Native. Frequent.

3. **C. obtusangula** Le Gall.
   2. 1877, Gunness, *Lees*.
   Recorded for Divs. 2, 5, 10, 11, 16, 17.

   Native. Distribution unknown, not fully worked; peat-loving.

4. **C. intermedia** Hoffm.
   5. 1878, Bishopbridge, *Lees*. 1933, E. Stockwith, *Willoughby-Smith*.
   6. 1963, Lincoln Race Course, *E. J. Gibbons*.
   11. 1894, Welton, *Mason*.
   15. 1958, Corby, *J. H. Chandler*.
   16. 1895, Bourne, *Mason*.†

   Native. Distribution unknown.

6. **C. truncata** Guss.
   6. 1970. Nr. Lincoln, *B. Howitt*.

LORANTHACEAE
263. Viscum L.

1. **V. album** L. *Mistletoe*
   16. 1836, Bourne, *Dodsworth*.
   Recorded for Divs. 2, 3, 8, 9, 10, 11, 15, 16.

   Introduced, but established in Div. 16. On branches of apple, horse-chestnut, poplar and hawthorn.

SANTALACEAE
264. Thesium L.

1. **T. humifusum** DC. *Bastard Toadflax*
    13. 1865, Leadenham, *Burtt*.† Extinct. 1897, Fulbeck, *L.N.U. Meeting*.† 1894, Potterhanworth, *Mason*.†
    15. 1903, Holywell, *Trollope*. 1967, *J. Gibbons*.
    16. 1940, Aunby, *S. A. Taylor*. 1958, *J. Gibbons*.

    Native. Rare; semi-parasitic on herb roots; limestone grassland. At Northern limit.

CORNACEAE
265. Thelycrania (Dumort.) Fourr. [Swida Opiz.]

1. **T. sanguinea** (L.) Fourr. *Dogwood*
    10 1820, Horncastle, *Ward*.
    Not recorded for Divs. 12, 17, 18.

    Native. In old hedgerows and woods; particularly common on the lias.

266. Cornus L.

C. alba L.
   3. 1930, Brocklesby Woods, *E. J. Gibbons*.
   15. 1963, Twyford Forest, *E. J. Gibbons and J. H. Chandler*.

   Introduced. In large woods on estates.

ARALIACEAE
268. Hedera L.

1. **H. helix** L. *Ivy*
    6 or 13. 1810, Lincoln, *Stark*.
    Recorded for all Divs.
    Native. Most variable in form.

UMBELLIFERAE
269. Hydrocotyle L.

1. **H. vulgaris** L. *Pennywort*
    15. 1746, Honington, *Blackstone*. (Misidentified as *Sibthorpia*).
    Recorded for all Divs.
    Native. Locally common; damp, acid soils.

270. Sanicula L.

1. **S. europaea** L. *Sanicle*
    10. 1820, Tetford, *Ward*.
    Not recorded for Divs. 9, 12, 17, 18.
    Native. Limited distribution; calcareous woodland.

272. Eryngium L.

1. **E. maritimum** L. *Sea Holly*
    1. 1958, W. Butterwick, *L. Smith*. Waterborne. Extinct.
    4. 1892, Cleethorpes, *Sneath*.
    9. 1834, Mablethorpe, *Bayley*. 1895, N. Somercotes, *Allett*.†
    1903, Humberstone, *Larder*. 1960, G. *Newton*.
    11. 1860, Skegness, *Mason*. 1967, *Weston*. 1889, Chapel St. Leonards, *Burgess*.
    12. 1696, Near Boston? *Merrett*.
    Native. Rare; sandy shores.

273. Chaerophyllum L.

1. **C. temulentum** L. *Rough Chervil*
    11. 1847, Burgh, *Grantham*.
    Recorded for all Divs.
    Native. Common; on shady hedge-banks.

274. Anthriscus Pers.

1. **A. caucalis** Bieb. *Bur Chervil*
    2. 1856, Broughton, *Fowler*.
    Not recorded for Divs. 1, 17, 18.
    Native. Infrequent; dry, sandy places.

2. **A. sylvestris** (L.) Hoffm. *Cow Parsley, Keck, Humlick, Ewe Bennet, Motherdie*
    14. 1790, Threckingham *Cragg*.
    Recorded for all Divs.
    Native. Very common; in woods, meadows and on road-sides.

3. **A. cerefolium** (L.) Hoffm. *Chervil*
    3. 1894, Cadney, *Peacock*.†
    7. 1893, Wispington, *Alston*.†
    Introduced. An escape; garden relic?

THE COUNTY FLORA

275. Scandix L.

1. **S. pecten-veneris** L.   *Shepherd's Needle*
    10. 1820, Hemingby, *Ward*.
    Recorded for all Divs.
    Native. Decreasing; calcareous, arble land.

276. Myrrhis Mill.

1. **M. odorata** (L.) Scop.   *Sweet Cicely*
    2. 1875, Broughton, *Fowler*.
    3. 1918, Nr. Kirmington, *Frith*.†
    8. 1896, Louth, *Goodall*.†
    12. 1856, Boston, *Thompson*.
    15. 1896, Belton, *Woolward*.
    Introduced. Uncommon; near dwellings.

277. Torilis Adans.

1. **T. japonica** (Houtt.) DC.   *Upright Hedge-parsley*
    12 or 17. 1837, Boston, *Dodsworth*.
    Recorded for all Divs.
    Native. Very common; in hedgerows, small woods, etc.

2. **T. arvensis** (Huds.) Link   *Spreading Hedge-parsley*
    6 or 13. 1851, Lincoln, *Watson*.
    Not recorded for Divs. 1, 8, 11, 17.
    Native? Uncommon and decreasing; cornfields.

3. **T. nodosa** (L.) Gaertn.   *Knotted Hedge-parsley*
    5. 1842, Lea, *Miller*.
    Not recorded for Div. 1.
    Native. Scattered distribution; dry banks.

278. Caucalis L.

1. **C. platycarpos** L.   *Small Bur-parsley*
    14. 1789, Sleaford, *Gough*.
    Recorded for Divs. 2, 3, 4, 6, 12, 13, 14.
    Occasional.

2. **C latifolia** L.   *Great Bur-parsley*
    16. 1640, Witham-on-the-Hill, *Parkinson*.
    Recorded for Divs. 4, 6, 12, 13, 14, 15, 16.
    Introduced. Cornfield casual; occasional.

279. Coriandrum L.

C. sativum L. *Coriander*

16. 1805, Folkingham, *Turner and Dillwyn.* "Wild in uncultivated places about Folkingham, very plentiful and apparently indigenous." — 1805, *Botanists' Guide.*
17. 1959, nr.Fosdyke, *E. J. Gibbons.* (In cultivation).
Recorded for Divs. 3, 6, 8, 14, 16, 17.
Escape from cultivation? To be looked for.

280. Smyrnium L.

S. olusatrum L. *Alexanders*

16. 1932, Witham-on-the-Hill, *Ridlington.* 1954, Uffington *Chandler.*
Introduced, but naturalised in these localities on roadsides; absent from the coast.

282. Conium L.

C. maculatum L. *Hemlock*

12 or 17. 1820, Boston, *Thompson.*
Recorded for all Divs.
Native. Scattered distribution, but locally common; absent from acid soil.

283. Bupleurum L.

2. **B. rotundifolium L.** *Hare's-ear*

16. 1796, Carlby, *Woodward.* 1940, *S. A. Taylor.*
Recorded for Divs. 2, 3, 4, 6, 12, 13, 14, 16.
Cornfield casual; occasional.

4. **B. tenuissimum L.** *Slender Hare's-ear*

12. 1688, Boston, *Plukenet.* 1937, Wrangle, *J. P. M. Brenan.*
Recorded for Divs. 3, 4, 9, 11, 12, 17, 18.
Native. Rare; coastal mud.

285. Apium L.

1. **A. graveolens L.** *Wild Celery*

12 or 17. 1724, Near Boston, *Stukeley.*
Not recorded for Divs. 6, 7, 8, 10, 13, 14, 15, 16.
Native. Along the Rivers Trent and Witham; and coastal salt-marshes and inland as an indication of former tidal creeks.

2. **A. nodiflorum** (L.) Lag. *Fool's Watercress*
   16. 1836, Bourne, *Dodsworth*.
   Recorded for all Divs.
   Native. Abundant in ditches, streams and ponds.
3. **A. repens** (Jacq.) Lag.
   1. 1895, Isle of Axholme, *Fowler*.
   Native. Rare; possibly overlooked.
4. **A. inundatum** (L.) Reichb. f. *Least Marshwort*
   5. 1840, Laughton, *Miller*.
   Not recorded for Divs. 8, 15, 17, 18.
   Native. Uncommon, peaty dykes.

   A. inundatum x nodiflorum — A. moorei (Syme) Druce.
   5. 1893, Walkerith, *Lees*.†
   6. 1893, Torksey, *Mills*.†
   Possibly overlooked.

286. Petroselinum Hill

1. **P. crispum** (Mill.) Nyman *Sheep's Parsley*
   8. 1856, Louth, *Bogg*.
   Not recorded for Divs. 5, 6, 10, 13, 15, 16, 17.
   Introduced. Escape of cultivation; sown with grasses and clover, but well established in chalk quarries, exposed roadsides and near the sea.
2. **P. segetum** (L.) Koch *Corn Caraway*
   11. 1840, Burgh, *Grantham*.
   Not recorded for Divs. 1, 2, 15, 18.
   Native. Northern limit; coastal and other dry places. Rare.

287. Sison L.

1. **S. amomum** L. *Stone Parsley*
   7. 1849, Newball, *Carrington*.
   Not recorded for Divs. 2, 12, 18.
   Native. Northern limit; uncommon, but frequent in the Sleaford area; clay soil.

288. Cicuta L.

1. **C. virosa** L. *Cowbane*
   12. 1799, East Fen, *Young*.
   16. 1838, Bourne, *Dodsworth*.
   6 or **13.** 1849, Near Lincoln, *Carrington*. (*Bot. Gaz.*, 1849, p. 323-4).
   Native; extinct. Needs very wet peat.

290. Falcaria Bernh.

1. **F. vulgaris** Bernh. *Long Leaf*
   - 4. 1934, Grimsby, *Cox.*†
   - 5. 1905, Grayingham, *J. G. Nicholson.*†
   - 6. 1895, Lincoln, *Higginbottom.*† 1959, Greetwell, *C. V. Sutton.* 1967, Lincoln, *Weston.*
   - 10. 1963, Tattershall, *Porter.*
   - 12. 1912, Boston, *B. Reynolds.*

   Introduced. Established in Div. 6.

291. Carum L.

2. **C. carvi** L. *Caraway*
   - 12. 1661, Boston, *Ray.* 1937, Frieston, *A. J. Wilmott* (spec. B.M.)
   - 15. 1726, Swayfield Dale, *V. Bacon.*

   Not recorded for Divs. 1, 4, 6, 7, 8, 9, 14.

   It appears definitely native around the Wash; probably introduced elsewhere. Old pastures.

293. Conopodium Koch

1. **C. majus** (Gouan) Loret *Pignut, Earthnut*
   - 16. 1836, Bourne, *Dodsworth.*

   Not recorded for Divs. 12, 17, 18.

   Native. Common; in woods and undisturbed pastures.

294. Pimpinella L.

1. **P. saxifraga** L. *Burnet Saxifrage*
   - 13. 1830, Canwick, *J. F. Wray.*

   Not recorded for Divs. 12, 17, 18.

   Native. Locally common; on dry chalky banks and roadsides.

2. **P. major** (L.) Huds. *Greater Burnet Saxifrage*
   - 5. 1842, Morton, *J. K. Miller.*

   Not recorded for Divs. 9, 17.

   Native. Scarce in extreme north of county, but frequent elsewhere; roadsides, woods and shady places.

295. Aegopodium L.

1. **A. podagraria** L. *Ground Elder*
   - 16. 1836, Bourne, *Dodsworth.*

   Recorded for all Divs.

   Introduced? Possibly native in woods in Div. 3. Very common; in old gardens and dumps on roadsides.

296. Sium L.

1. **S. latifolium** L   *Great Water Parsnip*
   12. 1826, Freiston, *Howitt*.
   Not recorded for Divs. 2, 7, 15.

   Native. Uncommon, decreasing; fen dykes and pools. Two forms have been noticed, one with leaves much broader and with larger umbels, the other taller with narrow leaves.

297. Berula Koch

1. **B. erecta** (Huds.) Coville   *Narrow-leaved Water Parsnip*
   16. 1836, Bourne, *Dodsworth*.
   Recorded for all Divs.

   Native. Frequent; ditches and pools with basic water.

298. Crithmum L.

1. **C. maritimum** L.   *Rock Samphire*
   11. 1962, Gibraltar Point, *Weston*. (Extinct 1967).
   Recorded by Stukeley (1724), but probably a mistake for *Salicornia sp.* locally called Samphire.

   Casual, perhaps seed washed up.

300. Oenanthe L.

1. **O. fistulosa** L.   *Water Dropwort*
   14. 1797, Threckingham, *Cragg*.
   Recorded for all Divs.

   Native. Not very common; in pools and marshy places.

4. **O. lachenalii** C. C. Gmel.   *Parsley Water Dropwort*
   18. 1688, Whaplode, *Plukenet* (Recorded as *O. pimpinelloides* L.).
   Not recorded for Divs. 2, 3, 10, 16.

   Native. Chiefly near the sea and occasionally inland; damp, clay soils. Local, decreasing.

5. **O crocata** L.   *Hemlock Water Dropwort*
   2. 1970, Whitton, *J. Gibbons*.
   6. 1870, Marton, *Lees*.† Extinct.
   7. 1877, North Willingham, *Lees*.
   10. 1851, Kirkstead, *Watson*.
   13. 1850, Boultham, *Carrington*.
   15. 1851, Grantham, *J. G. Baker*. (From train window).

   Native. Rare in E. Midlands.

6. **O. aquatica** (L.) Poir. *Fine-leaved Water Dropwort*
    1790, Bridge End, *Cragg*.
    Not recorded for Divs. 4, 8.
    Native. Scattered distribution; pools and stagnant water.

7. **O. fluviatilis** (Bab.) Colem.
    14. 1894, Leasingham, *Mason*.† 1895, Sleaford, *Mason*. 1908, Haverholme, *Mason*. 1960, E. J. Gibbons.
    16. 1884, Deeping, *Beeby*. 1895, Dunsby Fen, *Mason*.† 1905, Stamford, *Peacock*.† 1954, Uffington, E. J. Gibbons.
    17. 1895, Gosberton, *Mason*.
    Native. Locally frequent in Rivers Slea and Welland.

301. Aethusa L.

1. **A. cynapium** L. *Fool's Parsley*
    14. 1790, Threckingham, *Cragg*.
    Recorded for all Divs.
    Native. Common; especially in chalky cornfields.

302. Foeniculum Mill.

1. **F. vulgare** Mill. *Fennel*
    8. 1856, Louth, *Bogg*.
    Not recorded for Divs. 1, 17.
    Garden escape, but well established in many divisions; in waste places and along the coast.

303. Silaum Mill.

1. **S. silaus** (L.) Schinz & Thell. *Pepper Saxifrage*
    13. 1830, Branston, R. J. Bunch and J. F. Wray.
    Not recorded for Divs. 17, 18.
    Native. Scattered distribution; old pastures, chalk and clay meadows.

305. Selinum L.

1. **S. carvifolia** (L.) L.
    2. 1880, Broughton, *Fowler*. (*Bot. Rec. Club*, 1880, p. 156 and *Crit. Cat.*, Nov. 1894, p. 341).
    First British record.
    Native. Last seen 1931. Extinct?

307. Angelica L.

1. **A. sylvestris** L. *Wild Angelica*
   12. 1799, East Fen, *Young*.
   Recorded for all Divs.
   Native. Now rare in the Fens; locally common in woods and damp roadsides.
2. **A. archangelica** L. *Angelica*
   2. 1974, Burringham, *A. Frankish*.
   Washed down from Notts.

309. Peucedanum L.

2. **P. palustre** (L.) Moench. *Hog's Fennel, Milk Parsley*
   1. 1879, New Idle River near Sandtoft, *Fowler*.
   5. 1878-94, Laughton near Peacock Hole, *Fowler*.†
   1922-49, Waddingham, ½ mile N.E. of Brandy Wharf, *M. and E. J. Gibbons*. Extinct.
   12. 1789, East Fen, *Gough*. 1797, *Young*.
   Native. Extinct?

3. **P. ostruthium** (L.) Koch *Master-wort*
   1. 1933, Epworth, *A. Roebuck* (L.N.U. Meeting).
   18. 1958, Spalding, *J. H. Chandler*.
   Garden escape?

310. Pastinaca L.

1. **P. sativa** L. *Wild Parsnip*
   16. 1800, Stamford, *Hailstone*.
   Recorded for all Divs.
   Native. Scattered distribution; on calcareous soils.

311. Heracleum L.

1. **H. sphondylium** L. *Cow Parsnip, Hogweed*
   12. 1823, Wainfleet, *Sinclair*.
   Recorded for all Divs.
   Native. Very common; roadsides, hedgerows, woods and river banks.

314. Daucus L.

1. **D. carota** L. *Wild Carrot*
   14. 1790, Threckingham, *Cragg*.
   Recorded for all Divs.
   Native. Locally common; on calcareous soils.

CUCURBITACEAE
315. Bryonia L.
1. **B. dioica** Jacq. *White Bryony, Mandrake*
   14. 1790, Threckingham, *Cragg*.
   Recorded for all Divs.
   Native. Locally common; in hedgerows, but not on acid sand.

ARISTOLOCHIACEAE
316. Asarum L.
1. **A. europaeum** L. *Asarabacca*
   2. 1884, Manby Hall, *Woolward*. 1896, Broughton, *Peacock*.†
   1961, *Powell*.
   Garden relic.

317. Aristolochia L.
1. **A. clematitis** L. *Birthwort*
   5. 1838, Hemswell, *Irvine*.
   15. 1850, Harlaxton, *Skipworth*.
   Recorded for Divs. 2, 5, 10, 15.
   Alien. In old gardens.

EUPHORBIACEAE
318. Mercurialis L.
1. **M. perennis** L. *Dog Mercury*
   7. 1829, Bardney, *Rev. J. F. Wray*.
   Not recorded for Divs. 9, 12, 17, 18.
   Native. Locally abundant; in shady places, but not on acid soils.
2. **M. annua** L. *Annual Mercury*
   11. 1847, Burgh, *Grantham*.
   Recorded for Divs. 4, 6, 10, 11, 12, 15, 18.
   Casual. Rare; gardens and waste places; weed of cultivation.

319. Euphorbia L.
2. **E. lathyrus** L. *Caper Spurge*
   15. 1862, Great Gonerby, *Browne*.
   Not recorded for Divs. 1, 4, 5, 9, 13, 16, 17.
   Introduced. Scattered distribution; relic of cultivation.

THE COUNTY FLORA

7. **E. platyphyllos** L. *Broad Spurge*
   6. 1943, Welton, *M. Gibbons*.† Extinct. 1962, Torksey, *Howitt*.
   7. 1877, Middle Rasen, *Lees*.
   16. 1938, Carlby, *S. A. Taylor* (Herb. Leics.). 1953, *J. Gibbons*.
   Native? Very rare; cornfield weed.

9. **E. helioscopia** L. *Sun Spurge*
   7. 1829, Bardney, *Rev. J. F. Wray*.
   Recorded for all Divs.
   Native. Very common; cultivated ground.

10. **E. peplus** L. *Petty Spurge*
    16. 1836, Bourne, *Dodsworth*.
    Recorded for all Divs.
    Native. Very common; cultivated ground.

11. **E. exigua** L. *Dwarf Spurge*
    16. 1836, Witham-on-the-Hill, *Dodsworth*.
    Not recorded for Divs. 9, 17.
    Native. Common; on chalky, cultivated ground.

12. **E. portlandica** L. *Portland Spurge*
    11. Before 1897, Skegness, *Lees*.† (See *Naturalist*, 1899, p. 39, 1900, p. 86).
    Casual.

14. **E. uralensis** Fisch ex Link.
    3. 1968, Kirmington Airfield, *F. Lammiman*.
    Casual.

16. **E. cyparissias** L. *Cypress Spurge*
    3. 1918, Wrawby, *Frith*.†
    15. 1945, Bassingthorpe, Railway Bank, *Gibbons*.
    Garden escape.

17. **E. amygdaloides** L. *Wood Spurge*
    2. 1909, Black Holt, Brumby, *Hawkins*. (Doubtful).
    15. 1957, Little Bytham, Twyford Forest, *J. H. Chandler*.
    16. 1838, Bourne, *Dodsworth*. Also at Careby, Uffington, Edenham and Auster Wood, *J. H. Chandler*, 1954-60.
    Native. Rare, but fairly frequent in Division 16; calcareous woods. Northern limit.

POLYGONACEAE
320. Polygonum L.

1. **P. aviculare** L. *Knotgrass*
   14. 1790, Threckingham, *Cragg*.
   Native. Recorded for all divisions; common.

2. **P. raii** Bab. *Ray's Knotgrass*
   4. 1865, Cleethorpes, *Britten*. 1868, *Charters*.
   11. 1872, Gibraltar Point, *Streatfeild*. 1968, *L.N.U.*†
   Native. Rare. To be searched for.

5. **P. viviparum** L. *Alpine Bistort*
   13. 1746, Lincoln Heath, *Hill*.
   Presumably a mistake copied by others.

6. **P. bistorta** L. *Snake-root, Bistort*
   10. 1820, Horncastle, *Ward*.
   3. 1963, Limber, *R. May*.
   Not recorded for Divs. 1, 7, 9, 11, 17, 18.
   Doubtful native.

8. **P. amphibium** L. *Amphibious Bistort*
   12. 1799, East Fen, *Young*.
   Recorded for all Divs.
   Native.

9. **P. persicaria** L. *Persicaria, Red-leg, Willow Weed*
   16. 1836, Bourne, *Dodsworth*.
   Recorded for all Divs.
   Native.

10. **P. lapathifolium** L. *Pale Persicaria*
    6 or **13.** 1830, Lincoln, *R. J. Bunch*.
    Recorded for all Divs.
    Native.

11. **P. nodosum** Pers. *Knotted Persicaria*
    5. 1865, Gainsborough, *Charters*.
    Recorded for all Divs.
    Native.

12. **P. hydropiper** L. *Water-pepper*
    7. 1829, Bardney, *J. F. Wray*.
    Not recorded for Divs. 4, 9, 12, 17.
    Native. Acid and peaty soil.

13. **P. mite** Schrank.
    5 or 6. 1862, Near Gainsborough, *Stanwell.*
    6. 1956, Torksey, *Gibbons and Howitt*, confirmed *Lousley.*
    10. 1892, Coningsby, *F. Alston.*
    18. 1911, Cowbit Wash, *Druce.*†
    Native. Very rare; possibly overlooked.

14. **P. minus** Huds.
    18. 1911, Cowbit Wash, *Druce.*†
    Native. Very rare; possibly overlooked.

15. **P. convolvulus** L. Black Bindweed
    17. 1688, Kirton, *Holden.*
    Recorded for all Divs.
    Native.

18. **P. baldschuanicum** Regel [**P. aubertii** L. Henry]
    3. 1955, Nettleton, *S. A. Cox.*
    7. c. 1948, Snarford, *E. J. Gibbons.*
    16. 1959, Tallington, *J. H. Chandler.*
    Introduced. Planted and established in these localities.

19. **P. cuspidatum** Sieb. and Zucc. *Japanese Knotweed*
    1. 1957, Isle of Axholme, *Gibbons and Howitt.*
    2. 1950, Brumby, *B.S.B.I.* 1955, Ferriby Sluice, *Dunn.*
    12. 1957, North of Boston, *P. and J. Hall.*
    13. 1957, Sincil Dyke, Lincoln, *Gibbons.*
    15. 1958, Stainby, *Gibbons and Lowe.*
    16. 1959, Stamford Railway, *J. H. Chandler.*
    18. 1939, Spalding. 1957, Holbeach area, *P. and J. Hall.*
    Introduced. Becoming naturalised.

20. **P. sachalinense** F. Schmidt
    14. 1962, Aswarby Gorse, Green Hill, *Gibbons.*
    Introduced.

21. **P. polystachyum** Wall ex. Meisn.
    10. 1919, Woodhall Spa (chicken farm), *F. Alston.*
    Introduced.

321. Fagopyrum Mill.

1. **F. esculentum** Moench Buckwheat
    13. 1862, Doddington, *Cole.*†
    Recorded for Divs. 2, 3, 5, 6, 7, 12, 13, 16.
    Introduced. Pheasant food or relic of cultivation.

325. Rumex L.

1. **R. acetosella** agg. *Sheep's Sorrel*
   2. 1822, Appleby sandhills, *Strickland*.
   Recorded for all Divs.
   Native. Rare in the south.

   R. tenuifolius (Wallr.) Löve
   7. 1951, Holton le Moor, *E. J. Gibbons*, confirmed *J. E. Lousley*.
   Recorded for Divs. 1, 2, 3, 5, 7, 10, 13, 15.
   Native. On very acid soil. Locally frequent.

2. **R. acetosa** L. *Sorrel, Sour Sauce*
   14. 1790, Threckingham, *Cragg*.
   Recorded for all Divs.
   Native. Decreasing as old turf is ploughed up.

4. **R. hydrolapathum** Huds. *Great Water Dock*
   12. 1799, East Fen, *Young*.
   Not recorded for Div. 3.
   Native. Decreasing due to pollution and mechanised dredging.

11. **R. crispus** L. *Curled Dock*
    16. 1836, Bourne, *Dodsworth*.
    Recorded for all Divs.
    Native. Chiefly round farm buildings.

12. **R. obtusifolius** L. *Broad-leaved Dock, Dockens*
    16. 1836, Bourne, *Dodsworth*.
    Recorded for all Divs.
    Native, very common.

13. **R. pulcher** L. *Fiddle Dock*
    3. 1893, South Ferriby, *Miss Firbank*.
    6. 1959, Lincoln, *Gibbons*.
    11. 1910, Ingoldmells, *H. M. Nash*.
    13. 1851, Bracebridge, *Watson*.
    15. 1922, Harrowby, *Miss S. C. Stow*. 1955, Ancaster, *J. Gibbons*.
    16. 1934, Tallington, *Miss S. C. Stow*. 1956, Careby, *J. Gibbons*.
    Native. Rare, less so in south Lincolnshire.

14. **R. sanguineus** L. *Red-veined Dock*
    1851, *Watson*.
    Recorded for all Divs.
    Native. Not uncommon in shady places, hedges, etc.

15. **R. conglomeratus** Murr. *Clustered Dock*
    1851, *Watson*.
    Recorded for all Divs.
    Native. In wetter places than *R. sanguineus*, and rather more in the open.

    R. conglomeratus x palustris.
    1. 1949, Belton, *Gibbons*, det. *Lousley*.
    6. 1960, Lea, *Gibbons*, det. *Lousley*.

17. **R. palustris** Sm. *Marsh Dock*
    16. 1780, Crowland, *Sibthorp*.
    Not recorded for Divs. 4, 7, 8, 9, 15.
    Native. Scattered and appears in some years according to the weather and water level.

18. **R. maritimus** L. *Golden Dock*
    16. 1670, Crowland, *Ray*.
    Not recorded for Divs. 4, 9, 15.
    Native. Scattered; in drier seasons round ponds; not a maritime plant.

URTICACEAE
326. Parietaria L.

1. **P. diflusa** Mert. & Koch [**P. judaica** L.] *Pellitory-of-the-Wall*
   14. 1790, Threckingham, *Cragg*.
   Recorded for all Divs.
   Native. Formerly on many churches, walls of Boston waterways, etc., but decreasing due to repairs.

328. Urtica L.

1. **U. urens** L. *Small Nettle*
   7. 1829, Bardney, *Rev. J. F. Wray*.
   Recorded for all Divs.
   Native. Usually on chicken manure.

2. **U. dioica** L. *Stinging Nettle*
   12. 1799, Wainfleet, *Young*.
   Recorded for all Divs.
   Native. Abundant.

3. **U. pilulifera** L. *Roman Nettle*
    4. 1902, Grimsby, *A. Smith*.†
    7. 1640, Bardney, *Parkinson*.
    13. 1899, Boultham, *Peacock*.

    Introduced. Casual.

CANNABIACEAE
329. Humulus L.

1. **H. lupulus** L. *Hop*
    14. 1790, Threckingham, *Cragg*.
    Recorded for all Divs.

    Native. In hedges in many villages as a relic of cultivation; indigenous in wet thickets.

Cannabis sativa L. *Hemp*
    1666, "Plenty sowed between Bourne and Horncastle," *Lister*.
    12. 1799, East Fen, *Young*.
    Recorded for Divs. 1, 3, 4, 5, 7, 9, 13, 15, 18.

    Casual. Formerly cultivated for weaving and ropes; "Hemp pits" for retting are still surviving by name or field names. Plants are found on dumps, having originated from bird seed.

ULMACEAE
330. Ulmus L.

1. **U. glabra** Huds. *Wych Elm*
    1799, *Young*.
    Recorded for all Divs.

    Generally common in suitable habitats.

    U. glabra x plotti = U. x elegantissima Horwood.
    10. 1945, Carrington, *A. H. G. Alston and Simpson*.
    15. Swinstead, *J. H. Chandler*.
    16. Newstead, *J. H. Chandler*.
    18. Gedney Dyke, *J. H. Chandler*.

2. **U. procera** Salisb. *Common Elm*
    12. 1820, Boston, *Thompson*.
    Recorded for all Divs.

    Locally common, but not frequent.

THE COUNTY FLORA

3. **U. x sarniensis** (Loud.) Bancroft
    (A hybrid of *U. stricta*, the Cornish Elm.)
    16. 1959, Stamford, *J. H. Chandler*.
    Introduced about 1900? Recorded for divs. 6 and 13, and elsewhere on roadsides.

4. **U. coritana** Melville *East Anglian Elm*
    8. 1965, Wold Newton, *R. Melville*.
    U. coritana x glabra
    U. coritana x plotii
    U. plotii x coritana x glabra
    All these hybrids occur here and there, but not very commonly.

5. **U. carpinifolia** Gled.
    2. 1965, Scawby, *R. Melville*.
    5. 1962, Swinstead, *J. H. Chandler*.

    U. carpinifolia x glabra = U. x vegeta (Loud.) A. Ley.
    16. 1959, Stamford, *J. H. Chandler*.
    Distribution not yet known.

6. **U. plotii** Druce *Plot Elm*
    6. 1965, Marton, *R. Melville*.
    10. 1967, Roughton, *J. Gibbons*.
    12. 1945, Fishtoft, *A. H. G. Alston and N. D. Simpson*.
    15. 1959, Bulby, *J. H. Chandler*.
    16. 1959, Barholm, *J. H. Chandler*.
    Distributed here and there in many places.

MORACEAE
331. Ficus L.

1. **F. carica** L. *Fig*
    2. 1950, Sewage Dyke, Scunthorpe, *Gibbons*.
    9. 1901, Mablethorpe, *Allett*.
    Alien. Occasionally growing from seed.

JUGLANDACEAE
332. Juglans L.

1. **J. regia** L. *Walnut*
    6. 1836, Burton Park, *Simpson Collection*.
    Recorded for Divs. 1, 2, 3, 6, 11, 13, 15, 16.
    Introduced. Planted, and occasionally self set or by squirrels.

MYRICACEAE
333. Myrica L.

1. **M. gale** L.   *Bog Myrtle, Sweet Gale*
   - **1.** 1607, Axholme, *Camden.*
   - **2.** 1875, Crosby, *W. Fowler.* 1900, Frodingham. 1944, Twigmoor, Manton and Burringham. 1960, Brumby West Common, *E. J. Gibbons.*
   - **5.** 1935, Laughton and Scotter, *E. J. Gibbons.*
   - **7.** In check list but no locality known.
   - **10.** 1881, Kirkby Moor, *Fowler.* 1964, *E. J. Gibbons.*
   - **12.** 1696, Wainfleet, *C. Merrett.* Extinct.
   - **13.** 1835, Skellingthorpe, *Simpson Collection.*
   - **14.** In check list, but no locality known. (Gale Fen? N. Kyme).

   Native. Decreasing through drainage.

BETULACEAE
335. Betula L.

1. **B. pendula** Roth   *Birch*
   - **7.** 1877, Linwood, *Lees.*

   Recorded for all Divs.
   Native.

2. **B. pubescens** Ehrh. ssp. *pubescens*   *Downy Birch*
   - **7.** 1877, Linwood, *Lees.*

   Not recorded for Divs. 4, 9, 14, 17, 18.

   Native. Usually in the same localities as *B. pendula* but needs wetter soil.

   (Very many notes on depth in peat and level on moors in Peacock's Rock Soil Flora MS, Cambridge).

[3. **B. nana** L.   *Dwarf Birch*
   - **4.** 1894, Freshney Bog in interglacial peat, *Cordeaux.*

   Peacock's *Rock Soil Flora*, Cambridge. Native.]

336. Alnus Mill.

1. **A. glutinosa** (L.) Gaertn.   *Alder*
   - **18.** 1086, Spalding, *Domesday Book.*

   Recorded for all Divs.

   Native. Locally common in north Lincs.; in the Fens before drainage.

2. **A. incana** (L.) Moench   *Grey Alder*

   Introduced. Planted on Osgodby Moor (Div. 7) and hybridising with *A. glutinosa;* confirmed J. P. M. Brenan (1950). so at Louth water works (Div. 8), 1960, *J. Gibbons.*

CORYLACEAE
337. Carpinus L.
1. **C. betulus** L.   Hornbeam
   13. 1856, Doddington, *Cole*.
   Not recorded for Divs. 1, 9, 12, 16, 18.
   Not native. Planted as an ornamental tree near farms and occasionally on roadsides.

338. Corylus L.
1. **C. avellana** L.   Hazel, Cob-nut
   14. 1814, Digby, *Sir Joseph Banks*.
   Recorded for all Divs.
   Native. Besides being in coppiced woods, also in older hedges. Found in peat at Keadby Bridge in 1862.

FAGACEAE
339. Fagus L.
1. **F. sylvatica** L.   Beech
   12. 1799, Near Boston, *Arthur Young*.
   Recorded for all Divs.
   Possibly indigenous on the chalk.

340. Castanea Mill.
1. **C. sativa** Mill.   Sweet Chestnut, Spanish Chestnut
   12. 1820, Near Boston, *Thompson*.
   Not recorded for Divs. 1, 9, 18.
   Introduced. Always planted; some very big trees in parks; fruit rarely ripens in large quantities.

341. Quercus L.
1. **Q. cerris** L.   Turkey Oak
   Introduced. Nearly always planted; not common. Regenerating at Uffington (Div. 16).

2. **Q. ilex** L.   Evergreen Oak
   Introduced. Always planted; rare.

3. **Q. robur** L. *Common Oak, Pedunculate Oak*
    2. 1638, Broughton, *De la Pryme*.
    Recorded for all Divs.
    Native, though much planted. Large trunks found in peat in the Fens, Trent, Witham and Ancholme valleys and Isle of Axholme; also in submerged forest at Sutton-on-Sea.

4. **Q. petraea** (Mattuschka) Liebl. *Durmast Oak, Sessile Oak*
    7. 1877, North Willingham, *Lees*.
    Not recorded for Divs. 4, 6, 8, 9, 11, 18.
    Native. Scarce but present in old acid woods and a few hedges.

SALICACEAE
342. Populus L.

1. **P. alba** L. *White Poplar, Abele*
    6 or 13. 1841, Lincoln, *Stark*.
    Not recorded for Divs. 14, 16, 17, 18.
    Introduced.

2. **P. canescens** (Ait.) Sm. *Grey Poplar*
    13. 1862, Doddington and Boultham, *Cole*.
    Generally planted; distribution not yet worked out.

3. **P. tremula** L. *Aspen*
    6 or 13. 1851, Lincoln, *Watson*.
    Not recorded for Divs. 9, 17.
    Native. In old woods, but not generally common; also sometimes beside railway in dykes. Usually female; male noticed in Div. 11 at Monksthorpe and Div. 13 at Stapleford.

4. **P. nigra** L. *Black Poplar*
    6. 1841, Lea near Gainsborough, *Stark*.
    Possibly native at Stapleford near R. Witham, Div. 13, 1960; two hedgerow trees Kettleby and Searby Moor, 1960 or 70 (Div. 3); and at South Kelsey in 1962, trees female, definitely planted (Div. 7).

    P. italica (Duroi) Moench *Lombardy Poplar*
    6. 1841, Lea, *Stark*.
    Introduced. Distribution unknown; often planted as a quick growing shelter belt.

THE COUNTY FLORA

5. **P. x canadensis** Moench var. serotina (Hartig)
   Black Italian Poplar
   12. 1799, *Young*.
   Introduced. Distribution general except on Wolds; much planted; always male.

6. **P. gileadensis** Rouleau *Balsam Poplar, Balm of Gilead*
   2. 1876, Bottesford, *Peacock*.
   At Dunholme (Div. 6) and at South Elkington (Div. 8) and elsewhere, but distribution imperfectly known. Introduced. Always planted.

343. Salix L.

1. **S. pentandra** L. *Bay Willow*
   1. 1815, Axholme, *Peck*.
   Not recorded for Divs. 6. 9, 12, 17, 18.
   Native. Chiefly by Wold streams; Donnington-on-Bain, Thoresway.

2. **S. alba** L. *White Willow*
   1638, Fens, *Mercator*.
   Recorded for all Divs.
   Native. Scarce in some parts.

4. **S. fragilis** L. *Crack Willow*
   1638, Fens, *Mercator*.
   Recorded for all Divs.
   Native. Generally common.

5. **S. triandra** L. *Almond Willow*
   8. 1856, South Elkington, *Bogg*.†
   Recorded for all Divs.
   Doubtfully native except round Stamford.

6. **S. purpurea** L. *Purple Willow*
   8. 1856, Louth, *Bogg*.†
   Not recorded for Divs. 3, 5, 9, 10, 12, 16, 17.
   Native. Rather scarce.

9. **S. viminalis** L. *Common Osier*
   6. 1810, Lea, *Stark*.
   Recorded for all Divs.
   Native.

11. **S. caprea** L. *Great Sallow, Goat Willow*
   1638, Fens, *Mercator*.
   Recorded for all Divs.

   Native. Absent on some soils; less common than the next species.

12. **S. cinerea** L. *Common Sallow*
   subsp. **atrocinerea** (Brot.) Silva c Sobr.
   4. 1851, Grimsby, *Watson*.
   Recorded for all Divs.

   Native, very common.

13. **S. aurita** L. *Eared Sallow*
   13. 1851, Thorpe, *Watson*.
   Not recorded for Divs. 4, 6, 9, 12, 14, 15, 17, 18.

   Native. Acid soil; not common.

16. **S. repens** L. *Creeping Willow*
   12. 1799, East Fen, *Young*.
   Recorded for Divs. 1, 2, 3, 5, 7, 9, 10, 12, 13.

   Native. On sandy heaths; not known on maritime dunes until 1963 when 2 plants were seen on old dunes, North Somercotes. Two forms occur; true repens 15 in. and a taller form to 30 ins.

   Hybrids between many of these willows occur, especially **S. x geminata** (Forbes).

ERICACEAE

345. Rhododendron L.

1. **R. ponticum** L.
   7. 1894, North Willingham, *Lees*.
   Recorded for Divs. 2, 3, 6, 7, 10, 12, 13, 15, 16.

   Introduced. Planted in many woods on acid soil.

350. Andromeda L.

[1] **A. polifolia** L. *Marsh Andromeda, Bog Rosemary*
   1. 1840, near Wroot (Yorkshire?), *Revd. T. Owston.*†
   1893, Epworth, *Hudson.*† (See *Naturalist*, 1895, pp, 101 and 166; two separate localities and finders — *Fowler*).

   Native. Very rare and nearly extinct.

356. Calluna Salisb.

1. **C. vulgaris** (L.) Hull   *Ling, Heather*
   - **10.** 1724, Tattershall, *Stukeley.*
   - Not recorded for Divs. 4, 8, 9, 11, 12, 17, 18; presumably in 12, 17, 18 before drainage.
   - Native. Moors and woods, very rare in Divs. 6, 14, 15, 16.

357. Erica L.

1. **E. tetralix** L.   *Cross-leaved Heath*
   - **10.** 1724, Tattershall, *Stukeley.*
   - Recorded for Divs. 1, 2, 3, 5, 7, 10, 13, 16.
   - Native. Damp peat.

4. **E. cinerea** L.   *Bell Heather*
   - **10.** 1724, Tattershall, *Stukeley.*
   - Recorded for Divs. 1, 2, 3, 5, 7, 10, 13.
   - Native. Sand, decreasing.

358. Vaccinium L.

2. **V. myrtillus** L.   *Bilberry*
   - **2.** 1917, Broughton, *Canon Clayc.*†
   - **7.** 1929, Linwood, *Gibbons.*
   - Native. The absence of Bilberry from Lincolnshire caused comment by Lees in his *"Botany and Outline Flora of Lincolnshire"*, published in White's Lincolnshire Directory for 1892; also in *Naturalist* 1899, p. 336. Brogden's Glossary gives "Bilberry" but this is *Rubus caesius* — dune form.

4. **V. oxycoccos** L.   *Cranberry*
   - **1.** 1815, Axholme, *Peck.*
   - **2.** 1865, Brumby West Common, *Peacock.* 1917, Santon, *Peacock.* (L.N.U. Meeting).
   - **6.** 1842, Lea, *J. K. Miller.* (Doubtful).
   - **7.** 1877, Linwood Warren, *F. A. Lees.*†
   - **10.** 1820, Horncastle Moor, *Ward.*
   - **12.** 1799, Wainfleet (many acres in East Fen), *Arthur Young.* 1780, Fens, *Sibthorp.*
   - Native. Unfortunately drained out; presumably extinct.

PYROLACEAE

359. Pyrola L.

1. P. minor L. *Common Wintergreen*
- **1.** c. 1895, Axholme, *S. Hudson.*†
- **3.** 1872, North Kelsey Moor, *H. C. Brewster.* 1915, *Claye.*
- **5.** 1840, Laughton Common, *J. K. Miller;* recorded as *rotundifolia.*
- **7.** 1875, Osgodby Lane, *Bowstead.* 1877, Walesby, etc., *F. A. Lees.*† 1892, Usselby, *J. Brewster.* 1908, Holton le Moor, *M. Gibbons.* 1910, Tealby, *E. F. Lewin.* c. 1930, Linwood Warren, *G. Allison.*
- **10.** 1890, Woodhall Spa, *Miss Nash and Mrs. Jarvis.*† 1896, Coningsby, *F. S. Alston.*†
- **12.** 1856, East Fen, *Thompson.*
- **13.** 1890, Skellingthorpe, *Miss Nash.* 1894, South Hykeham, *C. H. Fearenside.*†

Native. Surviving only in division 7?

MONOTROPACEAE

362. Monotropa L.

1. M. hypopitys agg. *Yellow Bird's-nest*

Not examined by experts.

1. M. hypopitys L. (under conifers or on sandy warrens)
- **2.** 1945, Risby Warren, *R. May.* Ashby, *M. Barnes.*
- **4.** 1926, Swallow, *S. A. Cox.*† 1928, Croxby, *S. A. Cox.*
- **6.** 1805, Fillingham, *Dalton.* Firwood.

Native.

2. M. hypophegea Wallr. (under beech)
- **3.** 1932, Limber, *R. May.* 1959, South Ferriby, *A. Orchard and R. Smith.*

Native.

See *L.N.U. Trans.* (1934), pp. 214-216. S. A. Cox.

EMPETRACEAE

364. Empetrum L.

1. E. nigrum L. *Crowberry*
- **1.** 1815, Axholme, *Peck.* 1893, Lincs. Herb., *N. C. Marris.*†
- **2.** 1856, Frodingham, *W. Fowler.*
- **10.** 1830, Tattershall, *Allen and Saunders.*
- **12.** 1799, East Fen "cranberry ground", *Young.*

Native. No recent record.

PLUMBAGINACEAE
365. Limonium Mill.

1. **L. vulgare** Mill.   *Sea Lavender*
    17. 1688, Kirton, *Holden.*
    Recorded for Divs. 3, 4, 9, 11, 12, 17, 18.
    Native. Very variable in size and in time of flowering; *Spartina anglica* is threatening some of its colonies; scarce in the Humber area, grows as far up the estuary as Killingholme.

2. **L. humile** Mill.   *Lax-flowered Sea Lavender*
    4. 1919, Cleethorpes, *Peacock.*
    11. 1957, Gibraltar Point, *H. G. Baker.*
    No specimen; confirmation needed.

3. **L. bellidifolium** (Gouan) Dum.   *Matted Sea Lavender*
    11. 1952, Gibraltar Point, *M. Smith and another.* 1967, Extinct.?
    12. 1789, Freiston, *Gough* in "Camden's Britannia". 1805, *Banks.* 1826, *G. Howitt.* 1919, *Newman and Walworth.*
    17. 1789, Fosdyke, *Gough* in "Camden's Britannia". 1805, *Banks.*†
    18. 1805, Tydd, *Scrimshire.*
    Native. Extinct in VC 53, scarce in VC 54.

5. **L. binervosum** (G. E. Sm.) C. E. Salmon sensu lato
                                                  Rock Sea Lavender
    4. 1852, Cleethorpes, *S. M. Skipworth* (ffytche).
    9. 1854, Saltfleetby, *Fowler.* 1892, Humberstone, *Lees.*† 1960, *J. Gibbons.*
    11. 1954, Gibraltar Point, *J. Gibbons.*
    Native. Varying in quantity from year to year.

366. Armeria Willd.

1. **A. maritima** (Mill.) Willd.   *Thrift, Sea Pink*
    (*a*) ssp. *maritima.*
    17. 1688, Kirton, *Holden.*
    Recorded for Divs. 3, 4, 9, 11, 12, 17, 18.
    Native. Chiefly found in the north; varying in quantity.

    (*b*) ssp. *elongata* (Hoffm.) Bonnier   *Elongated Thrift*
    14. 1930, Wilsford, *H. Fisher.* (Wollaton Hall, Nottingham).
    15. 1726, Near Grantham, *V. Bacon.* 1896, Manthorpe, *L.N.U. Meeting.*† 1930, *E. E. Orchard.* 1953-54, Ancaster, *J. Gibbons* (confirmed Prof. H. G. Baker).
    See *Watsonia*, Vol. 4, p. 135.
    Native. Not satisfactorily identified till 1956.

PRIMULACEAE
367. Primula L.

3. **P. veris** L. *Cowslip, Pigseye*
   14. 1790, Threckingham, *Cragg.*
   Native. Found in all Divs. but decreasing and rare on some soils.

[4. **P. elatior** (L.) Hill *Oxlip*
   Recorded by Hawkins, 1906 in Div. 15, but incorrect identification. Specimens always prove to be *P. veris* x *vulgaris*.]

5. **P. vulgaris** Huds. *Primrose*
   7. 1829, Bardney, *J. F. Wray.*
   12. 1820, Near Boston, *Thompson.*
   Not recorded for Divs. 9, 18, though may have grown in the latter. (Primrose Holt, 1825 map).
   Native. Scattered and often plundered; chiefly on the east side of the Wolds.

368. Hottonia L.

1. **H. palustris** L. *Water Violet*
   14. 1790, Threckingham, *Cragg.*
   Recorded for all Divs.
   Native. Becoming scarce due to mechanical dredging.

369. Cyclamen L.

1. **C. hederifolium** Ait.
   1597, Gerarde's *Herbal* (See *Naturalist,* 1895, p. 102).
   Probably introduced. Last seen about 1920 by R. May, Div. 3.

370. Lysimachia L.

1. **L. nemorum** L. *Yellow Pimpernel*
   10. 1820, Roughton, *Ward.*
   Not recorded for Divs. 3, 4, 5, 9, 14, 15, 17, 18.
   Native. Scarce and local, chiefly east of Wolds.

2. **L. nummularia** L. *Creeping Jenny*
   14. 1790, Threckingham, *Cragg.*
   Recorded for all Divs.
   Native. Rather scarce.

3. **L. vulgaris** L. *Yellow Loosestrife*
    12. 1799, East Fen, *Young*.
    Recorded for all Divs.
    Native. Peaty soil, local.

4. **L. ciliata** L.
    3. 1849, Brocklesby, *Dr. B. Carrington;* also Rev. J. Mossop's paintings.
    Introduced. Garden escape or planted.

5. **L. punctata** L.
    7. 1950, Langton by Wragby, *M. Gibbons.*
    8. 1892, Redhill, Goulceby, *J. Larder.*†
    Introduced. Garden escape.

7. **L. thyrsiflora** L. *Tufted Loosestrife*
    1. 1840, between Keadby and Crowle Wharfe, *J. K. Miller.*
       1950, Epworth, *Bunting.* (Very little).
    12. 1842, near Boston, *Kippist and Woods.* (Dublin specimen).
    Native. Probably extinct.

372. Anagallis L.

1. **A. tenella** (L.) L. *Bog Pimpernel*
    1. 1815, Isle of Axholme, *Peck.*
    Not recorded for Divs. 8, 11, 12, 14, 17, 18.
    Native. Rare.

2. **A. arvensis** L. *Scarlet Pimpernel, Shepherds' Weather-glass*
    14. 1790, Threckingham, *Cragg.*
    Recorded for all Divs.
    Native.

3. **A. foemina** Mill. *Blue Pimpernel*
    14. 1829, S. Rauceby, *R. J. Bunch.*
    15. 1837, Holywell, *Dodsworth.* 1951, *L.N.U. Meeting.*
       1949, Ancaster, *L.N.U. Meeting.*
    Not recorded for Divs. 1, 6, 18.
    Native. Records probably include *A. arvensis,* blue form (see article by Peacock, *L.N.U. Transactions,* 1913).

4. **A minima** (L.) E. H. L. Krause *Chaffweed*
    7. 1877, between Osgodby and Middle Rasen, *Lees.*†
    The only record.
    Native.

373. Glaux L.
1. **G. maritima** L. *Sea Milkwort*
   4. 1800, Grimsby, *Salt*.
   Not recorded for Divs. 5, 6, 7, 8, 13, 14, 15, 16.
   Native.

374. Samolus L.
1. **S. valerandi** L. *Brookweed*
   16. 1829, Swanpool, Lincoln, *Rev. J. F. Wray*. 1836, Bourne, *Dodsworth*.
   Recorded for all Divs.
   Native. Scarce.

BUDDLEJACEAE
375. Buddleja L.
1. **B. davidii** Franch.
   17. 1957, near Gosberton, *Dr. Perring*.
   Introduced. Distribution not known; seeding from gardens.

OLEACEAE
376. Fraxinus L.
1. **F. excelsior** L. *Ash*
   1695, Lindsey, *De la Pryme*.
   Recorded for all Divs. (see *Naturalist*, 1896, p. 53).
   Native and planted.

377. Syringa L.
1. **S. vulgaris** L. *Lilac*
   15. 1960, Stainby near Grantham, *J. Gibbons*.
   Introduced. Appearing naturalised in a few places.

378. Ligustrum L.
1. **L. vulgare** L. *Privet*
   14. 1790, Threckingham, *Cragg*.
   Recorded for all Divs.
   Native. Often planted as cover for game.

APOCYNACEAE
379. Vinca L.

1. **V. minor** L. *Lesser Periwinkle*
   12. 1856, near Boston, *Thompson*.
   Recorded for Divs. 3, 8, 11, 12, 13, 15, 16.
   Possibly wild in the south, but doubtful elsewhere.

2. **V. major** L. *Greater Periwinkle*
   15. 1805, Woolsthorpe, *Crabbe*.
   Not recorded for Divs. 1, 6, 7, 12, 14, 17, 18.
   Introduced. Garden escape.

GENTIANACEAE
380. Cicendia Adans.

1. **C. filiformis** (L.) Delarb.
   12. 1887, between Friskney and Wainfleet on the Roman Bank, *J. Abbott*. ("Specimens brought to me" — F. A. Lees).
   14. 1790, Newton, *Cragg*.
   Native. To be searched for.

382. Centaurium Hill

1. **C. pulchellum** (Sw.) Druce *Slender Centaury*
   11. 1847, Skegness, *Dr. Grantham*.
   Recorded for Divs. 4, 9, 11, 17.
   Native.

4. **C. erythraea** Rafn. *Centaury*
   14. 1790, Osbournby, *Cragg*.
   Not recorded for Divs. 17, 18.
   Native. Not common.

6. **C. littorale** (D. Turner) Gilmour
   9. 1830-70, Saltfleet, *Revd. J. Mossop's* paintings.
   Native?

383. Blackstonia Huds.

1. **B. perfoliata** (L.) Huds. *Yellow-wort*
   8. 1666, Burwell, *Lister*.
   Not recorded for Divs. 17, 18.
   Native. Scarce.

384. Gentiana L.

1. **G. pneumonanthe** L. *Marsh Gentian*
   3. 1633, Nettleton Moor, *Johnson*.
   Recorded for Divs. 2, 3, 5, 7, 10, 12, 13.
   Native. Extinct in Divs. 3, 12, 13.

385. Gentianella Moench

1. **G. campestris** (L.) Borner *Field Gentian*
   13. 1829, Canwick Common, *Loughborough Collection*, R. J. Bunch.
   13. 1890, South Common, Lincoln, *J. S. Sneath*.†
   15. 1780, near Grantham, *Sibthorp*.
   Other specimens proved to be *G. amarella*. Native.

3. **G. amarella** (L.) Borner *Autumn Gentian, Felwort*
   8. 1666, Burwell, *Lister*.
   Not recorded for Divs. 1, 7, 12, 17, 18.
   Native.

4. **G. anglica** (Pugsl.) E. F. Warb. *Early Felwort*
   3. 1886, Thornton Curtis, Mrs. *A. Flowers*.
   14. 1937, Rauceby, *E. J. Gibbons*. Extinct.
   15. 1774, near Grantham, *T. G. Cullum* (Botanists' Guide). 1894, Ancaster, *Stow*.† 1954, Holywell, *J. H. Chandler*. (1951, L.N.U. Meeting). Conf. N. Pritchard.
   16. 1933, Grimsthorpe, *Fisher*.
   Native. Rare.

MENYANTHACEAE
386. Menyanthes L.

1. **M. trifoliata** L. *Bogbean*
   12. 1799, East Fen, *Young*.
   Not recorded for Divs. 16, 17, 18.
   Native. Rare and decreasing.

387. Nymphoides Hill

1. **N. peltata** (S. G. Gmel.) Kuntze *Fringed Waterlily*
   12. 1971, Boston, *F. Brasier*.
   13. 1909, Nocton, *Mason*.†
   15. 1860's, Grantham and Nottingham Canal, *E. M. Browne*. 1895, Syston Lake, *M. and S. Craster*.†
   18. 1890, Cowbit Wash, *C. Hufton*. 1958, *J. H. Chandler*. 1959, Clough's Cross, *E. J. Gibbons*.
   Native, or bird sown.

POLEMONIACEAE
388. Polemonium L.
1. **P. caeruleum** L. *Jacob's Ladder*
   - 2. 1856, Broughton, *Fowler.*†
   - 8. 1834, Louth, *Bayley.*
   - 15. 1780, Near Stoke Rochford, *Sibthorp.*
     1888, Great Ponton, *Fisher.*

   Also recorded for Divs. 9, 11; possibly native in 2, 15.

BORAGINACEAE
389. Cynoglossum L.
1. **C. officinale** L. *Hound's-tongue*
   - 10. 1820, Martin, *Ward.*

   Not recorded for Divs. 4, 5, 12, 17, 18.
   Native.

391. Asperugo L.
1. **A. procumbens** L. *Madwort*
   - 2. 1695-1701, Broughton, *A. De la Pryme* in a letter to Sloane, 1701, "I got it plentifully in the garth of Richard Robinson of Broughton amongst the corn." (*J. of B.*, 1915, p. 310).
     1876, Burton Stather, *Peacock.*
   - 3. 1915, Brigg, *Claye.* 1932, Limber, *R. May* (Pheasant food).
   - 4. 1900, Grimsby Dock, *Parker and A. Smith.*†
   - 6. 1893, Lincoln, *Goodall.*†
   - 12. 1907, Boston Dock, *Hurst.*

   Alien. Docks and dumps.

392. Symphytum L.
1. **S. officinale** L. *Comfrey*
   - 12. 1784, Boston, *Nicholls.*

   ssp. *ochroleucum* DC.
   - 6. 1968, Reepham, *I. Weston.*
   - 10. 1966, Ashby Puerorum, *M. N. Read.*
   - 12. 1945, Boston, *A. H. Alston and N. D. Simpson.*
   - 16. 1966, Nr. Crowland, *F. H. Perring.*
   - 17. 1960, Spalding, *E. J. Gibbons.*

   Native, wet fen banks.

   ssp. *purpureum* Pers.
   - 6. 1968, Nettleham, *I. Weston.*
   - 13. 1969, Potterhanworth, *E. J. Gibbons and F. H. Perring.*

   ssp. *ochroleucum x purpureum.*
   - 14. 1969, Haverholme, *E. J. Gibbons and F. H. Perring.*

   All recs. conf. F. H. Perring.

2. **S. asperum** Lepech. *Rough Comfrey*
Not so far identified or recorded.

S. asperum x officinale = S. x uplandicum Nyman.
Probably in all divisions; has not been properly distinguished from *S. officinale* in the past.
- **3.** 1895, Bigby, *E. P. Field* (first record).
  1968, Nettleton, *E. J. Gibbons*.
- **5.** 1968, Laughton, *E. J. Gibbons*.
- **6.** 1968, Sudbrooke, *I. Weston*.
- **9.** 1968, Skidbrooke, *E. J. Gibbons*.
- **10.** 1968, Dalderby, *M. N. Read*.

Introduced. All recs. conf. F. H. Perring.

3. **S. orientale** L.
- **10.** 1956, Scrivelsby, *E. J. Gibbons*.

Recorded for Divs. 4, 6, 7, 8, 10, 13, 16.
Introduced. Relic of old gardens.

6. **S. tuberosum** L. *Tuberous Comfrey*
1796, Fens, *Woodward*.
- **2.** 1961, Broughton, *D. McClintock and E. J. Gibbons*.
- **5.** 1892, Blyton or Blyborough, *Mabel Peacock*.† 1964, Thonock, *D. Wright*.
- **12.** 1856, Skirbeck, *Thompson*.
- **16.** 1836, Bourne (Falkners Decoy), *Dodsworth*.
- **17.** 1860, Quadring, *M. E. Dixon*.

Native. Rare; formerly in small quantity among *S. officinale* in the fens.

393. Borago L.

1. **B. officinalis** L. *Borage*
1835, No locality, *Miss Skipworth*.
Not recorded for Divs. 1, 3, 5, 7, 14.
Garden escape.

Amsinckia

A. lycopsioides Lehm. *Orange Bugloss, Fiddleneck*
- **13.** 1895, Mere, *A. Pears*.

Recorded for Divs. 2, 3, 4, 5, 7, 8, 10, 13.
Alien, not established.

395. Pentaglottis Tausch

1. **P. sempervirens** (L.) Tausch *Alkanet*
   11. 1890, Skegness, *Miss Lane-Claypon*.
   Not recorded for Divs. 1, 2, 9, 13, 17.
   Garden escape, becoming naturalised.

396. Anchusa L.

1. **A. officinalis** L.
   8. 1900, Keddington, *Kiddall*.
   11. 1960 Skegness *J. A. Lowe*.
   17. 1963, Wigtoft, *C. J. Allerton*.
   Alien.

397. Lycopsis L.

1. **L. arvensis** L. *Bugloss*
   10. 1836, Woodhall Spa, *Simpson collection*.†
   Recorded for all Divs. except 17, 18.
   Native.

400. Myosotis L.

1. **M. scorpioides** L. *Water Forget-me-not*
   12. 1799, East Fen, *Young*.
   Recorded for all Divs.
   Native. Fairly frequent.

2. **M. secunda** A. Murr.
   6. 1907, Newton-on-Trent, *Mason*.†
   7. 1877, Holton le Moor, *Lees*. (Doubtful) *M. caespitosa* frequent.
   9. 1907, Marshchapel, *Mason*. *M. caespitosa* frequent.
   13. 1856, Doddington, *Cole*.†
   Very rare, if correctly named.

4. **M. caespitosa** K. F. Schultz *Tufted Forget-me-not*
   8. 1850, Louth, *M. E. Dixon*.
   Not recorded for Divs. 17, 18.
   Native. Frequent.

7. **M. sylvatica** Hoffm. *Wood Forget-me-not*
   7. 1854, Claxby, *M. E. Dixon*.
   Not recorded for Divs. 1, 5, 9, 14, 17, 18.
   Native. Locally abundant except in the west.

8. **M. arvensis** (L.) Hill *Common Forget-me-not*
   14. 1790, Threckingham, *Cragg*.
   Recorded for all Divs. in woods and in stubble fields.
   Native.
9. **M. discolor** Pers. *Yellow and Blue Forget-me-not*
   5. 1840, Laughton, *Miller*.
   Not recorded for Divs. 1, 8, 17, 18.
   Native. Chiefly on dry acid sand but also in damp meadows.
10. **M. ramosissima** Rochel *Early Blue Forget-me-not*
    1851, *Watson*.
    Not recorded for Div. 1.
    Native. On coastal dunes and sand inland.

401. Lithospermum L.

2. **L. officinale** L. *Gromwell*
   2. 1822, Appleby Woods, *Strickland*.
   Not recorded for Divs. 1, 9, 10, 17, 18.
   Native. Rare and occasional except in the north.
3. **L. arvense** L. *Corn Gromwell*
   15. 1757, Near Grantham, *Pulteney*.
   Not recorded for Divs. 1, 17.
   Native. Occasional; chiefly on limestone.

402. Mertensia Roth

1. **M. maritima** (L.) Gray *Oyster Plant*
   4. 1884, Cleethorpes, *E. M. Browne*.
   Fleeting visitor.

403. Echium L.

1. **E. vulgare** L. *Viper's Bugloss*
   10. 1820, West Ashby, *Ward*.
   Recorded for all Divs., sometimes as a casual.
   Native.
2. **E. lycopsis** L. *Purple Viper's Bugloss*
   3. 1961, South Killingholme, *R. Parker*. (Pigeon food).
   4. 1897, Grimsby, *Wood*.†
   6. *Peacock's Check List*.
   13. 1898, Lincoln, *Sneath and Peacock*.†
   Casual alien.

CONVOLVULACEAE

405. Convolvulus L.

1. **C. arvensis** L.   *Bindweed, Cornbine*
   7. 1835, South Kelsey, *S. Skipworth.*
   Recorded for all Divs.
   Native.

406. Calystegia R. Br.

1. **C. sepium** (L.) R. Br.   *Larger Bindweed, (Columbine)*
   12. 1799, East Fen, *Young.*
   Recorded for all Divs.
   Native in wet fen and carr, rooting to a depth of six feet; a pink form in Divs. 1 and 12.

2. **C. pulchra** Brummitt and Heywood   *Pink Bindweed*
   12. 1945, Wainfleet, *A. H. G. Alston and N. D. Simpson.*
   Not recorded for Divs. 2, 8, 9, 11, 14, 15, 18.
   Introduced? Usually near houses.

3. **C. silvatica** (Kit.) Griseb.   *American Bellbine*
   10. 1945, Woodhall Spa, *A. H. G. Alston and N. D. Simpson.*
   Recorded for all Divs.
   Introduced? Usually near habitation.

4. **C. soldanella** (L.) R. Br.   *Sea Bindweed*
   4. 1950, Cleethorpes, *Watkinson.*
   9. 1834, Mablethorpe, *Bayley.* 1856, *Bogg.*†
   11. 1869, Skegness, *Mason.* 1953, Ingoldmells, *E. J. Gibbons.* 1903, Sutton-on-Sea, *Peacock.*† 1921-29, *E. J. Gibbons.* extinct 1953. 1960, Gibraltar Point, *M. Smith.*
   Recorded for Divs. 4, 9, 11 only.
   Native. Disappearing due to sea defences and collecting.

407. Cuscuta L.

1. **C. europaea** L.   *Large Dodder*
   15. 14th June, 1884, Grantham Journal, *E. M. Browne:* "in hedges from crops of tares or clover".
   Identification very doubtful, probably *C. trifolii.*

2. **C. epilinum** Weihe *Flax Dodder*
    1. 1850-60, Epworth, *S. Hudson*.
    2. 1853, Bottesford, *E. Peacock*.
    Once a nuisance on flax; not known now.

3. **C. epithymum** (L.) L. *Common Dodder*
    2. 1950, Frodingham, *R. Howitt*. On carrot.
    4. 1954, Thoresway, *E. J. Gibbons*. On medick. *M. lupulina*.
    7. 1878, Nova Scotia near Market Rasen, *Lees*. "Very little on heather". 1968, Bardney, *L.N.U. Meeting*, on medick.
    10. c. 1920, Woodhall Spa, *Alston*. 1948, *D. Marsden*. On heather.
    14. 1894, Billinghay, *Mrs. Walker*.† 1895, Sleaford, *Larder and Peacock*.†? *C. trifolii*.
    15. 1923, between Great Ponton and Stoke Rochford, *S. C. Stow*.
    Native. Rare.

C. trifolii Bab. *Clover Dodder*
    8. 1856, Burwell, *Bogg*.†
    Recorded for Divs. 1, 2, 3, 7, 8, 11, 12, 13, 14, 15.
    Once a nuisance on clover, but now extinct — no record since 1918.

SOLANACEAE
409. Lycium L.

1. **L. halimifolium** Mill [**L. barbarum** L.] *Duke of Argyll's Tea-plant*
    2. 1866, Bottesford, *Peacock*.
    Recorded for all Divs.
    Introduced. Generally common in hedges near villages; not fruiting.

2. **L. chinense** Mill.
    9. 1958, Mablethorpe, *E. J. Gibbons*.
    11. 1961, Chapel, *E. J. Gibbons and M. N. Read*.
    1963, Skegness, *J. Gibbons*.
    Introduced. Less common; coastal thickets.

410. Atropa L.

1. **A. bella-donna** L. *Deadly Nightshade*
    17 or 18. 1597, Holland (Lincs.), *Gerarde*.
    Recorded for Divs. 6, 8, 9, 10, 12, 13, 14, 15, 16, 18.
    Native in the south, casual in some localities, established in Divs. 8 and 16.

411. Hyoscyamus L.

1. **H. niger** L.   *Henbane*
   14. 1790, Threckingham, *Cragg*.
   Recorded for all Divs.
   Casual.

413. Solanum L.

1. **S. dulcamara** L.   *Bittersweet, Woody Nightshade*
   14. 1790, Threckingham, *Cragg*.
   Recorded for all Divs.
   Native.

3. **S. nigrum** L.   *Black Nightshade*
   13. 1830, Canwick, *Rev. J. Wray*.
   Recorded for all Divs.
   Native.

4. **S. sarrachoides** Sendtn.
   3. 1970, Nettleton, *S. A. Cox*. In potatoes.
   6. 1970, Newton-on-Trent, *Howitt*. In carrots.
   9. 1956, Humberstone, *Gibbons*.
   17. 1897, Wyberton, *Miss Lane-Claypon*.
   Casual.

415. Datura L.

1. **D. stramonium** L.   *Thorn-apple*
   1724, *Stukeley*. 1805, Near Gainsborough, *Salt*.
   Not recorded for Div. 1.
   Introduced. Seeds mature in hot summers and come up in disturbed ground.

SCROPHULARIACEAE
416. Verbascum L.

1. **V. thapsus** L.   *Aaron's Rod, Great Mullein*
   10. 1820, Tetford, *Ward*.
   Not recorded for Divs. 4, 6.
   Casual on waste ground.

3. **V. phlomoides** L.
   11. 1957, Skegness, *Howitt*.
   16. 1961, Uffington, *J. H. Chandler*.
   Casual.

4. **V. lychnitis** L. *White Mullein*
   5. 1850, Gainsborough, *Lowe.*
   13. 1875, Mere, *Miss A. Pears.*†
   Casual.

   V. lychnitis x thapsus — V. x thapsi I.
   3. 1909, Cadney, *T. W. Peacock.*†

   V. lychnitis x nigrum — V. x schiedeanum Koch.
   8. 1908, Muckton, *Larder.*†

7. **V. nigrum** L. *Dark Mullein*
   13. 1830, Washingborough, *Rev. J. F. Wray.*
   Recorded for Divs. 2, 5, 6, 10, 12, 13, 15, 16.
   Native. Rare; commoner in the south.

   V. nigrum x blattaria.
   11. 1909, Ingoldmells, *B. Reynold (J. of B.,* Feb., 1910).

8. **V. chaixii** Vill.
   10. 1919, Woodhall Spa (as *V. nigrum*), *F. S. Alston;* also 1964, *W. Heath* (det. Lousley).
   Chicken food, persisting.

9. **V. blattaria** L. *Moth Mullein*
   10. 1879, West Keal, *Burgess.*
   Recorded for Divs. 2, 3, 4, 10, 11, 13, 14.
   Casual.

10. **V. virgatum** Stokes *Twiggy Mullein*
    3. 1937, Melton Ross, *A. Malins Smith.*
    13. 1835, near Lincoln, *J. Nicholson* (Hooker's Flora, 1835). (See *The Naturalist,* 1896, p. 56).
    Casual.

417. Misopates Raf.

1. **M. orontium** (L.) Raf. *Weasel's Snout*
   3. *c.* 1920, Brigg, *E. Wright.* (Casual).
   4. 1901, Grimsby, *A. Smith.*†
   5. 1865, Gainsborough, *Charters.*†
   11. 1915, Skendleby, *R. Hamond.* 1933, *M. Gibbons.*
   12. 1856, Skirbeck, *P. Thompson.*
   13. 1890, Stapleford, *H. Fisher.*
   14. 1855, Brauncewell, *J. Lowe.*
   15. 1805, Allington, *Crabbe.* (May be this — Botanist's Guide). 1950, Barrowby, *A. C. Footiit.*
   Native or casual; weed of cornfields; rare.

418. Antirrhinum L.

1. **A. majus** L. *Snapdragon*
    6. 1805, Lincoln, *Stovin*. 1894, *Sneath*.†
    Recorded for Divs. 2, 6, 8, 11, 13, 15, 16.
    Introduced. Established on old walls and in quarries.

420. Linaria Mill.

2. **L. purpurea** (L.) Mill. *Purple Toadflax*
    3. 1958, Wootton, *E. J. Gibbons*.
    7. 1960, Walesby, *J. A. Lowe*. (On tip).
    Introduced. Not fully established outside gardens elsewhere.

3. **L. repens** (L.) Mill. *Pale Toadflax*
    1. 1962, Greenholme Bank, *E. J. Gibbons*.
    13. 1971, Eagle, *Mrs. E. M. Pearce*.
    15. 1960, Stainby, *M. Lowe*.
    ? Introduced; naturalised.

4. **L. vulgaris** Mill. *Toadflax*
    10. 1820, Horncastle, *Ward*.
    Not recorded for Div. 17.
    Native.

421. Chaenorhinum (DC.) Reichb.

1. **C. minus** (L.) Lange *Small Toadflax*
    15. 1805, Allington, *Crabbe*. (Possibly *Misopates* — see p. 202).
    Recorded for all Divs.
    Native on chalk and limestone; on railways almost everywhere.

422. Kickxia Dumort.

1. **K. spuria** (L.) Dumort. *Fluellen, Round-leaved Toadflax*
    8. 1666, Burwell, *Lister*.
    Not recorded for Divs. 1, 2, 9, 10, 12, 17, 18.
    Native. Scattered and rather scarce.

2. **K. elatine** (L.) Dumort. *Pointed-leaved Toadflax*
    8. 1666, Burwell, *Lister*.
    Not recorded for Divs. 1, 2, 5, 9, 12, 17, 18.
    Native. Scattered and rather scarce but commoner than the last on the Wolds.

423. Cymbalaria Hill

1. **C. muralis** Gaertn., Mey & Scherb. *Mother of Thousands, Ivy-leaved Toadflax*
    1835, Simpson collection, Lincoln.†
    Not recorded for Div. 1.
    Introduced. Established on walls in villages.

424. Scrophularia L.

1. **S. nodosa** L. *Figwort*
    6. Lincoln, Loughborough Collection, *Rev. R. J. Bunch.*
    Not recorded for Divs. 9, 12.
    Native. Rather uncommon.

2. **S. aquatica** L. [**S. auriculata** L.] *Water Figwort, Fiddles*
    10. 1820, Hemingby, *Ward.*
    Recorded for all Divs.
    Native.

3. **S. umbrosa** Dumort.
    13. 1908, Mere, *Mason.*†
    16. 1909, Auster Wood, Bourne, *W. H. Daubney.*
    Native. Distribution unknown.

5. **S. vernalis** L. *Yellow Figwort*
    15. 1954, Stubton, *Beaney and Baines.*
    Garden relic.

425. Mimulus L.

1. **M. guttatus** DC. *Monkey-flower*
    15. 1864, Hough, *Charters.*†
    Recorded for Divs. 2, 5, 6, 7, 8, 13, 15.
    Introduced.

426. Limosella L.

1. **L. aquatica** L. *Mudwort*
    12. 1856, near Boston, *Thompson.*
    16. 1836, Bourne, *Dodsworth.*
    Native. Extinct.

427. Sibthorpia L.

1. **S. europaea** L.   *Cornish Moneywort*
   15. 1763, Honington, *Hill or Blackstone*.
   Mistake for *Hydrocotyle* (*Naturalist* 1896, p. 57).

429. Digitalis L.

1. **D. purpurea** L.   *Foxglove*
   10. 1820, Holbeck, *Ward*.
   Not recorded for Divs. 4, 9, 12, 14, 17.
   Very scarce except on west side of the county and near Spilsby. Some records are of escapes or purposely planted. (*Phyt.* 1857, p. 303, W. Fowler).

430. Veronica L.

1. **V. beccabunga** L.   *Brooklime*
   2. 1822, Appleby, *Strickland*.
   Not recorded for Div. 17.
   Native. Ponds and streams.

2. **V. anagallis-aquatica** L.   *Water Speedwell*
   7. 1835, South Kelsey, *Skipworth*.
   Recorded for all Divs.
   Native.

3. **V. catenata** Pennell   *Pink Water Speedwell*
   11. 1905, Skegness, *F. M. Robinson* (Nott.).
   12. 1945, Baker's Bridge near Boston, *A. H. G. Alston and N. D. Simpson*.
   Recorded for all Divs.
   Native. The distribution of this species and the previous one would repay study.

4. **V. scutellata** L.   *Marsh Speedwell*
   13. 1829, Swanpool, Lincoln, *Rev. J. F. Wray*.
   Not recorded for Divs. 8, 12, 17, 18.
   Native. Uncommon.

5. **V. officinalis** L.   *Common Speedwell*
   2. 1822, Appleby, *Strickland*.
   Not recorded for Divs. 17, 18.
   Native. Dry banks of woods and pastures.

6. **V. montana** L. *Wood Speedwell*
    2. 1857, Broughton, *Fowler*.
    Not recorded for Divs. 1, 5, 6, 9, 12, 14, 17, 18.
    Native. Chiefly in woods on east side of Wolds; rare elsewhere.

7. **V. chamaedrys** L. *Germander Speedwell*
    7. 1835, South Kelsey, *Skipworth*.
    Recorded for all Divs.
    Native. Woods and banks.

13. **V. serpyllifolia** L. *Thyme-leaved Speedwell*
    16. 1838, Bourne, *Dodsworth*.†
    Recorded for all Divs.
    Native, old pastures and wood rides.

15. **V. arvensis** L. *Wall Speedwell*
    7. 1835, South Kelsey, *Skipworth*.
    Recorded for all Divs.
    Native. Walls and dry places.

20. **V. hederifolia** L. *Ivy Speedwell*
    16. 1836, Bourne and Denton, *Dodsworth*.†
    Recorded for all Divs.
    Native, arable land and gardens.

21. **V. persica** Poir. *Buxbaum's Speedwell*
    2. 1857, Winterton, *Fowler*.
    Recorded for all Divs.
    Introduced after 1850; frequent.

22. **V. polita** Fr. *Grey Speedwell*
    16. 1836, Bourne, *Dodsworth*.†
    Not recorded for Div. 5.
    Native. Rather scarce; not on acid soil.

23. **V. agrestis** L. *Field Speedwell*
    7. 1835, South Kelsey, *Skipworth*.
    Recorded for all Divs.
    Native. Chiefly in garden ground.

24. **V. filiformis** Sm. *Slender Creeping Speedwell*
    12. 1951, Boston, *E. J. Gibbons*.
    Recorded for Divs. 2, 5, 7, 10, 11, 12, 13, 16, 18.
    Introduced into gardens and is spreading as an escape in turf.

432. Pedicularis L.

1. **P. palustris** L. *Red Rattle*
    12. 1799, East Fen, *Young*.
    Not recorded for Divs. 3, 15, 17, 18.
    Native. Decreasing, becoming extinct. Last seen in Divs. 6, 10, before 1960.

2. **P. sylvatica** L. *Lousewort*
    13. 1807, Stapleford, *Ordoyno*.
    Not recorded for Divs. 12, 15, 17, 18.
    Native. Heaths.

433. Rhinanthus L.

1. **R. serotinus** (Schonh.) Oborny *Great Yellow Rattle*
    5. 1838, Hemswell, *Irvine*.
    Not recorded for Divs. 3, 4, 8, 16, 18.
    Native. Decreasing and only recorded 1950-60 in 3 places in Div. 1,

2. **R. minor** L. *Yellow Rattle*
    14. 1790, Threckingham, *Cragg*.
    Recorded for all divs.
    Native. Old pastures.

434. Melampyrum L.

1. **M. cristatum** L. *Crested Cow-wheat*
    15. 1896, Ropsley, *Woolward*. 1925, Holywell, *L. Bond*.†
    16. 1667, N. of Stamford, *Merrett*. 1880, Careby, *W. Fowler*.
    Recorded for Divs. 15, 16 only.
    Native. Very rare; last seen 1958. J. H. Chandler.

2. **M. arvense** L. *Field Cow-wheat*
    4. 1868, Scartho, *W. H. Daubney*.
    7. c. 1900, South Kelsey, *W. Holmes*. (In foreign wheat).

3. **M. pratense** L. *Common Cow-wheat*
    10. 1820, Tetford, *Ward*.
    Not recorded for Divs. 9, 12, 14, 17, 18.
    Native. Scattered and in small quantity generally in old woods.

435. Euphrasia L.

1. **E. officinalis** L. agg.   *Eyebright*
   - **14.** 1790, Threckingham, *Cragg.*
   - Not recorded for Divs. 17, 18.
   - Native, local.

   13. **E. nemorosa** (Pers.) Wallr.
       - **2.** 1893, Broughton, *Peacock.*†
       - **5.** 1892, Grayingham, *J. G. Nicholson.*†
       - **7.** 1905, Moortown, *Peacock.*†
       - **8.** 1856, Haugham, *Bogg.*†
       - **9.** 1931, Humberstone, *A. A. and B. E. Bullock.* (Kew).
       - **11.** 1888, Well, *J. Burtt Davy.*
       - **14.** 1896, Rauceby, *Stow.*†
       - **15.** 1903, Sapperton, *Stow.*†   1913, Easton, *Fisher.*
       - **16.** 1925, Lincs. Gate, *Fisher.*

   16. **E. pseudokerneri** Pugsl.
       - **7.** 1963, Walesby Top, *E. J. Gibbons.*†
       - **13.** 1943, Ermine St., 12 miles S. of Lincoln, *H.F.D.* (Kew).
       - **15.** 1950, Holywell, *T. G. Tutin.*

   18. **E. brevipila** Burnat and Gremli
       - **4.** 1892, Cleethorpes, *F. A. Lees.*†
       - **13.** 1856, Doddington, *Cole.*†

   22. **E. anglica** Pugsl.
       - **5.** 1892, Scotter Common, *M., G. and A. Peacock.*†
       - **7.** 1952, Hatton, *E. J. Gibbons,* det. E. F. Warburg.
       - **10.** 1903, Woodhall Spa, *Peacock.*†   1961, *J. Gibbons.*
       - **15.** 1934, W. Willoughby, *Fisher,* det. Pugsley.

       All det. P. F. Yeo.

436. Odontites Ludw.

1. **O. verna** (Bellardi) Dumort.   *Red Bartsia*
   - 1636, *Johnson.*
   - Not recorded for Div. 17.
   - Native. Wood rides and cultivated land.

437. Parentucellia Viv.

1. **P. viscosa** (L.) Caruel   *Yellow Bartsia*
   - **12.** 1957, Wainfleet, *P. and J. Hall.* (Two localities).
   - Casual in grass seed on sea bank.

OROBANCHACEAE

439. Lathraea L.

1. **L. squamaria** L. *Toothwort*
    2. 1895, Broughton, *Revd. W. Wyatt.*
    4. 1958, Croxby, *R. Smith.* On Sycamore.
    7. 1895, Claxby, *Peacock.*†
    8. 1901, Burwell, *C. S. Carter.* 1969, *L.N.U. Meeting.*
    11. 1884, Well Vale, *J. E. Mason.* 1888, Welton Wood, *W. W. Mason.*† 1971, *L.N.U. Meeting.*
    15. Before 1896, Wood Nook, near Grantham.
    16. 1840, Witham-on-the-Hill, *Dodsworth.*† 1936, Grimsthorpe, *J. H. Chandler.*

    Native. Frequent in division 11; rare elsewhere.

440. Orobanche L.

2. **O. purpurea** Jacq. *Blue Broomrape*
    3. 1972, Killingholme, *C. J. Potts.*
    4. 1929, Immingham Docks, *S. A. Cox.*† 1938, *L.N.U. Meeting.*

3. **O. rapum-genistae** Thuill. *Greater Broomrape*
    5. 1887, near Gainsborough, *Stanwell.*† On *Ulex.*
    6. 1894, Laughterton, *F. Mills.*† On *Sarothamnus.*
    1969, *I. Weston.*

    Recorded by Dodsworth in Div. 16, but possibly *O. elatior.*
    Native. Very rare.

4. **O. alba** Steph. ex Willd. *Red Broomrape*
    8. 1897, Cawthorpe Wood, *E. Lewin.* On *Thymus.* (Ref. *Journal of Botany,* 1903, p, 461).

6. **O. elatior** Sutton *Tall Broomrape*
    6. 1800, Fillingham, *Revd. J. Dalton.*
    Recorded for Divs. 2, 5, 6, 13, 14, 15, 16.
    Native. Only found on the limestone.

8. **O. minor** Sm. *Lesser Broomrape*
    16. 1836, Witham-on-the-Hill, *Dodsworth.*
    Not recorded for Divs. 1, 4, 9, 11, 12, 14.
    Native. No record since 1956.

LENTIBULARIACEAE

441. Pinguicula L.

3. **P. vulgaris** L. *Common Butterwort*
    10. 1724, Tattershall, *Stukeley.*
    Recorded for Divs. 1, 2, 3, 5, 6, 7, 10, 15, 16.
    Native. Peacock says 43 records in "*Rock Soil Flora*"; now very scarce.

442. Utricularia L.

1. **U. vulgaris** L.   *Greater Bladderwort*
   13. 1829, Canwick Common, *R. J. Bunch.*
   Not recorded for Divs. 8, 15, 17, 18.
   Native. Very scarce now.

3. **U. intermedia** Hayne   *Intermediate Bladderwort*
   No definite records — see Peacock's "*Rock Soil Flora*".

4. **U. minor** L.   *Lesser Bladderwort*
   5. 1838, Scotter, *Irvine.* 1843, Laughton, *T. V. Wollaston.*
   10. *c.* 1900, Woodhall Spa, *F. S. Alston.*
   13. 1961, Potterhanworth Fen, *E. H. Clifton.*
   14. 1895, Billinghay, *Revd. E. R. Walker.*
   16. 1838, Bourne, *Dodsworth.*†
   Native. Probably extinct.

ACANTHACEAE

443. Acanthus L.

1. **A. mollis** L.   *Bear's Breech*
   11. 1959, Trusthorpe, *R. Smith.*
   Alien, established.

VERBENACEAE

444. Verbena L.

1. **V. officinalis** L.   *Vervain*
   10. 1820, West Ashby, *Ward.*
   Not recorded for Divs. 1, 7, 9, 18.
   Native. Rare and elusive.

LABIATAE

445. Mentha L.

2. **M. pulegium** L.   *Penny-royal*
   12. 1856, near Boston, *Thompson.*
   Recorded for Divs. 4, 10, 12.
   Probably as a casual only.

3. **M. arvensis** L.   *Corn Mint*
   1851, *Watson.*
   Not recorded for Div. 9.
   Native. Common in suitable habitats as a marsh plant; as a cornfield weed occasional.

4. **M. aquatica** L. *Water Mint*
    12. 1799, East Fen, *Young*.
    Recorded for all Divs.
    Native. General; more usual by running water.

    x smithiana R. A. Graham [**M. rubra** Sm.]
    13. 1865, Skellingthorpe, *Cole*.†
    Distribution not known; to be investigated. Grows in the valleys near Louth and in Div. 16?

    x piperita L.
    7. 1905, South Kelsey, *Peacock*.†
    11. 1894, Saleby, *H. Charman*.†
    13. 1862, Doddington, *Cole*.†
    15. 1959, Gunby, *J. H. Chandler*.

6. **M. longifolia** (L.) Huds. *Horsemint*
    8. 1666, Burwell, *Lister*.
    Recorded for Divs. 1, 4, 8, 10, 11, 12, 15, 16.
    Garden escape.

7. **M. rotundifolia** (L.) Huds. *Apple-scented Mint*
    10. 1900, E. and W. Keal, *Mason*.
    11. 1911 (*Journal of Botany*), *Reynolds*.
    Garden escape.

446. Lycopus L.

1. **L. europaeus** L. *Gipsy-wort*
    10. 1820, Horncastle, *Ward*.
    Recorded for all Divs.
    Native. Frequent in damp spots.

447. Origanum L.

1. **O. vulgare** L. *Marjoram*
    3. 1840, Bigby, *Elwes*.
    15. 1819, Ropsley, *Dodsworth*.
    Not recorded for Divs. 4, 5, 9, 17, 18.
    Native. Locally abundant but not common.

448. Thymus L.

1. **T. pulegioides** L. *Larger Wild Thyme*
    4. 1953, Thoresway, *Hope-Simpson*.
    15. 1958, near Grantham, *Stirling*.
    Native. Rare or unnoticed.

3. **T. drucei** Ronn. *Wild Thyme*
   2. 1822, Appleby, *Strickland.*
   Not recorded for Divs. 1, 9, 12, 18.
   Native. Rather scarce; chiefly on limestone and chalk; shows a liking for ironstone on wold side. Not on the coast. The Grimsthorpe Thyme flowers earlier and is bushier in habit.

451. Calamintha Mill.

2. **C. ascendens** Jord. *Common Calamint*
   14. 1790, Threckingham, *Cragg.*
   Recorded for Divs. 3, 5, 6, 8, 10, 11, 13, 14, 16.
   Native. Rather rare. Decreasing.

452. Acinos Mill.

1. **A. arvensis** (Lam.) Dandy *Basil Thyme*
   15. 1726, near Grantham, *Bacon.*
   Not recorded for Divs. 3, 9, 12, 17, 18.
   Native. Uncommon.

453. Clinopodium L.

1. **C. vulgare** L. *Wild Basil*
   13. 1829, Canwick, *Rev. J. F. Wray.*
   Not recorded for Divs. 9, 12, 18.
   Native. Uncommon.

454. Melissa L.

1. **M. officinalis** L. *Lemon Balm*
   10. 1893, Coningsby, *Revd. F. S. Alston.*†
   16. 1958, Uffington gravel pit, *J. H. Chandler.*
   Introduced. Garden excape.

455. Salvia L.

1. **S. verticillata** L.
   4. 1901, Grimsby Docks, *A. Smith.*†
   12. 1911, Boston Docks, *S. Hurst and B. Reynolds.*†
   15. 1895, Ancaster railway, *Mason.*†
   16. 1961-2, Little Bytham, *A. Rasell.*
   Introduced. Casual.

2. **S. pratensis** L. *Meadow Clary*
   11. 1888, Partney, *Burgess.*
   14. 1939-54, Rauceby, *E. J. Gibbons.*
   15. 1896, Belton, *Woolward.*†
   Casual? Extinct?

THE COUNTY FLORA

4. **S. horminoides** Pourr. *Wild Clary*
   10. 1820, Horncastle, *Ward*.
   Not recorded for Divs. 1, 2, 17, 18.
   Native. Uncommon.

457. Prunella L.

1. **P. vulgaris** L. *Self-heal*
   12. 1820, near Boston *Thompson*.
   Recorded for all Divs.
   Native. Frequent.

2. **P. laciniata** (L.) L.
   10. 1945, Tower on the Moor, Woodhall Spa, *N. D. Simpson*.†
   Native. The only record.

458. Betonica L.

1. **B. officinalis** L. *Wood Betony*
   14. 1790, Threckingham, *Cragg*.
   Not recorded for Divs. 12, 17, 18.
   Native, old pastures and woods. Decreasing.

459. Stachys L.

3. **S. arvensis** (L.) L. *Field Woundwort*
   15. 1726, near Grantham, *Bacon*.
   Not recorded for Divs. 1, 6, 9, 12, 17, 18.
   Native, arable weed. Uncommon.

4. **S. germanica** L. *Downy Woundwort, Base Horehound*
   4. 1903, Grimsby Docks, *A. Smith*.†
   15. 1727, opposite Easton, *Dr. Richardson*. 1762, *W. Hudson*. 1794-6, *Botanist Guide*, "In profusion a little wide of Colsterworth", *Turner and Dillwyn*. 1800, Stoke Rochford, Old Herbal (G. C. Druce) 1805, Colsterworth, *Revd. J. Davies*. (Specimen at Kew).
   16. 1840, Thurlby, *Dodsworth*.
   See *Naturalist* 1896, p. 181 and 1897, p. 170.
   Presumably extinct.

6. **S. palustris** L. *Marsh Woundwort*
   12. 1825, Wainfleet, *Sinclair*.
   Recorded for all Divs.
   Native. Decreasing.

7. **S. sylvatica** L. *Hedge Woundwort*
    7. 1829, Bardney, *Rev. J. F. Wray.*
    Recorded for all divs.
    Native. Common.

460. Ballota L.

1. **B. nigra** L. *Black Horehound*
    14. 1790, Threckingham, *Cragg.*
    Recorded for all Divs.
    Native. Chiefly in villages.

461. Lamiastrum Adans. [Galeobdolon Heist. ex Fabr.]

1. **Lamiastrum galeobdolon** *Yellow Archangel*
    ssp. *galeobdolon* (S. Wegmuller).
    8. 1969, Burwell, *M. Smith.*
    11. 1969, Welton Wood, *M. Smith.*
    Rare. First British record.

    ssp. *montana.*
    10. Langton by Horncastle, *Ward.*
    Not recorded for Divs. 4, 9, 12, 17, 18.
    Native, uncommon.

462. Lamium L.

1. **L. amplexicaule** L. *Henbit*
    5. 1840, Morton, *Miller.*
    Not recorded for Divs. 1, 12.
    Native. Chiefly arable land and gardens.

3. **L. hybridum** Vill. *Cutleaved Dead-nettle*
    6. *c.* 1836, Greetwell, *Simpson Collection.*†
    Not recorded for Divs. 4, 8, 14.
    Native. Uncommon on sandy peat.

4. **L. purpureum** L. *Red Dead-nettle*
    3. 1835, Wootton, *E. J. Nicholson.*
    Recorded for all Divs.
    Native. Abundant.

5. **L. album** L. *White Dead-nettle*
    14. 1790, Threckingham, *Cragg.*
    Recorded for all Divs.
    Native. Frequent.

6. **L. maculatum** L. *Spotted Dead-nettle*
    2. 1876, Bottesford, *Peacock*.†
    Recorded for Divs. 2, 4, 10, 11, 15.
    Introduced, a garden escape.

463. Leonurus L.

1. **L. cardiaca** L. *Motherwort*
    4. 1897, Grimsby, *A. Smith*.†
    7. Check List.
    9. 1893, Mablethorpe, *Miss Mackinder*.†
    10. 1918, Woodhall Spa, *J. S. Sneath*.†
    11. 1913, Ingoldmells, *Peacock*.†, 1936, *L.N.U. Meeting*. Extinct.
    12. 1856, near Boston, *Thompson*.
    Relic of herb gardens or dock alien.

465. Galeopsis L.

1. **G. angustifolia** Ehrh. ex Hoffm. *Red Hemp-nettle*
    16. 1837, Bowthorpe, *Dodsworth*.
    Not recorded for Divs. 17, 18.
    Native or casual, calcareous soil. Uncommon.

2. **G. bifida** Boenn.
    3. 1838, North Kelsey, *Skipworth*.
    Recorded for Divs. 2, 3, 5, 7, 10, 12. 13, 14, 15.

3. **G. segetum** Neck. *Downy Hemp-nettle*
    1. 1899, Carr Houses, *S. Hudson*.
    2. 1877, Twigmoor, *Fowler*.
    16. 1838, Bourne, *Dodsworth*.
    Native? Extinct.

4. **G. tetrahit** L. sensu lato *Common Hemp-nettle*
    15. 1763, near Grantham, *Martyn*.
    Not recorded for Div. 9.
    Casual. Arable land and manure heaps.

5. **G. speciosa** Mill. *Large-flowered Hemp-nettle*
    18. 1695, Spalding, *Ray*.
    Not recorded for Divs. 8, 9.
    Native. Locally abundant; peaty arable ground.

466. Nepeta L.

1. **N. cataria** L. *Wild Catmint*
    15. 1726, near Grantham, *V. Bacon*.
    Not recorded for Divs. 4, 9, 11, 12, 14, 17, 18.
    Native. Rare.

467. Glechoma L.

1. **G. hederacea** L. *Ground Ivy*
    14. 1790, Threckingham, *Cragg*.
    Recorded for all Divs.
    Native. Frequent.

468. Marrubium L.

1. **M. vulgare** L. *White Horehound*
    15. 1726, near Grantham, *V. Bacon*.
    Not recorded for Divs. 4, 7, 8.
    Native. Rare. Decreasing.

469. Scutellaria L.

1. **S. galericulata** L. *Skullcap*
    13. 1829, Bracebridge, *Rev. J. F. Wray*.
    Not recorded for Divs. 17, 18.
    Native. Scattered, locally abundant. Occasionally pink in Div. 2.

2. **S. minor** Huds. *Lesser Skullcap*
    2. 1876, Santon, *Fowler*. 1969, Twigmoor, *E. J. Gibbons*.
    5. 1840, Laughton, *Miller*. 1894, Scotter, *Mason*. 1926, *E. J. Gibbons*.
    16. 1836, River Glen, Bourne, *Dodsworth*.
    Native. Rare; perhaps extinct.

470. Teucrium L.

1. **T. chamaedrys** L. *Wall Germander*
    2. 1858, Broughton, *Fowler*.† 1967, *J. Gibbons*.
    7. 1877, Railway bank by second gatehouse, Middle Rasen, *Lees*. 1952, *J. Gibbons*.
    Introduced. Established and naturalised.

2. **T. scordium** L. *Water Germander*
    1602, Camden's *Britannia*.
    12. 1790, East Fen, *Sir J. Banks*.
    13. 1870, Washingborough Fen, *F. A. Lees*†
    18. 1911, Cowbit, *Druce*.† 1953, *Amner*.
    Native. Very rare; perhaps extinct.

4. **T. scorodonia** L. *Wood Sage*
    10. 1820, Holbeck, *Ward*.
    Recorded for Divs. 1, 2, 3, 5, 7, 10, 11, 13, 14.
    Native. Very scarce in south Lincs.; commoner in north-west.

471. Ajuga L.

2. **A. reptans** L. *Bugle*
   16. 1836, Bourne, *Dodsworth*.
   Not recorded for Divs. 9, 17, 18.
   Native. In most woods; also in boggy hollows.

PLANTAGINACEAE
472. Plantago L.

1. **P. major** L. *Great Plantain*
   14. 1790, Threckingham, *Cragg*.
   Recorded for all Divs.
   Native. Arable fields and paths.

2. **P. media** L. *Hoary Plantain*
   7. 1835, South Kelsey, *Skipworth*.
   Recorded for all Divs.
   Native. Mostly confined to calcareous soil.

3. **P. lanceolata** L. *Ribwort*
   16. 1836, Bourne, *Dodsworth*.
   Recorded for all Divs.
   Native. A very important grazing plant.

4. **P. maritima** L. *Sea Plantain*
   9. 1834, Mablethorpe, *Bayley*.
   Recorded for Divs. 2, 3, 4, 9, 11, 12, 17, 18.
   Native. On muddy shores all along the coast.

5. **P. coronopus** L. *Buck's-horn Plantain*
   12. 1666, Wainfleet, *Lister*.
   Not recorded for Divs. 8, 15, 16.
   Native. On sandy soil inland as well as all along the coast.

6. **P. indica** L.
   4. 1901, Grimsby, *A. Smith and G. W. Marris*.†
   6. 1899, Lincoln, *H. M. W. Hinchliff*.
   12. 1938, Boston, *F. T. Baker*.
   13. 1897, Boultham, *Sneath*.
   Alien. Occasional on dumps.

473. Littorella Berg.

1. **L. uniflora** (L.) Aschers. *Shore-weed*
    - 2. 1910, Crosby Warren, *W. D. Roebuck.*†
    - 5. 1846, Laughton, *Miller.* 1876, Scotter, *Fowler.*
    - 12. Check List.
    - 13. 1829, Swanpool, *R. J. Bunch.*
      1948, Lincoln ballast pits, *E. J. Gibbons.*

    Native. Very rare.

CAMPANULACEAE

475. Campanula L.

1. **C. latifolia** L. *Giant Campanula*
    - 10. 1820, Revesby, *Ward.*

    Not recorded for Divs. 4, 5, 9, 17, 18.

    Native. In calcareous woods chiefly in the north.

2. **C. trachelium** L. *Nettle-leaved Campanula*
    - 2. 1875, Broughton, *Fowler.* 1965, *J. Gibbons.*
    - (10. 1820, Revesby, *Ward.* (A mistake).)
    - 11. Check List.
    - 13. 1892, Potterhanworth, *Sneath.* 1969, Nocton, *J. Gibbons.*
    - 14. Check List.
    - 15. 1903, Sapperton, *Miss Stow.*† 1969, Pickworth, *J. Gibbons.*
    - 16. 1836, Bourne, *Dodsworth.* 1969, Careby, *J. Gibbons.*

    Definite records for Divs. 2, 13, 15, 16.

    Native. Rare in the north; in most woods in the south on the limestone and some roadsides; no definite record for the chalk.

3. **C. rapunculoides** L. *Creeping Campanula*
    - 14. 1855, Brauncewell, *Lowe.*

    Recorded for Divs. 2, 3, 6, 7, 8, 10, 11, 14, 15, 16.

    Possibly native. A very persistent garden weed, occasionally established elsewhere.

6. **C. glomerata** L. *Clustered Bellflower*
    - 15. 1726, near Grantham, *Bacon.*

    Not recorded for Divs. 4, 9, 12, 17, 18.

    Native. Chiefly on limestone, sparingly; very rare on chalk.

7. **C. rotundifolia** L. *Harebell*
    - 14. 1790, Osbournby, *Cragg.*

    Not recorded for Div. 18.

    Native. Scattered, locally abundant but decreasing.

8. **C. patula** L.
   11. 1893, Spilsby, *Burgess*. (Painting; wrongly identified; a spurious *C. rotundifolia*).

9. **C. rapunculus** L.   *Rampion*
   11. 1893, Spilsby, *Burgess*.
   Several plants in a clover field, one season.

476. Legousia Durande

1. **L. hybrida** (L.) Delarb.   *Venus's Looking-glass*
   16. 1840, Manthorpe, *Dodsworth*.†
   Not recorded for Divs. 5, 9, 12, 17, 18.
   Native. Not common, decreasing.

477. Trachelium L.

1. **T. caeruleum** L.
   10. 1820, Revesby, *Ward*. Mistake, or garden escape.
   12. 1938, Boston Dock, *F. T. Baker*.
   Alien.

479. Jasione L.

1. **J. montana** L.   *Sheep's-bit*
   2. 1853-97, Winterton, *Fowler*.
   3. 1900, Nettleton, *S. Allett*.†   1969, *J. Gibbons*.
   5. c. 1930, Scotter, *S. A. Cox*.
   6. 1909, Kettlethorpe, *F. Mills*.†   1945, Laughterton, *L.N.U.*
   7. 1877, Tealby, *Lees*.   1890, Holton to Claxby, *Brewster*. 1905, Linwood, *A. Smith*.†
   10. 1820, Horncastle, *Ward*.   1892, Coningsby, *F. S. Alston*. 1897, Hagworthingham, *S. Borrass*.
   13. 1858, Doddington, *Cole*.†   1905, North Hykeham, *Peacock*.† 1967, *J. Gibbons*.
   15. 1884, Little Ponton, *Browne*.†   1900, Great Ponton, *Mason*. 1930, Belton, *E. Orchard*.
   Native. Rare, decreasing.

RUBIACEAE

481. Sherardia L.

1. **S. arvensis** L.   *Field Madder*
   3. 1835, North Kelsey, *Skipworth*.
   Recorded for all Divs.
   Native, arable weed.

483. Asperula L.

2. **A. cynanchica** L. *Squinancy Wort*
   1835, no locality, *Skipworth*.
   2. 1856, Broughton, *Fowler*.
   5. 1953, Waddingham, *E. Rylatt*.
   6. Check List.
   13. 1893, Nocton, *Mason*. 1903, Fulbeck, *Stow*.†
   14. 1855, Wilsford, *Lowe*. 1895, Rauceby, *Mason*. 1937 and 1963, *E. J. Gibbons*.
   15. 1879, Ancaster, *Fowler*. 1963, Holywell, *J. H. Chandler*.
   16. Check List.
   Native. Rare and decreasing; on limestone only.

484. Cruciata Mill.

1. **C. chersonensis** (Willd.) Ehrend. [**C. laevipes** Opiz] *Crosswort*
   14. 1790, Threckingham, *Cragg*.
   Not recorded for Div. 18.
   Native. Probably decreasing.

485. Galium L.

1. **G. odoratum** (L.) Scop. *Sweet Woodruff*
   14. 1790, Threckingham, *Cragg*.
   Not recorded for Divs. 4, 9, 12, 14, 17, 18.
   Native. Scarce and local.

3. **G. mollugo** L. *Great Hedge Bedstraw*
   16. 1836, Bourne, *Dodsworth*.
   Not recorded for Divs. 1, 11.
   Native. Rare and scattered except in Divs. 15 and 16.

   (*b*) ssp. *erectum* Syme *Erect Hedge Bedstraw*
   2. 1856, Broughton, *Fowler*.
   5. 1903, Redbourn, *Peacock*.
   6. 1961, Knaith, *E. J. Gibbons*.
   12. 1856, near Boston, *Thompson*.
   15. 1780, near Easton, *Sibthorp*.

4. **G. verum** L. *Lady's Bedstraw*
   14. 1790, Threckingham, *Cragg*.
   Recorded for all Divs.
   Native. Rare in south east.

5. **G. saxatile** L. *Heath Bedstraw*
   10. 1820, Moorby, *Ward*.
   Not recorded for Divs. 4, 9, 11, 12, 17, 18.
   Native. Locally abundant on sandy soil.

6. **G. pumilum** Murr. *Slender Bedstraw*
    3. 1949, Nettleton, *E. J. Gibbons.*
    8. 1952, Walmsgate, *J. Hope Simpson.* Both records confirmed by K. Goodway.
    Native. Very rare on chalk.

8. **G. palustre** L. *Marsh Bedstraw*
    12. 1820, East Fen, *Thompson.*
    Recorded for all Divs.
    Native. Frequent.

    b. ssp. *elongatum* (C. Presl) Lange.
    Distribution not worked up. To be looked for.

10. **G. uliginosum** L. *Bog Bedstraw*
    12. 1829, Wainfleet, *Oldfield.*
    Not recorded for Divs. 1, 17, 18.
    Native. Widespread, particularly in wold valleys.

11. **G. tricornutum** Dandy *Rough Corn Bedstraw*
    16. 1778, Stamford, *Hudson.*
    Recorded for Divs. 2, 3, 5, 6, 7, 12, 13, 15, 16.
    Native. Very scarce and decreasing. Last seen Careby 1953, J. Gibbons.

12. **G. aparine** L. *Goosegrass, Hairif, Sweethearts*
    14. 1790, Threckingham, *Cragg.*
    Recorded for all Divs.
    Native. There are two forms of this: a slenderer, earlier flowering form in woods, not uncommon; the other, in hedges mainly and arable fields, much coarser, very abundant.

14. **G. parisiense** L. *Wall Bedstraw*
    12. 1836, Boston (on a wall), *Dodsworth.* The only record.

CAPRIFOLIACEAE
487. Sambucus L.

1. **S. ebulus** L. *Danewort*
    1. 1895, Haxey, *Fowler.*†
    7. 1890, Sotby, *Mrs. Jarvis.*†
    12. 1688, Fishtoft, *Plukenet.*
    13. 1868, Welbourne, *G. W. Burtt.*†
    14. 1947, Pickworth, *E. J. Gibbons.*
    15. 1903, Ropsley, *S. C. Stow.*† 1960, *J. H. Chandler.*
    17. 1688, Kirton, *Plukenet.*
    ?Native or relic of cultivation. Established.

2. **S. nigra** L. *Elder*
 12. 1696, Wainfleet, *C. Merrett, Jnr.*
 Recorded for all Divs.
 Native, very common.

488. Viburnum L.
1. **V. lantana** L. *Wayfaring Tree*
 3. 1901, Cabourn, *Mason*. (Planted).
 4. 1892, Rothwell, *W. J. LeTall*. (Planted. Nat. 1892).†
 5. 1893, Morton, *Lees*. (Planted in hedge).
 11. 1890, Ailby, *J. W. Chandler*. (Probably wrongly identified — Dogwood).
 13. 1856, Doddington, *Cole*.
 14. *Check List*.
 15. 1903, Holywell, *Peacock*.† 1953, *J. Gibbons*.
 16. 1836, Obthorpe, *Dodsworth*. 1831, Market Deeping (?Planted) *R. J. Bunch*.
 Native. Doubtfully wild except on limestone in the southwest.

3. **V. opulus** L. *Guelder Rose, Dogberry*
 14. 1790, Threckingham, *Cragg*.
 Not recorded for Divs. 9, 12, 17, 18.
 Native. Fairly general but not common. Woods and hedges.

489. Symphoricarpos Duham.
1. **S. rivularis** Suksd. *Snowberry*
 2. 1864, Bottesford, *Peacock*.
 Not recorded for Div. 17.
 Introduced; planted in many woods and copses and naturalised.

491. Lonicera L.
1. **L. xylosteum** L. *Fly Honeysuckle*
 15. 1896, Belton, *Woolward*.†
 Not native.

3. **L. periclymenum** L. *Honeysuckle, Woodbine*
 10. 1820, Holbeck, *Ward*.
 Recorded for all Divs.
 Native. Abundant in woods and old hedges.

ADOXACEAE
493. Adoxa L.
1. **A. moschatellina** L.   Moschatel, Townhall Clock
    7. 1829, Bardney,† *Revd. J. F. Wray and R. J. Bunch.*
    Not recorded for Divs. 1, 9; not confirmed for Divs. 12, 17, 18.
    Native. Scattered and not common.

VALERIANACEAE
494. Valerianella Mill.
1. **V. locusta** (L.) Betcke   Corn Salad, Lamb's Lettuce
    2. 1822, Appleby, *Strickland.*
    Not recorded for Divs. 1, 14, 15, 18.
    Native. Not very common except on sand dunes, perhaps ssp. *dunense.*
    Casual.

2. **V. carinata** Lois.
    4. 1902, Grimsby, *A. Smith.*†
    Casual.

3. **V. rimosa** Bast.
    10. 1882, Woodhall Spa, *J. C. Melvill.*
    Native. Possibly overlooked.

5. **V. dentata** (L.) Poll.
    7. 1836, South Kelsey, *Skipworth.* 1971, Sotby, *L.N.U.*
    16. 1836, Bowthorpe, *Dodsworth.*
    Not recorded for Divs. 9, 12, 17.
    Native or colonist, chiefly in calcareous arable fields.

495. Valeriana L.
1. **V. officinalis** L.   *Valerian*
    10. 1724, Tattershall, *Stukeley.*
    Not recorded for Div. 18.
    Native. Decreasing owing to River Board spraying.

3. **V. dioica** L.   *Marsh Valerian*
    8. 1834, near Louth, *Bayley.*
    Not recorded for Divs. 17, 18.
    Native. Widespread but local.

496. Centranthus DC.

1. **C. ruber** (L.) DC. *Red Valerian*
    3. 1835, ?Thornton Abbey, *E. J. Nicholson*.
    Recorded for Divs. 2, 3, 5, 8, 15, 16.
    Introduced. Established on walls and quarries.

DIPSACACEAE
497. Dipsacus L.

1. **D. fullonum** L. ssp. **sylvestris** (Huds.) *Teasel*
    10. 1820, Horncastle, *Ward*.
    Recorded for all Divs.
    Native. Common, abundant along the Humber bank and waste places.

2. **D. pilosus** L. *Small Teasel*
    4. 1936, Stainton-le-Vale, *E. J. Gibbons*.
    6. 1950, Fiskerton, *R. Hull*.
    7. 1877, Claxby, *Lees*.†
    11. 1890, Claythorpe, *J. W. Chandler*.†
    12. 1829, Wainfleet, *Oldfield*. Extinct.
    13. 1949, Potterhanworth, *E. J. Gibbons*.
    15. 1780, Near Easton, *Sibthorp*.
    1963, Twyford Forest, *E. J. Gibbons*.
    16. 1836, Billingborough, *Dodsworth*. 1960, Kirkby Underwood, *E. J. Gibbons*.
    Native. Calcareous woods. Uncommon.

498. Knautia L.

1. **K. arvensis** (L.) Coult. *Field Scabious*
    13. 1829, Canwick, *Rev. J. F. Wray*.†
    Recorded for all Divs.
    Native. Absent from acid soils.

499. Scabiosa L.

1. **S. columbaria** L. *Small Scabious*
    6 or 13. 1800, Lincoln, *Salt*.
    Not recorded for Divs. 1, 7, 9, 17, 18.
    Native. Uncommon, on calcareous soil only.

500. Succisa Haller

1. **S. pratensis** Moench   *Devil's-bit Scabious*
    7. 1836, South Kelsey, *Skipworth*.
    16. 1836, Bourne, *Dodsworth*.
    Not recorded for Divs. 12, 17, 18.
    Native. Decreasing.

COMPOSITAE
502. Bidens L.

1. **B. cernua** L.   *Nodding Bur-Marigold*
    12. 1799, East Fen, *Young*.
    Not recorded for Divs. 4, 8, 16, 17, 18.
    Native. Rare and decreasing.

2. **B. tripartita** L.   *Threecleft Bur-Marigold*
    17 or 18. 1666, Fens, *Lister*.
    Not recorded for Divs. 4, 8, 9.
    Native. Not common.

503. Galinsoga Ruiz & Pav.

1. **G. parviflora** Cav.   *Gallant Soldier*
    13. 1968, Lincoln, *R. J. Burton*.
    Introduced.

2. **G. ciliata** (Raf.) Blake   *Shaggy Soldier*
    11. 1967, Skegness, *M. N. Read*,
    14. 1969, Rauceby, *F. Lamnniman*.
    16. 1961, Stamford, *J. H. Chandler*.
    Introduced.

504. Ambrosia L.

1. **A. artemisiifolia** L.
    7. 1878, Rasen, *Lees*.
    10. 1917, Woodhall Spa, *F. S. Alston*.†
    Casual.

505. Xanthium L.

2. **X. spinosum** L.   *Spiny Cocklebur*
    4. 1967, Immingham.
    17. 1937, Donington, *F. L. Kirk*.
    Alien.

506. Senecio L.

1. **S. jacobaea** L. *Ragwort*
    12. 1799, East Fen, *Young*.
    Recorded for all Divs.

    Native. Locally abundant on sand.

2. **S. aquaticus** Hill *Water Ragwort*
    13. 1836, Boultham, *Simpson Collection*.†
    Recorded for all Divs.

    Native. Uncommon. Decreasing.

3. **S. erucifolius** L. *Hoary Ragwort*
    8. 1850, Louth, *M. E. Dixon*.
    Recorded for all Divs.

    Native. Replaces *S. jacobaea* on stiff clay; not very common.

4. **S. squalidus** L. *Oxford Ragwort*
    2. 1936, Scunthorpe Slag Heaps, *J. Gibbons*.
    14. 1855, Anwick, *Lowe*.
    Recorded for all Divs.

    Has spread to Scunthorpe slag heaps, Barton quarries, Brigg sugar factory, the ironstone quarries south of Grantham and to Lincoln railway sidings since 1930. Infrequent, not established in the east and south east.

6. **S. sylvaticus** L. *Wood Groundsel*
    5. 1840, Morton, *Miller*.
    Not recorded for Divs. 17 and 18.

    Native. Frequent on sandy and gravelly soils.

7. **S. viscosus** L. *Sticky Groundsel*
    15. 1884, Grantham Canal, *Browne*.
    Not recorded for Divs. 17, 18.

    Doubtfully native. Spread from gravel pits.

8. **S. vulgaris** L. *Groundsel*
    14. 1790, Threckingham, *Cragg*.
    Recorded for all Divs.

    Native, abundant. A rayed form is spreading along railways.

THE COUNTY FLORA

11. **S. paludosus** L. *Great Fen Ragwort*
    13. 1797, Brayford, Lincoln, *H. Wollaston*. (Specimen in Edinburgh Herbarium given by W. J. Hooker, or probably by Dr. John Nicholson, to Prof. J. Hutton, 1838).
    1805, Near Hare Booth, Metheringham, *Sir J. Banks*.
    Div.? 8, J. Bogg, not dated or localised (*Naturalist* 1895, page 96). c. 1820.
    Extinct.

13. **S. fluviatilis** Wallr. *Broad-leaved Ragwort*
    15. 1880, Little Ponton, *Browne*.†
    Introduced? Extinct.

16. **S. palustris** (L.) Hook. *Marsh Fleawort*
    7. 1820, Great Sturton, *J. Ward*. Probably *S. aquaticus* as the habitat is hardly suitable.
    12. 1789, East Fen, *Gough*. "In the East Fen in some years in vast abundance; in others very scarce" — Sir J. Banks (*Botanists Guide*, 1805).
    Formerly native. Extinct.

17. **S. integrifolius** (L.) Clairv. sensu lato *Field Fleawort*
    15. 1873, Ancaster, *Streatfield*. (Extinct c. 1930, *H. Fisher*).†
    1780, Near Grantham, *Sibthorp*.
    16. 1796, Stamford, *R. A. Salisbury* (Peacock).
    Native. Always very rare. Extinct.

507. Doronicum L.

1. **D. pardalianches** L. *Great Leopard's Bane*
    2. 1962, Scawby Park, *L.N.U.*
    7. 1878, Kirkby-cum-Osgodby, *Lees*.
    10. 1941, Woodhall Spa, *D. Marsden*.
    11. 1894, Gunby, *Mason*.†
    15. 1962, West Willoughby, *E. J. Gibbons*.
    16. 1957, Dowsby, *R. and B. Howitt*.
    Introduced and established.

2. **D. plantagineum** L. *Leopard's Bane*
    15. 1898, Saltersford, *Miss Woolward*.
    Introduced.

508. Tussilago L.

1. **T. farfara** L. *Coltsfoot*
    14. 1790, Threckingham, *Cragg*.
    Recorded for all Divs.
    Native. Not on acid sand, except by accident.

228   THE FLORA OF LINCOLNSHIRE

509. Petasites Mill.

1. **P. hybridus** (L.) Gaertn., Mey & Scherb.　　　*Butterbur*
    8. 1805, Hemingby, *Relhan.*
    Not recorded for Divs. 13, 18.
    Native. Female without exception on the wold streams; male and female occur on the Trent banks; male on Ancholme Canal, Brigg, and in Kesteven.

2. **P. albus** (L.) Gaertn.　　　*White Butterbur*
    2. 1896, Broughton, *Peacock.*†
    Introduced.

4. **P. fragrans** (Vill.) C. Presl　　　*Winter Heliotrope*
    11. 1879, Spilsby, *Burgess.*
    Not recorded for Divs. 1, 2, 9, 13, 17.
    Introduced, naturalised usually near houses.

512. Inula L.

1. **I. helenium** L.　　　*Elecampane*
    3. 1918, Wrawby, *Frith.*†
    5. 1840, Walkerith, *Miller.*
    6. 1892, Hardwick, *Paddison.*†
    7. 1890, Goltho, *Sneath.* 1969, Campney lane, *N. Read.*
    8. 1666, Burwell, *Lister.*
    12. 1926, Boston Dock, *M. E. Stewart.* 1931, *S. Hurst (L.N.U Trans.).*
    15. 1893, Dry Doddington, *Stow.*† 1960, *E. J. Gibbons.*
    17. 1962, Surfleet Fen, *E. J. Gibbons and F. H. Perring.*
    18. 1685, Whaplode, *Plukenet.*
    Introduced.

4. **I. conyza** DC.　　　*Ploughman's Spikenard*
    9. 1853, Saltfleetby, *Cordeaux.*
    Not recorded for Divs. 6, 11, 12, 17, 18.
    Native. Uncommon and local.

5. **I. crithmoides** L.　　　*Golden Samphire*
    4. 1886, Cleethorpes, *Browne.*
    12. 1856, Near Boston, *Thompson.*
        1886, Frieston, *Lane-Claypon* (Painting).
    18. 1861, Coast, *M. Walcott.* c. 1930, Gedney, *Ian Hepburn.*
    Dr. Burgess' painting is not *I. crithmoides* but *Aster tripolium* var. *discoideus.*
    Native. No specimen exists.

THE COUNTY FLORA

513. Pulicaria Gaertn.

1. **P. dysenterica** (L.) Bernh. *Fleabane*
    16. *c.* 1822, Uffington, *Twopeny.*
    14. (Var. glabra, 1908, Haverholme, *Mason.†*).
    Recorded for all Divs.
    Native.

514. Filago L.

1. **F. germanica** (L.) L. [**F. vulgaris** Lam.] *Cudweed*
    8. 1666, Burwell, *Lister.*
    16. 1790, Threckingham, *Cragg.*
    Not recorded for Div. 18.
    Native. Decreasing.

2. **F. apiculata** G. E. Sm. [**F. lutescens** Jord.]
    Red-tipped Cudweed
    6. 1911, Reepham, *N. Y. Sandwith.* (Sp. Kew).
    Native.

3. **F. spathulata** C. Presl [**F. pyramidata** L.]
    Spathulate Cudweed
    15. 1882, Corby, *W. Fowler.*
    Native. Other specimens in the County Herbarium are wrongly identified.

5. **F. minima** (Sm.) Pers. *Slender Cudweed*
    11. 1847, Burgh, *Dr. Grantham.*
    Not recorded for Divs. 4, 12, 15, 16, 17, 18.
    Native, on very dry sand.

515. Gnaphalium L.

1. **G. sylvaticum** L. *Wood Cudweed*
    8. 1666, Burwell, *Lister.*
    Not recorded for Divs. 12, 14, 16, 17, 18.
    Native, scarce.

4. **G. uliginosum** L. *Marsh Cudweed*
    13. 1836, Doddington, *Simpson Collection.*
    Recorded for all Divs.
    Native. Decreasing.

5. **G. luteoalbum** L. *Jersey Cudweed*
    12. 1945, Boston Dock, *N. D. Simpson and A. H. Alston.*
    Dock alien.

516. Anaphalis DC.

1. **A. margaritacea** (L.) Benth.   *Pearly Everlasting*
    1. 1877, Haxey, *W. Fowler.*
    3. 1835, Wootton, *E. J. Nicholson.*
    Introduced.

517. Antennaria Gaertn.

1. **A. dioica** (L.) Gaertn.   *Cat's-foot*
    2. 1875, Broughton, *Fowler*
    7. 1877, Middle Rasen, Nova Scotia, *Lees.*
    15. 1780, Near Easton, *Sibthorp.* 1805, Grantham Heath, *D. Turner.*
    Native. No recent record; probably extinct.

518. Solidago L.

1. **S. virgaurea** L.   *Golden-rod*
    2. 1894, Broughton. 1969, *E. J. Gibbons.*
    3. 1909, North Kelsey, *Peacock.*†
    5. *Check List.*
    7. 1877, Market Rasen, *Lees.* 1892, South Kelsey, *Miss J. Brewster.*† 1970, Osgodby, *E. J. Gibbons.*
    10. 1724, Tattershall, *Stukeley.* 1967, Woodhall Spa, *E. J. Gibbons.*
    13. 1851, Swinderby, *Watson.* 1960, Thurlby, *J. Gibbons.*
    Recorded for Divs. 2, 3, 4, 5, 7, 10, 13.
    Native. Rather scarce and local.

2. **S. canadensis** L.
    11. 1964, Chapel St. Leonards, *J. Gibbons and N. Read.*
    Introduced.

3. **S. gigantea** Ait.
    2. 1959, Broughton Woods, *E. J. Gibbons.*
    5. 1962, Snitterby, *E. J. Gibbons.*
    15. 1958, Castle Bytham, *J. H. Chandler.*
    16. 1961, Uffington, *J. H. Chandler.*
    Introduced. Spreading on tips and roadsides.

4. **S. graminifolia** (L.) Salisb.
    6. 1963, Greetwell, *E. J. Gibbons.*
    Introduced.

519. Aster L.

1. **A. tripolium** L.   *Sea Aster*
    17. 1688, Kirton, *Holden.*
    Not recorded for Divs. 6, 7, 8, 10, 13, 14, 15, 16.
    Native. On mud all round the coast, including R. Trent, abundantly by R. Humber.

    var. *discoideus* Reichb. f.
    12 or 17. 1838, Boston, *Dodsworth.*†
    Proc. B.S.B.I., 1966, vol. 6, p. 274, A. J. Gray.
    Frequent. More often on lower zones of salt marshes.

6. **A. novi-belgii** L.   *Michaelmas Daisy*
    13. 1950, Boultham, *E. J. Gibbons.*
    16. 1955, Stamford, *J. H. Chandler.*
    Occurring as a garden escape on roadsides and dumps. Distribution not yet known.

521. Erigeron L.

1. **E. acer** L.   *Blue Fleabane*
    7. 1836, Claxby, *Skipworth.*
    Recorded for all Divs.
    Native. Not common; locally abundant.

522. Conyza Less.

1. **C. canadensis** (L.) Cronq.   *Canadian Fleabane*
    11. 1909, Skegness, *Reynolds.*
    Recorded for all Divs.
    Introduced. Persisting, chiefly on railways.

524. Bellis L.

1. **B. perennis** L.   *Daisy*
    14. 1790, Threckingham, *Cragg.*
    Recorded for all Divs.
    Native. "Not on peat" — *Peacock.*

525. Eupatorium L.

1. **E. cannabinum** L.   *Hemp Agrimony*
    12. 1799, East Fen, *Young.*
    Not recorded for Div. 17.
    Native. Scarce and local.

526. Anthemis L.

1. **A. tinctoria** L. *Yellow Chamomile*
    2. 1892, Winterton, *Fowler*.†
    Recorded for Divs. 2, 8, 11, 13.
    Casual.

2. **A. cotula** L. *Stinking Mayweed*
    7. 1829, Bardney, *Revd. J. F .Wray*.†
    Not recorded for Divs. 1, 9, 17.
    Native.

3. **A. arvensis** L. *Corn Chamomile*
    8. 1855, Donington-on-Bain, *Bogg*.†
    Not recorded for Divs. 12, 16, 17, 18.
    Native.

527. Chamaemelum Mill.

1. **C. nobile** (L.) All. *Chamomile*
    Very doubtful; Peacock says "Garden escape", but gives seven records in Check List. Not native.

528. Achillea L.

1. **A. millefolium** L. *Yarrow, Milfoil*
    14. 1790, Threckingham, *Cragg*.
    Recorded for all Divs.
    Native. Decreasing.

3. **A. ptarmica** L. *Sneezewort*
    8. 1666, Muckton, *Lister*.
    Recorded for all Divs.
    Native. Not common and decreasing.

531. Tripleurospermum Schultz Bip.

1. **T. maritimum** (L.) Koch [**Matricaria inodora** L.]
    Scentless Mayweed
    16. 1836, Bourne, *Dodsworth*.
    Recorded for all Divs.
    Native. Weed of cultivation.

532. Matricaria L.

1. **M. recutita L.** *Wild Chamomile*
    16. 1836, Bourne, *Dodsworth*.
    Not recorded for Divs. 13, 15; probably overlooked.
    Native.

2. **M. matricarioides** (Less.) Porter *Rayless Mayweed*
    Pineapple Weed
    1895, Check List, *Peacock*.
    Recorded for all Divs.
    Casual, but since 1930 has become widespread. Carried on rubber tyres and boots into arable fields, where it is established.

533. Chrysanthemum L.

1. **C. segetum L.** *Corn Marigold*
    14. 1790, Osbournby, *Cragg*.
    Not recorded for Divs. 17, 18.
    Native. Decreasing. Locally abundant.

2. **C. leucanthemum L. [Leucanthemum vulgare Lam.]**
    Ox-eye Daisy
    14. 1790, Threckingham, *Cragg*.
    Recorded for all Divs.
    Native. Old pasture and railway banks.

3. **C. maximum** Ramond **[Leucanthemum maximum** (Ramond) DC.] *Shasta Daisy*
    7. 1950, Osgodby, *E. J. Gibbons*.
    Occurring as a garden escape on roadsides, etc.; no definite records yet.

4. **C. parthenium** (L.) Bernh. **[Tanacetum parthenium** (L.) Schultz Bip.] *Feverfew*
    10. 1820, Wilksby, *Ward*.
    Recorded for all Divs.
    Introduced and established near buildings.

5. **C. vulgare** (L.) Bernh. **[Tanacetum vulgare L.]** *Tansy*
    14. 1790, Threckingham, *Cragg*.
    Recorded for all Divs.
    Native. Distribution very variable; not common.

535. Artemisia L.

1. **A. vulgaris** L. — *Mugwort*
   1. 1790, Folkingham, *Cragg*.
   Recorded for all Divs.
   Native. Distribution variable.

6. **A. absinthium** L. — *Wormwood*
   10. 1820, Tetford, *Ward*.
   Not recorded for Divs. 1, 14, 18.
   Introduced, scarce.

7. **A. maritima** L. — *Sea Wormwood*
   12. 1723, Boston, *Blair*.
   Recorded for Divs. 3, 4, 9, 11, 12, 17, 18.
   Native, rather scarce.

537. Carlina L.

1. **C. vulgaris** L. — *Carline Thistle*
   9. 1851, Saltfleetby, *Watson*.
   Not recorded for Divs. 1, 17.
   Native, dry banks and quarries.

538. Arctium L.

1. **A. lappa** L. — *Great Burdock*
   3. 1877, Brigg, *Fowler*.
   Not recorded for Divs. 4, 8, 9, 17, 18.
   Native. Peaty ground near rivers and in woods; not very common.

2. **A. nemorosum** Lejeune — *Common Burdock*
   11. 1882, Scremby, *Fowler*.
   Recorded for Divs. 2, 3, 6, 7, 11, 14, 15, 16.
   Native. The commonest type on dry open ground, roadsides, etc.

4. **A minus** Bernh. — *Lesser Burdock*
   16. 1836, Bourne, *Dodsworth*.
   Recorded for all Divs.
   Native. In woods, widespread but not everywhere. (Hybrids between *nemorosum* and *minus* are frequent).

539. Carduus L.

1. **C. tenuiflorus** Curt. *Slender Thistle*
   - 3. 1957, Killingholme, *E. J. Gibbons.*
   - 4. 1892, Cleethorpes, *Lees.*†
   - 13. 1892, Boultham, *Goodall.*†

   Doubtfully native.

3. **C. nutans** L. *Musk Thistle*
   - 8. 1851, Louth, *Watson.*

   Recorded for all Divs.
   Native. In loose soil on banks and in arable fields.

4. **C. acanthoides** L. *Welted Thistle*
   - 8. 1851, Yarburgh, *Watson.*

   Not recorded for Div. 9.
   Native. Not common; absent from acid soil.

540. Cirsium Mill.

1. **C. eriophorum** (L.) Scop. *Woolly Thistle*
   - 3. 1805, Near Barton, *Winch* (*Botanists' Guide*).

   Recorded for Divs. 1, 3, 6, 8, 11, 12, 13, 14, 15, 16.
   Native. Very rare in the north; locally common in the south.

2. **C. vulgare** (Savi) Ten. *Spear Thistle, Buck Thistle*
   - 13. 1851, Doddington, *Watson.*

   Recorded for all Divs.
   Native. Increasing.

3. **C. palustre** (L.) Scop. *Marsh Thistle*
   - 12. 1799, East Fen, *Young.*

   Recorded for all Divs.
   Native. Decreasing, now very rare in the fens.

4. **C. arvense** (L.) Scop. *Creeping Thistle*
   - 12. 1799, East Fen, *Young.*

   Recorded for all Divs.
   Native. Increasing.

5. **C. oleraceum** (L.) Scop.
   - 16. 1816, between Crowland and Deeping, Mr. Oldham (nurseryman of Sheffield), who gave a specimen to Mrs. M. Stovin from his garden. (*Phyt.* II, 1845, pp. 53 and 115). c. 1823, between Crowland and Deeping, Mr. Cole of Bourne.

   This is a strange record for the Fens and has never been satisfactorily explained, doubtfully native.

6. **C. acaulon** (L.) Scop. [**C. acaule** Scop.] *Stemless Thistle*
    6. 1800, Fillingham, *Dalton*.
    Recorded for all Divs.
    Native. Decreasing and unaccountably rare in Div. 2.

8. **C. dissectum** (L.) Hill *Meadow Thistle*
    5. 1840, Laughton, *Miller*.
    Not recorded for Divs. 4, 8, 9, 11, 16, 17, 18.
    Native. Decreasing; very rare in the south.

541. Silybum Adans.

1. **S. marianum** (L.) Gaertn. *Milk Thistle*
    10. 1820, Stovin Wood (Edlington), *Ward*.
    Recorded for Divs. 3, 6, 7, 9, 10, 11, 12, 13.
    Established alien.

542. Onopordum L.

1. **O. acanthium** L. *Scotch Thistle*
    5. 1840, Thonock, *Miller*.
    Not recorded for Divs. 4, 8, 17.
    Only as a casual.

544. Centaurea L.

1. **C. scabiosa** L. *Greater Knapweed*
    10. 1820, Horncastle, *Ward*.
    Not recorded for Divs. 9, 18.
    Native on chalk and limestone; occasionally casual.

3. **C. cyanus** L. *Cornflower*
    14. 1790, Threckingham, *Cragg*.
    Recorded for all Divs.
    Native. Decreasing and rare; formerly locally abundant.

6. **C. nigra** L. *Lesser Knapweed, Hardheads, Horseknobs*
    14. 1790, Threckingham, *Cragg*.
    Recorded for all Divs.
    Native. Common.

7. **C. nemoralis** Jord.
    14. 1945, Sleaford, *N. D. Simpson and A. H. Alston*.
    Native. Distribution not fully worked; preferring calcareous soil.

9. **C. calcitrapa** L. *Star Thistle*
 10. 1890, Low Toynton, *Bayldon*. (Burgess painting).
 Recorded for Divs. 4, 6, 10, 13.
 Casual in foreign seed.

10. **C. solstitialis** L. *St. Barnaby's Thistle*
 6. 1892, Lincoln, *R. J. Owston*.† (In foreign seed).
 Not recorded for Divs. 1, 2, 7, 9, 11, 14, 16, 18.

 C. diluta, salmantica, melitensis, intybaca, axillaris, spinosa, pallescens and *iberica* have also occurred as casuals on dumps.

545. Serratula L.

1. **S. tinctoria** L. *Saw-wort*
 8. 1666, Muckton, *Lister*.
 Not recorded for Divs. 1, 12, 17, 18.
 Native. Scarce on calcareous clay chiefly.

546. Cichorium L.

1. **C. intybus** L. *Chicory*
 10. 1820, Revesby, *Ward*.
 Recorded for all Divs.
 Sown with grass seed or for fodder.

547. Lapsana L.

1. **L. communis** L. *Nipplewort*
 3. 1835, Wootton, *E. J. Nicholson*.
 Recorded for all Divs.
 Native, common.

548. Arnoseris Gaertn.

1. **A. minima** (L.) Schweigg. & Koerte *Lamb's Succory*
 1. 1950, Epworth, *B.S.B.I.*
 3. 1862, Wrawby, *Britten*.
 4. 1898, Grimsby.
 5. 1882, Laughton, *Lees*.†
 7. 1944, Middle Rasen, *E. J. Gibbons*.
  1958, Holton le Moor, *S. W. Parker*.
 Native. Rare on acid sand. Perhaps extinct.

549. Hypochoeris L.

1. **H. radicata** L.   *Cat's Ear*
    16. 1836, Bourne, *Dodsworth*.
    Recorded for all Divs.
    Native. Locally abundant.

2. **H. glabra** L.   *Smooth Cat's Ear*
    10. 1882, Woodhall Spa, *Melville*.
    Not recorded for Divs. 1, 4, 8, 12, 15, 16, 17, 18.
    Native. Often overlooked.

3. **H. maculata** L.   *Spotted Cat's Ear*
    13. 1896, Fulbeck, *Burtt*.† (See *Naturalist* 1897, p. 169).
    Recorded for Div. 13 only.
    Native.

550. Leontodon L.

1. **L. autumnalis** L.   *Autumnal Hawkbit*
    13. 1856, Doddington, *Cole*.†
    Recorded for all Divs.
    Native, locally abundant.

2. **L. hispidus** L.   *Rough Hawkbit*
    3. 1835, Wootton, *E. J. Nicholson*.
    Recorded for all Divs.
    Native. In the fens an indication of old pasture.

3. **L. taraxacoides** (Vill.) Merat   *Hairy Hawkbit*
    8. 1857, Louth, *Bogg*.†
    Recorded for all Divs.
    Native. Less common than the two preceding, particularly on slopes on old turf.

551. Picris L.

1. **P. echioides** L.   *Bristly Ox-tongue*
    16. 1837, Dunsby, *Dodsworth*.
    Not recorded for Div. 1.
    Native. Clay ditch banks.

2. **P. hieracioides** L.   *Hawkweed Ox-tongue*
    16. 1837, Carlby, *Dodsworth*.
    Not recorded for Divs. 1, 4, 9, 12, 17, 18.
    Native. Scarce; old quarries and roadsides on chalk and limestone.

552. Tragopogon L.

1. **T. pratensis** L. *Goat's Beard, Jack-go-to-bed-at-noon*
   14. 1790, Threckingham, *Cragg.*
   Recorded for all Divs.
   Native.

2. **T. porrifolius** L. *Salsify*
   3. 1893, Barton-on-Humber, *Miss Firbank.*†
   4. 1938, Immingham, *L.N.U.*
   10. 1894, Woodhall Spa, *Mackinder.*†
   16. 1953, Stamford, *J. H. Chandler.*
   17. 1938, Donington, *Kirk.*
   Casual. Waste ground.

554. Lactuca L.

1. **L. serriola** L. *Prickly Lettuce*
   2. 1950, Scunthorpe, *B.S.B.I.*
   6. 1952, Lincoln, *E. J. Gibbons.*
   13. 1957, Temple Bruer, *Howitt.*
   16. 1956, Tallington, *J. H. Chandler.*
   Casual. About roadsides and quarries.

2. **L. virosa** L.
   5. 1840, Corringham, *Miller.*
   Not recorded for Divs. 2, 18.
   Probably native. On sand dunes and waste places; there may be some confusion between this and the last.

555. Mycelis Cass.

1. **M. muralis** (L.) Dumort. *Wall Lettuce*
   2. 1858, Broughton, *Fowler.*
   Recorded for Divs. 2, 3, 5, 7, 10, 14, 16, 17.
   Native in some places, and colonizing near houses.

556. Sonchus L.

1. **S. palustris** L. *Marsh Sow-Thistle*
   3. 1930, Horkstow, *E. Dunn.* Extinct.
   7. 1942, Bardney, *F. Gough.* Extinct.
   10 or 12. *c.* 1820, East Fen, *Bloxam* (Specimen Herb. Dublin).
   12. 1799, East Fen, *Young.* Extinct.
   Native. Extinct.

2. **S. arvensis** L. *Corn Sow-Thistle*
   13. 1830, Branston, *Rev. R. J. Bunch.*
   Recorded for all Divs.
   Native. Common.

3. **S. oleraceus** L. *Milk- or Sow-Thistle*
   3. 1850, Caistor, *M. E. Dixon.*
   Recorded for all Divs.
   Native. Common.

4. **S. asper** (L.) Hill. *Spiny Milk- or Sow-Thistle*
   13. 1851, Boultham, *Watson.*
   Recorded for all Divs.
   Native. Common.

   S. asper x oleraceus?
   Hybrid or species; entire leaves.
   In garden ground. Distribution not worked up.

557. Cicerbita Wallr.

3. **C. macrophylla** (Willd.) Wallr.
   Purple Lettuce, Blue Sow Thistle
   3. 1956, Brocklesby, *R. May.*
   8. 1960, Keddington, *N. Read.* 1963, Binbrook and Kelstern, *E. J. Gibbons.*
   Garden escape; colonizing roadsides.

558. Hieracium L.

1. **H. murorum** L. sensu lato (All determined P. Sell & C. West).†
   (Sect. **Vulgata** F. N. Williams).

   98. **H. exotericum** Jord ex Bor. agg *Hawkweed*
       7. 1953, Sturton Hall, *E. J. Gibbons.*
       15. 1957, Easton, *E. J. Gibbons.*
       Garden weed with Mycelis. Colonist.

   149. **H. vulgatum** Fr. *Common Hawkweed*
        13. 1847, Doddington, *Cole.*†
        Recorded for Divs. 1, 2, 3, 4, 6, 13, 15.

   154. **H. maculatum** Sm. *Spotted Hawkweed*
        6. 1964, Normanby-by-Spital, *E. J. Gibbons.*
        15. 1960, Gunby, *E. J. Gibbons.*
        Colonist.

155. **H. submutabile** (Zahn) Pugsl.
    2. 1953, Broughton, *E. J. Gibbons*.
Native?

157. **H. diaphanoides** Lindeb.
    6. 1953, Lincoln Cathedral Walls, *E. J. Gibbons*.
Colonist.

159. **H. diaphanum** Fries (incl. **H. anglorum** (A. Ley) Pugsl.)
    1. 1958, Crowle, *E. J. Gibbons*.
    2. 1959, Brumby West Common, *E. J. Gibbons*.
    3. 1959, Melton Chalk Pits, *E. J. Gibbons*.
    6. 1954, Lincoln Cathedral, *E. J. Gibbons*.
    7. 1857, Benniworth, *Bogg*.† 1950, Holton, *E. J. Gibbons*. 1954, Tealby and Osgodby, *E. J. Gibbons*.
    10. 1956, Woodhall Spa, *E. J. Gibbons*.
    13. 1958, Whisby, *E. J. Gibbons*.
    14. 1954, Ancaster-Wilsford Railway, *E. J. Gibbons*.
Native.

163. **H. strumosum** (W. R. Linton) A. Ley
    3. 1952, Nettleton, *E. J. Gibbons*.
    4. 1960, Swallow, *E. J. Gibbons*.
    10. 1959, Harrington, *E. J. Gibbons*.
    14. 1954, Haydor, *E. J. Gibbons*.
Native on chalk, etc.; rare.

164. **H. lachenalii** C. C. Gmel.
    2. 1952, Brumby West Common, *E. J. Gibbons*.
    7. 1953, Bardney, *E. J. Gibbons*.
    13. 1847, Doddington, *Cole*.†
    15. 1954, Ancaster, *E. J. Gibbons*.
Native. Colonist?

(Sect. **Tridentata** F. N. Williams).

203. **H. eboracense** Pugsl.
    2. 1964, Holme Lane, *E. J. Gibbons*.
    7. 1951, Holton-le-Moor and South Kelsey, *E. J. Gibbons*.
Native.

204. **H. calcaricola** (F. J. Hamb.) Roffey
    3. 1905, Broughton, *Peacock*.†
    6. 1951, Dunholme, *E. J. Gibbons*.
    7. 1951, Holton-le-Moor, *E. J. Gibbons*.
    13. 1964, Doddington, *E. J. Gibbons*.
Native.

(Sect. **Umbellata**, F. N. Williams).

> **217. H. umbellatum** L.
> **13.** 1855, Doddington, *Cole*.†
> Not recorded for Divs. 12, 16, 18.
> Three forms, *H. umbellatum* and var. *coronopifolium* and var. *commune*.
> Locally frequent as a native on acid sand, and as a casual of gravel pits, railway sidings and at Grimsby docks.

(Sect. **Sabauda**, F. N. Williams).

> **219. H. perpropinquum** (Zahn) Druce
> **8.** 1666, Burwell, *Lister*. 1965, *E. J. Gibbons* (confirmed P. Sell.).
> Not recorded for Divs. 9, 15, 17, 18.
> Native. Less common than H. *vagum*. In woods on clay, etc.

> **221. H. rigens** Jord.
> **13.** 1952, Stapleford, *E. J. Gibbons*.
> Native.

> **223. H. vagum** Jord.
> **7.** 1895, Newball, *Raynor* (L.N.U.).†
> Not recorded for Divs. 15, 18.
> Native and colonist.

2. **H. pilosella** L. sensu lato.

> 1. **H. pilosella** L.   *Mouse-ear Hawkweed*
> **13.** 1856, Doddington, *Cole*.†
> Recorded for all Divs.
> Native. Dry banks.

> 7. **H. aurantiacum** L.   *Fox and Cubs*
> **9.** 1893, Somercotes, *Crow*.†
> **10.** 1961, Roughton, *E. J. Gibbons*.
> Recorded for Divs. 4, 7, 8, 9, 10, 13, 15.
> Colonist. Some of these records may be the following.

> 8. **H. brunneocroceum** Pugsl.
> **2.** 1950, Scunthorpe, *R. Lewis*.
> Colonist.

559. Crepis L.

2. **C. vesicaria** L. *Beaked Hawk's-beard*
    7. 1890, Hatton, *Jarvis*.
    Not recorded for Div. 11.
    Colonist. Roadsides and railways.

3. **C. setosa** Haller f. *Bristly Hawk's-beard*
    4. 1905, Grimsby, *Smith and Parker*.
    Dock alien.

5. **C. biennis** L. *Rough Hawk's-beard*
    10. 1820, Thornton, *Ward*.
    Recorded for Divs. 1, 3, 5, 7, 8, 10, 12, 15, 16.
    Native. Frequent in south-west; scarce elsewhere.

6. **C. capillaris** (L.) Wallr. *Smooth Hawk's-beard*
    13. 1855, Doddington, *Cole*.
    Recorded for all Divs.
    Native. Very common and variable.

8. **C. paludosa** (L.) Moench *Marsh Hawk's-beard*
    3. 1958, Elsham, *E. J. Gibbons*.
    4. 1939, Thoresway, *E. J. Gibbons*.
    7. 1878, Tealby, *Lees*. 1893, Hatton, *Jarvis*.
    Native. Rare. A northern plant in calcareous swamps.

560. Taraxacum Weber

1. **T. officinale** Weber *Common Dandelion*
    12. 1820, Boston, *Thompson*.
    Recorded for all Divs.
    Native.

2. **T. palustre** (Lyons) DC.
    12. 1856, Boston, *Thompson*.
    Recorded for Divs. 2, 3, 8, 12.
    Native.

3. **T. spectabile** Dahlst.
    15. 1961, Lobthorpe, *J. H. Chandler*.
    Native. Probably widespread.

4. **T. laevigatum** (Willd.) DC.   *Lesser Dandelion*
   16. 1883, Deeping, *Beeby*.
   Recorded for Divs. 3, 15, 16.
   Native.

   T. erythrospermum Andrz ex Bess.
   2. 1876, Bottesford, *Peacock*.
   Recorded for Divs. 2, 3, 5, 8, 10, 13.

MONOCOTYLEDONES
ALISMATACEAE
561. Baldellia Parl.

1. **B. ranunculoides** (L.) Parl.   *Lesser Water-Plantain*
   15. 1780, Fens, *Sibthorp*.
   16. 1838, Edenham, *Dodsworth*.†
   Not recorded for Divs. 4, 17.
   Native. Rare and decreasing.

562. Luronium Raf.

2. **L. natans** Raf.   *Floating Water-Plantain*
   Grown in Mr. F. M. Burton's garden, Gainsborough; said to have been brought from Scotter Common Div. 5.
   The only record; pre-1900.

563. Alisma L.

1. **A. plantago-aquatica** L.   *Water-Plantain*
   12. 1799, East Fen, *Young*.
   Recorded for all Divs.
   Native.

2. **A. lanceolatum** With.   *Narrow-leaved Water-Plantain*
   9. 1856, Conisholme, *Bogg*.
   Not recorded for Divs. 3, 5, 6, 8, 10, 11, 12.
   Native. Much less rare in the south. Peaty water.

3. **A. gramineum** Lejeune   *Ribbon-leaved Water-Plantain*
   17. 1955, Surfleet, *L.N.U.*
   (*B.S.B.I. Proc.*, 1957, 346, 1956, 75).
   Native. Abundant 1956. Decreasing owing to dredging. Rare, 1970.

564. Damasonium Mill.

1. **D. alisma** Mill. *Starfruit*
    - 11. 1923, Willoughby, *Miss Farmery* (not confirmed).
    - 16. 1836, Bourne, *Dodsworth*. (*Naturalist*, 1896, 309).
    Native. Extinct.

565. Sagittaria L.

1. **S. sagittifolia** L. *Arrowhead*
    - 10. 1820, Thornton, *Ward*.
    Recorded for all Divs.
    Native. Not common; decreasing.

BUTOMACEAE
566. Butomus L.

1. **B. umbellatus** L. *Flowering-Rush*
    - 16. 1790, Bridge End, *Cragg*.
    Recorded for all Divs.
    Native. Decreasing, becoming rare.

HYDROCHARITACEAE
567. Hydrocharis L.

1. **H. morsus-ranae** L. *Frogbit*
    - 14. 1790, Threckingham, *Cragg*.
    Not recorded for Divs. 2, 7.
    Native. Scarce and decreasing.

568. Stratiotes L.

1. **S. aloides** L. *Water Soldier*
    1633, Fens, *Johnson*.
    Not recorded for Divs. 1, 3, 4, 8, 13, 18; recently only found in three Divs. — 5, 7, 11.
    Native.

570. Elodea Michx.

1. **E. canadensis** Michx. *Canadian Waterweed*
    - 13. 1849, Boultham, *Carrington*.
    Recorded for all Divs.
    Colonist.

3. **E. nutallii** (Planch) St. John [**Hydrilla verticillata** Dandy]
    - 16. West Deeping, *J. H. Chandler*.
    Probably introduced.

JUNCAGINACEAE
574. Triglochin L.

1. **T. palustris** L. *Marsh Arrow-grass*
   1. 1815, Axholme, *Peck.*
   Recorded for all Divs.
   Native. Not very common; calcareous bogs, near springs and also coastal.

2. **T. maritima** L. *Sea Arrow-grass*
   12. 1688, Boston, *Plukenet.*
   Recorded for Divs. 2, 3, 4, 9, 11, 12, 17, 18.
   Native. Common on coastal mud.

ZOSTERACEAE
576. Zostera L.

1. **Z. marina** L. *Eel-grass*
   4. Cleethorpes. No definite record; Peacock's MS says — "Washed up in 1912."
   12. 1856, Near Boston, *Thompson.* 1896, Freiston, *Peacock·* 1919, between Boston and Wainfleet, *Newman and Walworth.* (*J. of E.*, 1919, 205-210).
   17. *Check List.*
   18. 1974, Holbeach, *S. M. Coles.*
   Native. Below low water mark. Specimen wanted.

3. **Z. noltii** Hornem. *Dwarf Eel-grass*
   4. 1884, Cleethorpes, *H. Searle.* (Specimen BM).
   9. 1889, North Somercotes, *J. Cordeaux* (BM).
   1956, Tetney, *A. J. Gray.*
   12. 1919, between Boston and Wainfleet, *Newman and Walworth.*
   1960, Freiston, *E. Seppings.*
   18. 1974, Gedney, *S. M. Coles.*
   Native. On mud above low water mark.

POTAMOGETONACEAE
(Records marked* have been checked by J. E. Dandy.)
577. Potamogeton L.

1. **P. natans** L. *Broad-leaved Pondweed*
   1. 1815, Axholme, *Peck.*
   Recorded for all Divs.
   Native. Common in still water.

2. **P. polygonifolius** Pourr. *Bog Pondweed*
   - **2.** 1893, Crosby Warren, *Fowler.** 1969, Twigmoor, *J. Gibbons.**
   - **3.** 1917, Near Brigg, *Peacock.**
   - **5.** 1894, Scotton, *Mason and Peacock.**
   - **7.** 1904, Linwood, *Peacock**. 1908, Holton-le-Moor, *Peacock.**
   - **10.** 1970, Roughton,*J.Blackwood.**
   - **13.** 1851, Skellingthorpe, *Watson.*
     1969, Stapleford, *E. J. Gibbons and B. Howitt.*

   Native. Decreasing. Peaty dykes.

3. **P. coloratus** Hornem. *Fen Pondweed*
   - **2.** 1856, Roxby, *Fowler.*
   - **3.** 1878, Barnetby, *Lees.** 1894, Bigby, *Mason.**
   - **10.** 1945, Woodhall Spa, *Alston and Simpson.**
   - **12.** 1896, Wainfleet, *Mason.**
   - **13.** 1907, Blankney, *Mason.**
   - **14.** 1955, North Kyme, *E. J. Gibbons and R. C. L. Howitt.**
   - **16.** 1911, Crowland, *Druce.**

   Native. Rare.

5. **P. lucens** L. *Shining Pondweed*
   - **13.** 1830, Lincoln, *Bunch.**

   Not recorded for Div. 4.

   Native. Abundant in fairly still waters.

   P. lucens x perfoliatus = P. x salicifolius Wolfg. (**P. decipiens** Nolte ex Koch).
   - **14.** 1955, North Kyme, *E. J. Gibbons and R. C. L. Howitt.**
   - **16.** 1883, Deeping Fen, *Beeby.** 1957, Dowsby, *R. C. L. Howitt.**

   Native. Rare.

6. **P. gramineus** L. *Various-leaved Pondweed*
   - **1.** 1878, Crowle, *Fowler.** 1894, Haxey, *Fowler.**
     1941, Belton, *Sledge.** 1943, *G. Taylor.**
     1943, Epworth and Wroot, *J. M. Taylor.**
   - **3.** 1893, Cadney, *Fowler and Peacock.**
   - **9.** 1856, Saltfleetby, *Fowler* (as P. polygonifolius).*
   - **13.** 1862, Bassingham, *Carrington.** 1954, Branston, *Gibbons.**
     1955, Boultham Ballast Dyke, Lincoln, *Gibbons.**
   - **14.** 1855, Anwick, *Lowe.**
     1955, North Kyme, *E. J. Gibbons and B. Howitt.**
   - **16.** 1895, Dowsby, *Mason and Peacock.**

   Native. Not common.

   P. gramineus x lucens = P. x zizii Koch ex Roth.
   - **1.** 1939, Haxey, *C. I. and N. Y. Sandwith.** 1943, Belton, Epworth, Wroot and Crowle, *J. M. Taylor.**
   - **13.** 1955, Boultham Ballast Dyke, Lincoln and Pyewipe Drain, Skellingthorpe, *E. J. Gibbons.**
   - **14.** 1955, North Kyme, *E. J. Gibbons and B. Howitt.*
   - **16.** 1895, Dunsby, *Mason and Peacock.** 1957, Dowsby, *R. C. L. Howitt.**

   Native. Rare.

P. gramineus x perfoliatus = P. x nitens Weber
    1. 1937, Wroot, *Brenan and C. I. Sandwith.** 1939, Haxey, *C. I. and N. Y. Sandwith.** 1941, Belton, *G. Taylor.** 1943, Crowle and Epworth, *J. M. Taylor.**
    3. 1959, Cadney, *Gibbons.**
    4. 1930, Little Coates, *Mason.**
    14. 1955, North Kyme, *E. J. Gibbons and B. Howitt.**
    16. 1883, Deeping Fen, *Beeby.**

Native. Scarce.

7. P. alpinus Balb.   *Reddish Pondweed*
    1. 1944, Wroot, *J. M. Taylor.**
    4. 1881, Cleethorpes, *Searle.**
    8. (1910, South Cockerington, *Mason*).
    9. 1856, Saltfleetby, *Bogg.**
    11. 1959, Irby in the Marsh, *Gibbons.**
    13. 1849, Near Boultham, Lincoln, *Carrington.**
    14. (1896, Billinghay, *Walker*).

Native. Rare.

8. P. praelongus Wulf.   *Long-stalked Pondweed*
    1. 1939, Wroot and Epworth, *C. I. and N. Y. Sandwith.** 1941, Belton, *G. Taylor.** 1943, Crowle, *J. M. Taylor.*
    6. 1949, Lincoln (near Bishop's Bridge), *G. Taylor.**
    13. 1849, Near Boultham, Lincoln, *Carrington.** 1862, Skellingthorpe, *Cole.** 1878, Five Mile House Station, Heighington, *Fowler.**
    16. (1940, Market Deeping, *H. Burchnall*).

Native. Rare.

9. P. perfoliatus L.   *Perfoliate Pondweed*
    6 or 13. 1829, Fossdyke, *Bunch.**
Not recorded for Div. 4.

Native. Grows in quantity in many slow rivers and drains.

P. perfoliatus x praelongus = P. x cognatus Aschers. & Graebn.
    1. 1943, Crowle and Belton, *J. M. Taylor.**
Recorded only from the Double Rivers and North Idle Drain.

Native. Very rare.

11. P. friesii Rupr.   *Flat-stalked Pondweed*
    8. 1857, Alvingham, *Bogg.**
Not recorded for Divs. 2, 5, 7.

Native. Frequent.

13. P. pusillus L.   *Lesser Pondweed*
    4. 1851, Near Grimsby, *Watson.**
Not recorded for Divs. 5, 7, 12, 14, 18.

Native. Frequent.

THE COUNTY FLORA 249

14. **P. obtusifolius** Mert. & Koch   *Blunt-leaved Pondweed*
    1. 1963, Wroot, *Bowden and Hillman.**
    4. 1881, Cleethorpes, *Searle.**
    10. 1942, Tattershall Thorpe, *G. Taylor.**
    13. 1849, Lincoln, *Carrington.**
   Native. Rare.

15. **P. berchtoldii** Fieb.   *Small Pondweed*
    4. 1881, Cleethorpes, *Searle.**
   Specimens for Divs. 1, 4, 5, 8, 10, 11, 13, 14, 15, 16 have been confirmed.
   Native. Frequent. Specimens for Holland not available.

16. **P. trichoides** Cham. & Schlecht.   *Hair-like Pondweed*
    12. 1891, Wainfleet, *J. B. Davy.**
    16. 1883, Crowland, *Beeby.**
   Native. Very rare.

17. **P. compressus** L.   *Grass-wrack Pondweed*
    1. 1965, Haxey, *E. J. Gibbons and B. Howitt.**
    6. 1829, Lincoln (drain beyond racecourse), *Bunch.**
    15. 1894, Grantham, *Stow.**
        1954, Denton, *E. J. Gibbons and B. Howitt.**
   Native. Rare.

18. **P. acutifolius** Link.   *Sharp-leaved Pondweed*
    10. About 1795, Northdike Bridge between Boston and Spilsby, *Banks.**
   Native. Very rare or extinct. The above is the only record.

19. **P. crispus** L.   *Curled Pondweed*
   1780, *Sibthorp.*
   Recorded for all Divs.
   Native. Common in ponds and running water.

21. **P. pectinatus** L.   *Fennel Pondweed*
    6 or 13. 1829, Fossdyke, *Bunch.**
   Recorded for all Divs.
   Native. In brackish dykes as well as fresh water inland.

578. **Groenlandia** Gay

1. **G. densa** (L.) Fourr [**Potamogeton densus** L.]
                                     Opposite-leaved Pondweed
    13. 1829, Canwick, *Bunch.**
   Not recorded for Div. 12.
   Native. Frequent.

RUPPIACEAE
(Records checked by J. E. Dandy.)*

579. Ruppia L.

1. **R. spiralis** L. ex Dumort. [**R. cirrhosa** (Petagne) Grande]
   Spiral Tasselweed
   12. *c.* 1795, Freiston, *Banks.* 1937, Wrangle, *Williams and Wilmott.* 1958, Butterwick, *Gibbons.*

   Native. Brackish pools. Local.

2. **R. maritima** L. *Beaked Tasselweed*
   9. 1954, North Coates, *E. J. Gibbons.*
   12. Before 1850, Boston, (*Herb. Brit. Mus.*). "Bottom of the Common, Boston, Charles Street near the Ropewalk."

   Native. Brackish pools. Rare.

ZANNICHELLIACEAE

580. Zannichellia L.

1. **Z. palustris** L. *Horned-pondweed*
   4. 1851, Grimsby, *Watson.*
   Not recorded for Div. 2.

   Native. Ponds and streams. Common.

LILIACEAE

584. Narthecium Huds.

1. **N. ossifragum** (L.) Huds. *Bog Asphodel*
   10. 1724, Tattershall, *Stukeley.*
   Recorded for Divs. 1, 2, 3, 5, 6, 7, 10; Div. 14 not confirmed. Extinct in all but 2 Divs.

   Native. Dying out through drainage and cultivation.

588. Convallaria L.

1. **C. majalis** L. *Lily of the Valley*
   2. 1697, Broughton, *de la Pryme.*
   Recorded for Divs. 2, 6, 7, 10, 13, 15, 16; records for Divs. 3, 5, 8 are planted.

   Native. Locally abundant in a number of large old woods; scarce in others. Decreasing owing to oaks being replaced by conifers.

589. Polygonatum Mill.

3. **P. multiflorum** (L.) All. *Solomon's Seal*
   - **7.** 1856, Benniworth, *Bogg.*†
   - **14.** 1905, Bloxholme, *Mason.*† 1959, *J. Gibbons.*

   Not native.

590. Maianthemum Weber

1. **M. bifolium** (L.) Schmidt *May Lily*
   - **10.** 1895, Fulsby, *Miss F. Rawnsley.*†
   - **11.** 1927, Welton Wood, *H. Carlton.*

   Native.

591. Asparagus L.

1. **A. officinalis** L. *Asparagus*
   (a) ssp. *officinalis* var. *altilis* L.
   - **9.** 1698, Mablethorpe, *M. Lister.* (*Nat.*, 1891, p. 42). 1922, *Mason.*
   - **18.** 1597, Holbeach and Moulton, *Gerarde.* (*Nat.*, 1896, p. 249).

   In check list, recorded for Divs. 3, 5, 8, 9, 11, 12, 18. These may be bird sown from gardens.

   Native. To be searched for.

592. Ruscus L.

1. **R. aculeatus** L. *Butcher's Broom*
   - **8.** 1855, Donington-on-Bain, *Bogg.*†

   Recorded for Divs. 2, 3, 4, 5, 6, 8, 9, 10, 11, 12.

   Always introduced.

594. Fritillaria L.

1. **F. meleagris** L. *Snake's Head, Fritillary*
   - **5.** 1837, Grayingham, *M. Nicholson.*†

   Recorded for Divs. 2, 5, 6, 7, 14, 15.

   Always introduced. No recent record.

597. Gagea Salisb.

1. **G. lutea** (L.) Ker-Gawl. *Yellow Star of Bethlehem*
   - **2.** 1959, Broughton, *T. Stones.*
   - **16.** 1956, Careby Wood, *J. H. Chandler.*

   Native. May be overlooked elsewhere, not seen recently.

598. Ornithogalum L.

1. **O. umbellatum** L. *Star of Bethlehem*
    13. 1890, Stapleford, *Fisher*.
    Recorded for Divs. 2, 3, 8, 9, 10, 13, 18.
    Doubtful native; usually near houses.

2. **O. nutans** L. *Drooping Star of Bethlehem*
    10. 1876, Somersby, *Brooks*.
    Recorded for Divs. 6, 10, 11, 12.
    Introduced.

599. Scilla L.

1. **S. verna** Huds. *Spring Squill*
    11. 1879, Old Garden, Eresby Hall, *Burgess*. (See *Naturalist* 1893, p. 332).
    Not native. Extinct.

600. Endymion Dumort.

1. **E. non-scriptus** (L.) Garcke *Bluebell, Wild Hyacinth*
    10. 1820, Thornton, *Ward*.
    Not recorded for Divs. 12, 17, 18.
    Native. Scarce in several parts of the county.

602. Colchicum L.

1. **C. autumnale** L. *Meadow Saffron*
    1. 1896, Axholme, *S. Hudson*.†
    2. 1865, Brumby, *Moore;* Flixboro, *Fowler* (Native).
    3. 1892, Bigby, *J. Field*.† (Introduced).
    5. 1877, Glentham, *F. A. Lees*.
    6. *Check List*.
    No recent record; dying out, always rare, and not always native.

603. Paris L.

1. **P. quadrifolia** L. *Herb Paris*
    10. 1820, Tetford, *Ward*.
    Not recorded for Divs. 1, 4, 9, 12, 17, 18.
    Native.

JUNCACEAE
605. Juncus L.

1. **J. squarrosus** L.   *Heath Rush*
   13. 1855, Doddington, *Cole*.†
   Recorded for Divs. 1, 2, 3, 5, 6, 7, 10, 13.
   Native. Locally common on acid soil. Decreasing.

4. **J. compressus** Jacq.   *Round-fruited Rush*
   5. 1868, Gainsborough, *Charters*.†
   Not recorded for Divs. 1, 2, 4, 8, 9.
   Native. Thinly scattered, chiefly in the south.

5. **J. gerardii** Lois.   *Saltmarsh Rush*
   12 or 17. 1836, Boston, *Dodsworth*.†
   Recorded for Divs. 1, 2, 3, 4, 9, 11, 12, 17, 18.
   Native. Locally frequent.

7. **J. bufonius** L.   *Toad Rush*
   16. 1836, Bourne, *Dodsworth*.
   Recorded for all Divs.
   Native. Common on waterlogged ground.

8. **J. inflexus** L.   *Hard Rush*
   16. 1836, Bourne, *Dodsworth*.†
   Recorded for all Divs.
   Native. Abundant on basic soil.

9. **J. effusus** L.   *Soft Rush*
   12. 1799, East Fen, *Young*.
   Recorded for all Divs.
   Native. Common on acid soil.

10. **J. conglomeratus** L. [**J. subuliflorus** Drejer]   *Compact Rush*
    1. 1815, Axholme, *Peck*.
    Recorded for all Divs.
    Native. Frequent on acid soil.

14. **J. maritimus** Lam.   *Sea Rush*
    9. 1851, Humberston, *Watson*.
    Recorded for Divs. 4, 9, 11.
    Native. Scarce in dune slacks.

15. **J. acutus** L.   *Sharp Rush*
    12 or 17. 1856, Near Boston, *Thompson*.
    Doubtful.

17. **J. subnodulosus** Schrank   *Blunt-flowered Rush*
     13. 1851, Doddington, *Watson*.
     Not recorded for Divs. 12, 17.
     Native. Frequent in wold valleys.

18. **J. acutiflorus** Ehrh. ex Hoffm.   *Sharp-flowered Rush*
     8. 1857, Louth, *Bogg*.†
     Not recorded for Divs. 12, 18.
     Native on acid soil.

19. **J. articulatus** L.   *Jointed Rush*
     1851, *Watson*.
     Recorded for all Divs.
     Native. Frequent on basic soil.

22. **J. bulbosus** L. [including **J. kochii** F. W. Schultz]
                                                   Bulbous Rush
     13. 1855, Doddington, *Cole*.†
     Not recorded for Divs. 4, 12, 15, 17, 18.
     Native. Acid heaths in wet peat.

606. Luzula DC.

1. **L. pilosa** (L.) Willd.   *Hairy Woodrush*
     16. 1836, Thurlby, *Dodsworth*.†
     Not recorded for Divs. 1, 5, 9, 12, 17, 18.
     Native. Rather scarce but not infrequent in old woods.

3. **L. sylvatica** (Huds.) Gaudin   *Great Woodrush*
     11. 1847, Near Burgh, *Dr. Grantham*.
     Not recorded for Divs. 1, 5, 6, 9, 12, 14, 17, 18.
     Native. Less frequent than *L. pilosa*.

8. **L. campestris** (L.) DC.   *Field Woodrush*
     1. 1815, Axholme, *Peck*.
     Recorded for all Divs.
     Native. Common in old pastures.

9. **L. multiflora** (Retz.) Lej.   *Many-headed Woodrush*
     5. 1833, Scotton, *C. M. Cautley*.
     Not recorded for Divs. 8, 9, 12, 17, 18.
     Native. Usually the var. *congesta* on acid soil.

10. **L. pallescens** Sw.
     1. 1958, Haxey, *E. J. Gibbons*. (Cambridge Botany School Herbarium). Not confirmed.
     Native, one plant only — probably extinct.

AMARYLLIDACEAE
607. Allium L.

3. **A. scorodoprasum** L. *Sand Leek*
    1. 1893, Axholme, *S. Hudson.*†
    2. 1895, Broughton, *Peacock.*† 1952, Scawby, *E. J. Gibbons.*
    5. 1842, Gainsborough, *Miller.* 1953, Morton, *E. J. Gibbons.* (Extinct).
    14 or 15. 1949, Ancaster, *L.N.U.* (S. Hopwood and R. E. Taylor). Specimen seen by E. J. Gibbons.

    Native. Noticeably near Roman roads.

5. **A. vineale** L. *Crow Garlic*
    5. 1842, Gainsborough, *Miller.*
    Not recorded for Divs. 14, 17, 18.

    Native. On sand-dunes on the coast and also inland on roadsides. Form with bulbils is commoner than the form with flowers and bulbils. Formerly in pastures, causing tainting of milk.

6. **A. oleraceum** L. *Field Garlic*
    5. 1842, Gainsborough, *Miller.*
    Recorded for Divs. 2, 5, 6, 7, 9, 15.

    Native. No recent records.

7. **A. carinatum** L.
    5. 1893, Walkerith, *F. A. Lees.*
    15. 1883, Dry Doddington, *Rev. W. S. Hampson.* 1894, *Stow.*†

    Doubtfully native?

8. **A. schoenoprasum** L. *Chives*
    12. 1856, Boston, *Thompson.*
    13. 1899, Boultham, *Peacock.*

    Not native.

12. **A. ursinum** L. *Ramsons*
    10. 1820, Tetford, *Ward.*
    Not recorded for Divs. 4, 9, 12, 14, 17, 18.

    Native. Locally dominant.

611. Leucojum L.

1. **L. vernum** L. *Spring Snowflake*
    15. 1959, Casewick Park, *J. H. Chandler.*
    Naturalised.

2. **L. aestivum** L. *Loddon Lily*
    8. 1897, Calcethorpe, *Lewin.*†
    Garden escape.

612. Galanthus L.

1. **G. nivalis** L. *Snowdrop*
    1863, Yaddlethorpe, *Woodruffe Peacock*.
    Recorded for Divs. 2, 3, 4, 7, 8, 10, 11, 12, 13, 15, 16.
    Naturalised and in some places appearing wild. Probably planted over 200 years ago near Holy Wells and Spas.

614. Narcissus L.

1. **N. pseudonarcissus** L. *Wild Daffodil*
    2. 1858, Broughton, *Fowler*.
    Recorded for Divs. 2, 3, 4, 6, 7, 8, 10, 11, 13, 15, 16.
    Doubtfully native.

IRIDACEAE
616. Iris L.

1. **I. spuria** L. *Butterfly Iris*
    11. 1896, Huttoft, *F. M. Burton*.
    16. 1836, Bourne, *Dodsworth*.
    Recorded for Divs. 9(?), 11, 12 (?), 16 (?).
    Native. Very rare. Seven distinct locations, only one remaining.

3. **I. foetidissima** L. *Gladdon, Stinking Iris*
    16. 1884, Careby Wood, *W. Fowler*.
    Recorded for Divs. 3, 4, 7, 11, 14, 16.
    Doubtfully wild; possibly extinct.

4. **I. pseudacorus** L. *Yellow Flag*
    12. 1799, East Fen, *Young*.
    Recorded for all Divs.
    Native. In wold valleys, dykes and wet places in woods.

618. Crocus L.

1. **C. nudiflorus** Sm. *Autumnal Crocus*
    5. 1885, Lea, *Cockin*.
    Water borne from Notts? Peacock says "Not doubtful."

620. Crocosmia Planch.

1. **C. x crocosmiflora** (Lemoine) N.E. Br. *Montbretia*
    10. 1957, Woodhall Spa, *F. H. Perring*.
    12. 1957, Near Boston, *P. and J. Hall*.
    Introduced. Naturalised as a garden escape.

DIOSCOREACEAE
622. Tamus L.

1. **T. communis** L.   *Black Bryony, Womandrake*
   10. 1820, West Ashby, *Ward.*
   Recorded for all Divs.
   Native, clay soils chiefly.

ORCHIDACEAE
625. Epipactis Sw.

1. **E. palustris** (L.) Crantz   *Marsh Helleborine*
   1. 1974, Crowle, *A. Frankish & E. J. Gibbons.*
   2. 1900, Frodingham, *L.N.U.* (Mason). Extinct 1930.
      1909, Alkborough, *Claye.*
   4. 1867, Freshney Bog, *M. G. Watkins.*† Extinct 1910.
   5. 1882, Ferry Flash, *W. Fowler.* Extinct 1910.
   9. Saltfleetby St. Clements. (Mason MS 1930). Extinct 1953.
   11. 1936, Chapel St. Leonards, *Hind.*†
   15. *c.* 1930, Ancaster, *E. Orchard.*
   Native. Previously believed extinct. Re-introduced at Saltfleetby.

2. **E. helleborine** (L.) Crantz.   *Broad Helleborine*
   15. 1780, Near Easton, *Sibthorp.*
   Recorded for Divs. 4, 7, 8, 11, 13, 15, 16.
   Native. Rare and fluctuating.

3. **E. purpurata** Sm.   *Violet Helleborine*
   11. 1890, Aby, *J. W. Chandler.*†
   16. 1895, Dunsby, *Peacock.*†
   These records not confirmed.

627. Spiranthes Rich.

1. **S. spiralis** (L.) Chevall.   *Autumn Lady's Tresses*
   8. 1666, Burwell, *M. Lister.* Extinct.
   12. 1873, Freiston, *Revd. H. Disbrowe.* 1937, *B.E.C.*
   15. 1836, Ropsley, *Dr. Latham.*† 1932, Ancaster, *E. Orchard.*†
   16. 1860, Near Stamford (not Lincs?), *Miss Wingfield.*
   17. 1896-1900, Surfleet, *Dr. Perry.* 1970, *Z. Porter.*
   18. 1961, Tydd Gote, *L.N.U.*
   Native. Rare. To be searched for.

628. Listera R. Br.

1. **L. ovata** (L.) R. Br.   *Twayblade*
   10. 1820, Horncastle, *Ward.*
   Recorded for all Divs.
   Native. Not really common. Occasionally colonising.

629. Neottia Ludw.

1. **N. nidus-avis** (L.) Rich. *Bird's-nest Orchid*
   15. 1780, Near Easton, *Sibthorp*.
   Recorded for Divs. 2, 7, 8, 10, 11, 13, 15, 16.
   Native. Scarce.

631. Hammarbya Kuntze

1. **H. paludosa** (L.) Kuntze *Bog Orchid*
   10. 1820, Tattershall Moor, *Ward*.
   Native. Extinct.

632. Liparis Rich.

1. **L. loeselii** (L.) Rich. *Fen Orchid*
   3. Between 1865 and 1900, Nettleton Moor, *H. C. Brewster*.
   13. 1884, Near Lincoln, *Burbidge*. (*Gardeners Chronicle*, 1884, p. 144). See *Naturalist*, 1896, p. 246.
   "Liparis grew there 20 years ago." (*H.C.B. History of South Kelsey* MS in Lincoln Cathedral Library).
   Native. Nettleton Moor was more suitable for *Hammarbya*.

635. Coeloglossum Hartm.

1. **C. viride** (L.) Hartm. *Frog Orchid*
   6 or 13. 1831, Near Lincoln, *Drury*.
   Not recorded for Divs. 1, 5, 6, 9, 17, 18.
   Native. Uncommon and decreasing.

   C. viride x Dactyorchis fuchsii
   11. 1952, Welton Wood, *M. Smith*, confirmed V. S. Summerhays.

636. Gymnadenia R. Br.

1. **G. conopsea** (L.) R. Br. *Fragrant Orchid*
   2. 1822, Broughton Common, *Strickland*.
   Not recorded for Divs. 1, 9, 12, 13, 17, 18.
   Native. Certainly rare. In divs. 2, 5, 7, and 14 probably var. *densiflora*.

638. Platanthera Rich.

1. **P. chlorantha** (Custer) Reichb. *Greater Butterfly Orchid*
   6 or 13. 1831, Near Lincoln, *Drury*.
   Not recorded for Divs. 1, 3, 4, 9, 12, 17, 18.
   Native. In several old woods.

THE COUNTY FLORA

2. **P. bifolia** (L.) Rich. *Lesser Butterfly Orchid*
    - **1.** 1877, Haxey, *Fowler.* 1960, *E. J. Gibbons.*
    - **4.** 1888, Maude Hole, Healing, *J. Cordeaux.* Before 1896, Great Coates. (Extinct).
    - **6.** 1948, Hackthorn, *J. Hull.*
    - **7.** 1877, Nova Scotia, Middle Rasen, *Lees.*
    - **8.** 1933, Tathwell Greasy Field, *D. Marsden.* (Or *P. chlorantha*). 1930-40, Firebeacon, *Mrs. Williams.*
    - **11.** 1905, Ailby and Tothby, *F. A. Lees.*†

    Records for this and the preceding are sometimes mixed.

    Native. Disappearing.

640. Ophrys L.

1. **O. apifera** Huds. *Bee Orchid*
    - **18.** 1745, Moulton, *Dr. Green.* (Spalding Gentleman's Society).

    Not recorded for Divs. 1, 17.

    Native. Widespread but uncommon and fluctuating.

2. **O. fuciflora** (Crantz) Moench. *Late Spider Orchid*
    "Several reports of it but finders were always dead." Peacock MS. Ray's Catalogue, 1670, p. 224 (Northants).
    Most unlikely. Out of range of distribution.

3. **O. sphegodes** Mill. *Early Spider Orchid*
    Watson says N. or S. Lincs. (Northants probably).

4. **O. insectifera** L. *Fly Orchid*
    - **2.** 1842, Broughton, *J. K. Miller.* 1966, *E. J. Gibbons.*
    - **8.** 1864, Cawthorpe Wood, *Rev. J. H. Thompson.*
    - **10.** 1908, Tetford, *Mr. Dale.*
    - **11.** 1867, Welton Wood, *Mason.*† 1938, *D. Marsden.*

    Native. Records from south Lincs. not confirmed. Might reoccur.

641. Himantoglossum Spreng.

1. **H. hircinum** (L.) Spreng. *Lizard Orchid*
    - **3.** 1835, "On Chalk", *Miss E. J. Nicholson* of Wootton.
    - **10.** 1929, "On Chalk", *E. J. Gibbons;* also 1930, 1936, 1939, 1947-52.
    - **11.** 1939, Gibraltar Point, "on sand dunes", *E. Chapman* of Bexhill.
    - **15.** 1931, (on limestone), Near Ancaster, *Miss Stow.*
    - **16.** 1790, Dunsby, *J. Cragg.* "Beside the turnpike, near 3 feet high." 1952, Greatford, *Dr. Dale.* Extinct.

    Native. Not seen since 1952.

642. Orchis L.

4. **O. ustulata** L.   *Burnt Tip Orchid, Dwarf Orchid*
   15. 1780, Near Easton, *Sibthorp*.
   Recorded for Divs. 2, 3, 4, 5, 13, 15, 16(?).
   Native. Always rare and disappearing through ploughing up.

5. **O. morio** L.   *Green-winged Orchid*
   14. 1790, Threckingham, *Cragg*.
   Not recorded for Div. 18.
   Native. Becoming rare through ploughing of meadows.

7. **O. mascula** (L.) L.   *Early Purple Orchid*
   10. 1820, Tetford, *Ward*.
   Not recorded for Divs. 1, 9, 18.
   Native. In many woods on clay soils but not in all.

643. Dactylorchis (Klinge) Vermeul

1. **D. fuchsii** (Druce) Vermeul. (Soó)   *Spotted Orchid*
   10. 1820, Thornton, *Ward*.
   Native. Frequent on basic soils; woods, roadsides, meadows and marshes. The dry soil chalk form is smaller and less common.

2. **D. maculata** (L.) Vermeul.   *Heath Spotted Orchid*

2b. **D. maculata** ssp. **ericetorum** (E. F. Linton) Hunt & Summerhayes.
   2. 1917, Appleby, *L.N.U.*
   Recorded for Divs. 1, 2, 3, 4, 5, 7, 9, 10, 15.
   Native. Rather scarce.

3. **D. incarnata** (L.) Vermeul.   *Early Marsh Orchid*

3a. **D. incarnata** ssp. **incarnata** L.Soó.
   8. 1856, Near Louth, *Bogg*.†
   Not recorded for Divs. 3, 4, 12, 13, 17, 18.
   Native. Less common than *praetermissa*.

4. **D. praetermissa** (Druce) Vermeul.   *Common Marsh Orchid*
   2. 1891, Yaddlethorpe, *Peacock*.†
   Not recorded for Divs. 1, 17, 18.
   Native. Often in quantity and hybridising with the last and with *D. fuchsii*. Habitats becoming scarce through ploughing.

THE COUNTY FLORA

644. Aceras R. Br.

1. **A. anthropophorum** (L.) Ait. f.   *Man Orchid*
   - **15.** 1939, Ancaster, *E. Orchard.* 1954, Little Ponton, *S. Bond.*
   - **16.** 1950, Carlby, *Locke.*

   Native. Northern extremity of its range. Unconfirmed records Div. 3 — c. 1920, Barton-on-Humber, *D. Witty* and 1930, Ferriby, *P. Pape.*

645. Anacamptis Rich.

1. **A. pyramidalis** (L.) Rich.   *Pyramidal Orchid*
   - **6 or 13.** 1831, Near Lincoln, *Drury.*
   - Not recorded for Div. 1.

   Native. Found on coastal dunes as well as basic soils inland. Not really common.

ARACEAE

646. Acorus L.

1. **A. calamus** L.   *Sweet Flag*
   - **6.** 1956, Torksey, *B. M. Howitt* and *E. J. Gibbons.*
   - **11.** Check List.
   - **12.** 1856, Near Boston, *Thompson.*
     1896, Wainfleet, *Mason (T. Hodson).*
   - **16.** 1840, Bourne, *Dodsworth* (Planted).

   Introduced.

649. Arum L.

1. **A. maculatum** L.   *Cuckoo Pint*
   - **16.** 1790, Threckingham, *Cragg.*
   - Recorded for all Divs.

   Native. Frequent on clay and basic soil — not found on blown sand.

LEMNACEAE

650. Lemna L.

1. **L. polyrhiza** L.   *Great Duckweed*
   - **16.** 1836, Bourne, *Dodsworth.*
   - Recorded for Divs. 6, 7, 9, 10, 11, 12, 14, 16, 18.

   Native. Infrequent but not rare.

2. **L. trisulca** L.   *Ivy Duckweed*
   - **16.** 1836, Bourne, *Dodsworth.*
   - Recorded for all Divs.

   Native.

3. **L. minor** L. *Duckweed*
   16. 1790, Threckingham, *Cragg*.
   Recorded for all Divs.
   Native.

4. **L. gibba** L. *Gibbous Duckweed*
   16. 1836, Bourne, *Dodsworth*.
   Not recorded for Div. 2.
   Native. Brackish dykes especially; not common.

SPARGANIACEAE
652. Sparganium L.

1. **S. erectum** L. *Bur-reed*
   12. 1597, (Wainfleet), *Gerarde*.
   Recorded for all Divs.
   Native.

2. **S. emersum** Rehm. *Simple Bur-reed*
   1. 1597, (Althorpe), *Gerarde*.
   Recorded for all Divs.
   Native. Rather scarce.

[
3. **S. angustifolium** Michx. *Floating Bur-reed*
   5. 1838, Scotton Common, *Irvine*. In *Top. Bot.* Watson (2nd Edn.). Most unlikely. Probably No. 4.
]

4. **S. minimum** Wallr. *Small Bur-reed*
   1. 1879, New Idle River, *W. Fowler*. 1895, Haxey, *Fowler*.†
      1962, Epworth, *Allerton and Hurst*.
   5. Before 1896, Scotter. (*Check List*). Crit. Cat.
   7. Before 1896, (*Check List*). Crit. Cat.
   12. 1826, Frieston, *Howitt*.
   13. 1879, Nocton, *Fowler*.
   16. 1895, Dunsby, *Peacock and Mason*.†
   Native. Rare. To be looked for.

TYPHACEAE
653. Typha L.

1. **T. latifolia** L. *Great Reedmace*
   1636, Fens, *Johnson*.
   Recorded for all Divs.
   Native. Common.

2. **T. angustifolia** L. *Lesser Reedmace*
    16. 1836, Bourne, *Dodsworth*.
    Not recorded for Divs. 4, 5.
    Native. Not very common.

CYPERACEAE
654. Eriophorum L.

1. **E. angustifolium** Honck. *Common Cotton-grass*
    12. 1799, East Fen, *Young*.
    Not recorded for Divs. 11, 14, 17, 18.
    Native. Widespread but becoming rather scarce; wet acid and peaty localities. In two forms, broad-leaved and later flowering; and a commoner form, narrow leaved and earlier flowering.

3. **E. latifolium** Hoppe *Broad-leaved Cotton-grass*
    1. 1840, Crowle, *Miller*.
    13. 1839, Stapleford, *G. Howitt*. } Doubtful if correct name or species; more likely to be the stout broadleaved form of *angustifolium*.
    16. 1874, Nr. Stamford, *Berkeley*. (?Northants).
    Records for Divs. 1, 16 doubtful — county boundary. (See Peacock's Check List).
    Native.

4. **E. vaginatum** L. *Cotton-grass, Hare's-tail*
    1. 1815, Isle of Axholme, *Peck*. 1959, Epworth, *Gibbons*. 1958, Crowle, *E. J. Gibbons*.
    3. 1959, Nettleton, *S. W. Parker*. 1966, Elsham, *P. Wood*.
    4. 1898, Great Coates. (Extinct). (*Nat.*, 1898, no. sp.).
    5. 1903, Scotton, *Peacock and Mason*.†
    7. 1878, Osgodby, *Lees*† (Extinct).
    11. 1830-70, Skendleby, *Mossop*. (Extinct).
    13. 1959, Stapleford, *I. Antoine*.
    Native. Becoming very scarce; found in drier places than *E. angustifolium*.

655. Scirpus L.

2. **S. cespitosus** L. *Deer-grass*
    1. 1815, Isle of Axholme, *Peck*.
    2. 1950, Manton, *M. E. Gibbons*.
    5. 1905, Scotton, *Peacock*.†
    7. 1878, Osgodby, *Lees*. 1895, Linwood, *Lees*.† 1966, *J. Gibbons*.
    10. 1946, Woodhall, *E. J. Gibbons*.
    Native. Uncommon; acid peat, usually amongst *Erica tetralix*.

3. **S. maritimus** L. *Sea Club-rush*
   12 or 17. 1836, Boston, *Dodsworth*.
   Not recorded for Divs. 7, 13, 14, 15, 16.
   Native. Common near the coast and throughout the length of the Trent in Lincs. and growing some distance inland.

4. **S. sylvaticus** L. *Wood Club-rush*
   7. 1964, Wragby, *N. Read*.
   Possibly brought by birds. Not known elsewhere nearer than Notts.

8. **S. lacustris** L. *Bulrush*
   1. 1815, Isle of Axholme, *Peck*.
   Not recorded for Divs. 8, 12.
   Native. Sometimes introduced into lakes and ornamental waters.

9. **S. tabernaemontani** C. C. Gmel. *Glaucous Bulrush*
   1. *c.* 1880, Althorpe, *Fowler*.†
   4. 1895, Great Coates, *Cordeaux and Peacock*.†
   6. Before 1909, Newton-on-Trent. 1963, Barlings, *W. Heath and B. Howitt*.
   9. 1903, Mablethorpe, *Mason*.† 1950, Tetney, *E. J. Gibbons*.
   12. 1851, Freiston, *H. C. Watson*. 1878, Boston, *N. C. Watkins*.†
   16. 1957, Dowsby, *Howitt*. 1967, Baston Fen, *J. Gibbons and I. Weston*.
   18. 1957, Whaplode Drove, *Howitt*.
   Native. Brackish dykes; uncommon.

10. **S. setaceus** L. *Bristle Scirpus*
    12. 1856, Boston, *Thompson*.
    Not recorded for Divs. 9, 14, 17, 18.
    Native. Fairly widespread; damp grassy places.

12. **S. fluitans** L. *Floating Scirpus*
    13. 1849, Skellingthorpe, *Carrington*.
    Not recorded for Divs. 4, 8, 11, 15, 17,
    Native. Occasional in north and uncommon in the south· Peaty dykes.

656. Eleocharis R. Br.

2. **E. acicularis** (L.) Roem. & Schult. *Slender Spike-rush*
   16. 1838, Bourne, *Dodsworth*.
   Not recorded for Divs. 4, 8, 9, 11, 15, 17, 18.
   Native. Peaty dykes; not always flowering; occasional.

3. **E. quinqueflora** (F. X. Hartmann) Schwarz
                                                                         Few-flowered Spike-rush
   - 5. 1946, Waddingham, *E. J. Gibbons*, det. A. J. Wilmott. (Ref. B.E.C. Report, 1947). Extinct.
   - 6. 1954, Hackthorn, *E. J. Gibbons*, det. S. M. Walters.
   - 9. 1963, North Somercotes, *E. J. Gibbons and N. Read*.

   Native. Calcareous bogs; possibly overlooked.

4. **E. multicaulis** (Sm.) Sm.   *Many-stemmed Spike-rush*
   - 1. 1858, Haxey, *Fowler*.†
   - 2. 1876, Manton, *Fowler*.
   - 5. 1905, Scotton, *Peacock*.†
   - 7. 1878, Linwood, *Lees*.† and 1957, S. M. *Walters*.

   Native. Scarce.

5. **E. palustris** (L.) Roem. & Schult.   *Common Spike-rush*
   - 1. 1815, Isle of Axholme, *Peck*.

   Recorded for all Divs.

   Native. Common; wet places.

   (a) ssp. *microcarpa* Walters.
   - 17. 1955, Surfleet, *E. J. Gibbons*, det. S. M. Walters.

   Native. Possibly overlooked.

6. **E. uniglumis** (Link) Schult.
   - 6. 1954, Hackthorn, *R. Hull and E. J. Gibbons*. (Conf. S. M. Walters)
   - 11. 1961, Irby, *S. M. Walters*.
   - 12. 1894, Leverton, *W. W. Mason*.† (Conf. S. M. Walters).

   Native. Rare or overlooked.

657. Blysmus Panz.

1. **B. compressus** (L.) Panz. ex Link   *Broad Blysmus*
   - 2. 1879, Broughton, *Fowler*.
   - 4. 1878, Grimsby, *H. Searle*. Liverpool Univ. Herb. J. Whitelegg.
   - 6. 1953, Nettleham. 1954, Hackthorn, *E. J. Gibbons*.
   - 7. 1958, Great Sturton, *E. J. Gibbons*.
   - 9. 1889, North Somercotes, *Cordeaux*.
   - 10. 1958, West Ashby, *M. N. Read*.
   - 13. 1851, Bracebridge, *H. C. Watson*. 1972, *E. J. Gibbons*.
   - 15. 1904, Holywell, *Mason and Stow*.† 1905, Little Ponton, *Stow*.†

   Native. Uncommon; calcareous bogs.

2. **B. rufus** (Huds.) Link   *Narrow Blysmus*
   - 4. 1882, Grimsby, *J. S. Rouse*, Liverpool Univ. Herb. 1958, Cleethorpes, *D. D. Bartley*.
   - 5. 1868, Gainsborough, *R. H. Charters*.†
   - 9. 1876, Humberston, *Fowler*. 1856, Saltfleet, *Fowler*. 1973, Saltfleetby, *C. Walker*.
   - 11. *c.* 1877, Skegness, *Fowler*.

   Native. Rare; brackish pools.

658. Cyperus L.

1. **C. longus** L. *Galingale*
    - **6.** 1969, Burton, *J. Blackwood* ? Introduced.
    - **15.** 1961, Long Bennington, *E. J. Gibbons*.
    - **16.** 1836, Bourne (planted), *Dodsworth*.
        1966, Langtoft, *Z. Porter* and *E. J. Gibbons* (rubbish tip).

    Doubtful native.

659. Schoenus L.

1. **S. nigricans** L. *Bog-rush*
    - **1.** 1815, Isle of Axholme, *Peck*.
    - **2.** 1875, Crosby, *Fowler*. (*Naturalist*, 1900, 298).
    - **3.** 1894, Bigby, *Mason*.†
    - **4.** 1923, Aylesby, *L.N.U.*
    - **5.** 1887, Scotton, *Fowler*. 1937, Waddingham, *L.N.U.*
    - **16.** 1836, Bourne, *Dodsworth*.

    Native. Rare; bogs, usually calcareous.

660. Rhynchospora Vahl

1. **R. alba** (L.) Vahl *White Beak-sedge*
    - **1.** 1815, Epworth, *Peck*.
    - **2.** *Check List*.
    - **5.** 1920, Scotton, *Miss E. Fowler*.
    - **7.** 1878, Linwood, *Lees*. (BM).
    - **10.** 1877, Kirkby Moor, *Fowler*. 1918, Woodhall Spa, *L.N.U.*

    Native. Possibly extinct.

661. Cladium (L.) Pohl

1. **C. mariscus** (L.) Pohl *Fen Sedge, Twigrush, Star Thack*

    1650, W. How, *Phytologia Britannica*, London.
    - **1.** 1898, Epworth and Haxey, *Fowler and Peacock*.† 1950, B.S.B.I.
    - **2.** 1877, Manton, *Fowler*.
    - **5.** *Check List*.
    - **10.** 1932, Tattershall, *S. A. Cox.* (L.N.U.).
    - **12.** 1799, East Fen, *Young*.

    Native. Possibly extinct, except in the Isle of Axholme; evidently far more widespread before drainage as indicated by place names: Star Carr, Wrawby, Thatch Fen, Potterhanworth, Star Lode, Pinchbeck, Star Fen, Heckington. Used in thatching and known as "Star Thack" and "Sorgh-grass".

663. Carex L.

1. **C. laevigata** Sm. *Smooth Sedge*
    - **8.** 1904, Acthorpe, *C. S. Carter*. (Extinct).
    - **10.** 1915, Woodhall Spa, *A. R. Horwood*. 1960, *E. J. Gibbons*.

    Native. Rare. Damp woods.

2. **C. distans** L.   *Distant Sedge*
    9. 1876, Saltfleet, *Fowler*.
    Not recorded for Divs. 1, 7, 13, 15, 16.
    Native. Common in dune slacks and uncommon inland on heavy clay.

4. **C. hostiana** DC.   *Tawny Sedge*
    16. 1836, Bourne, *Dodsworth*.
    Recorded for Divs. 2, 5, 6, 7, 8, 13, 14, 16.
    Native. Rare and in small quantity. Bogs.

5. **C. binervis** Sm.   *Green Ribbed Sedge*
    13. 1855, Doddington, *Cole*.†
    Recorded for Divs. 1, 2, 3, 4, 5, 6, 7, 10, 13, 15.
    Native. Local on dry heaths.

7. **C. lepidocarpa** Tausch   *Tall Yellow Sedge*
    2. 1857, Broughton, *Fowler*.
    Not recorded for Divs. 1, 9, 11, 12, 17, 18.
    Native. Near calcareous springs.

8. **C. demissa** Hornem.   *Common Yellow Sedge*
    5. 1878, Scotton, *Fowler*.
    Not recorded for Divs. 8, 12, 14, 16, 17, 18.
    Native. Acid soils; not very common.

10. **C. serotina** Mérat   *Dwarf Yellow Sedge*
    5. 1951, Scotter, *E. J. Gibbons*. Conf. E. Nelmes.
    13. 1855, Boultham, *Lowe*.
    Recorded for Divs. 2, 3, 5, 10, 10, 13 — but open to doubt.
    Native. Rare. Often confused with *C. demissa* (*oederi*).

11. **C. extensa** Gooden.   *Long-bracted Sedge*
    4. 1892, Cleethorpes, *Lees*.†   1964, *J. Gibbons*.
    9. 1876, Humberstone, *Fowler*. 1965, Saltfleetby, *L.N.U.*
    11. 1909, Ingoldmells, *B. Reynolds*. 1961, Skegness, *E. J. Gibbons*.
    17. 1851, Fosdyke, *Watson*.
    Native. Rare. On coast.

12. **C. sylvatica** Huds.   *Wood Sedge*
    8. 1857, Kenwick, *Bogg*.†
    Not recorded for Divs. 9, 17, 18.
    Native. Common in many woods.

15. **C. pseudocyperus** L. *Cyperus Sedge*
   16. 1838, Bourne, *Dodsworth.*
   Recorded for Divs. 1, 5, 6, 10, 13, 15, 16, 18.
   Native. Uncommon. In water.

16. **C. rostrata** Stokes *Bottle Sedge, Beaked Sedge*
   1. 1815, Axholme, *Peck.*
   Not recorded for Divs. 9, 12, 14, 15, 16, 17, 18.
   Native. Uncommon in N. Lincs. very rare in S. In water.

17. **C. vesicaria** L. *Bladder Sedge*
   1. 1815, Axholme, *Peck.*
   Not recorded for Divisions 12, 14, 15, 16, 18.
   Native. Uncommon.

20. **C. riparia** Curt. *Great Pond Sedge*
   16. 1838, Bourne, *Dodsworth.*
   Recorded for all Divs.
   Native. Not as common as *acutiformis*. In water.

21. **C. acutiformis** Ehrh. *Lesser Pond Sedge*
   16. 1838, Bourne, *Dodsworth.*
   Recorded for all Divs.
   Native. Fairly common. Often in quantity in damp pasture or woods.

22. **C. pendula** Huds. *Pendulous Sedge*
   2. 1959, Flixborough, *E. J. Gibbons.* (5 plants). 1964, Normanby Hall, planted, naturalised. *L.N.U.*
   7. 1961, Hallbush Wood, *N. Read.* 1961, Bullington Short Wood, *W. Heath.*
   10. 1956, Fulsby, *N. Read.*
   15. 1959, Twyford Forest, *M. Lowe.* 1962, Irnham, *L.N.U.*
   16. 1960, Grimsthorpe, *E. J. Gibbons.*
   Native. Rare; perhaps overlooked elsewhere. Planted in gardens.

23. **C. strigosa** Huds. *Thin Spiked Sedge*
   7. 1952, Claxby and 1961, Bullington Spring Wood, *E. J. Gibbons.*
   15. 1962, Irnham Old Park Wood, *L.N.U.*
   16. 1961, Dunsby, *J. H. Chandler and E. J. Gibbons.*
   Native. Rare; perhaps overlooked elsewhere. Wet woods.

24. **C. pallescens** L. *Pale Sedge*
   2. 1857, Broughton, *Fowler.*†
   Not recorded for Divs. 3, 4, 9, 14, 17, 18.
   Native. Not very common; damp woods.

THE COUNTY FLORA

26. **C. panicea** L. *Carnation Grass*
    8. 1856, Louth, *Bogg*.†
    Not recorded for Divs. 12, 17, 18.
    Native. Local in bogs and wet pastures.

31. **C. flacca** Schreb. *Glaucous Sedge*
    16. 1836, Bourne, *Dodsworth*.†
    Recorded for all Divs.
    Native. One of the most common species and rather variable.

32. **C. hirta** L. *Hammer Sedge, Hairy Sedge*
    16. 1836, Bourne, *Dodsworth*.
    Not recorded for Div. 12.
    Native. Abundant; meadows, roadsides.

33. **C. lasiocarpa** Ehrh. *Slender Sedge*
    5. Scotton, *J. Dickson*. BM. Date possibly before 1820.
    1878, Laughton, *Fowler*. c. 1947, Scotter, *W. Sledge*.
    Native. Very rare; Fowler's habitat was lost through dumping before 1900.

34. **C. pilulifera** L. *Pill-headed Sedge*
    13. 1851, Thorpe, *Watson*.
    Not recorded for Divs. 4, 8, 9, 11, 12, 14, 17, 18.
    Native. Dry acid ground chiefly in woods.

35. **C. ericetorum** Poll. *Breckland Spring Sedge*
    2. 1951, Broughton, *E. J. Gibbons*.
    14. 1953, Ancaster Valley, *E. J. Gibbons and R. and B. Howitt*.
    Native. Very rare on limestone.

36. **C. caryophyllea** Latourr. *Vernal Sedge*
    6 or 13. 1851, Lincoln, *Watson*.
    Not recorded for Divs. 9, 12, 17.
    Native. Generally common in old turf. Decreasing.

46. **C. elata** All. *Tufted Sedge*
    5. 1868, Gainsborough, *Charters*.
    Not recorded for Divs. 9, 17, 18.
    Native. A robust sedge, often forming big tussocks in water.

47. **C. acuta** L. *Graceful or Acute Sedge*
    16. 1836, Bourne, *Dodsworth*.†
    Not recorded for Divs. 4, 9, 11, 12, 14, 17, 18.
    Native. Perhaps overlooked or scarce. In water.

50. **C. nigra** (L.) Reichard *Common Sedge*
    12. 1799, East Fen, *Young*.
    Not recorded for Divs. 17, 18.
    Native. Variable in size; sometimes dominant in moorland.

54. **C. paniculata** L. *Tussock Sedge*
    16. 1836, Bourne, *Dodsworth*.
    Not recorded for Divs. 6, 9, 11, 12, 14, 17, 18.
    Native. Forming high tussocks usually; not common.

56. **C. diandra** Schrank *Two-stamened Sedge*
    5. 1878, Laughton, *Fowler*. 1896, Scotton, *Peacock*.†
    10. 1900, Somersby, *Mason*.
    Native. Very rare. No specimen.

57. **C. otrubae** Podp. *False Fox Sedge*
    16. 1838, Bourne, *Dodsworth*.†
    Recorded for all Divs.
    Native. Common inland and near the coast.

60. **C. disticha** Huds. *Brown Sedge*
    16. 1836, Bourne, *Dodsworth*.†
    Not recorded for Divs. 14, 17, 18.
    Native. Prefers basic soils. Locally abundant in wold valleys.

61. **C. arenaria** L. *Sand Sedge*
    12. 1856, Boston, *Thompson*.
    Not recorded for Divs. 8, 15, 17, 18.
    Native. Locally dominant on some inland heaths as well as on the coastal dunes.

62. **C. divisa** Huds. *Divided Sedge*
    2. 1945, Burton Stather, *J. M. Taylor*.
    3. 1893, Barton, *Firbank*.† 1966, S. Ferriby, *E. J. Gibbons*.
    4. 1851, Grimsby, *Watson*.
    9. 1961, Thoresby Bridge, *E. J. Gibbons*.
    11. 1929, Trusthorpe, *E. J. Gibbons*.
    12. *Check List*.
    Native. Scarce near the coast.

⎡ 64. **C. maritima** Gunn. *Curved Sedge* ⎤
⎣   **3.** Kirmington Inter-glacial deposit (Peacock's MS). ⎦

65. **C. divulsa** Stokes *Grey Sedge*
    3. 1870, Brocklesby, *F. A. Lees*.†
    11. *Check List*.
    Native. Rare, some of the old records being wrongly identified.

66. **C. polyphylla** Kar. & Kir. [**C. muricata** ssp. **leersii** Aschers & Graeb].
   14. 1957, Haverholme. *E. J. Gibbons.*
   15. 1960, S. Witham, *J. H. Chandler.*
   16. 1903, Careby, *Peacock.*†
   Native. Rare, but fairly frequent in Div. 16. Open ground.

67. **C. spicata** Huds. *Spiked Sedge*
   16. 1836, Bourne, *Dodsworth.*
   Not recorded for Divs. 1, 2, 12, 18.
   Native. Open ground and banks on basic soils.

69. **C. elongata** L. *Elongated Sedge*
   2. *c.* 1920, Manton, *W. Frith.*
   5. 1881, Laughton, *Fowler.*
   Native. Very rare. No specimen.

70. **C. echinata** Murr. *Star Sedge*
   13. 1851, Boultham, *Watson.*
   Not recorded for Divs. 1, 6, 9, 12, 14, 17.
   Native. Not very common. Bogs.

71. **C. remota** L. *Remote Sedge*
   13. 1851, Skellingthorpe, *Watson.*
   Not recorded for Divs. 3, 9, 17, 18.
   Native. Damp woods.

72. **C. curta** Gooden *Whitish Sedge*
   1. 1958, Crowle Moor, *E. J. Gibbons.*
   2. 1950, Manton, *B. Morgan.* B.S.B.I.
   5. 1865, Laughton, *Charters.*† 1868, Scotter, *Fowler.*
   7. 1911, Linwood, *Fowler.* 1950, *J. Gibbons.*
   10. 1954, Woodhall Spa, *N. Read.*
   Native. Rare, but in some quantity in two localities. Wet peat.

74. **C. ovalis** Gooden. *Oval Sedge*
   2. 1857, Sawcliff, *Fowler.*†
   Not recorded for Divs. 9, 12, 14, 17, 18.
   Native. Generally common on acid soil in damp pastures.

80. **C. pulicaris** L. *Flea Sedge*
   12. 1856, Boston, *Thompson.*
   Recorded for Divs. 1, 2, 3, 4, 5, 7, 8, 12, 14.
   Native. Rare and in small quantity. Bogs. Decreasing.

81. **C. dioica** L. *Dioecious Sedge*
   8. 1933, Tathwell, *Marsden.* B.E.C. Rep. (Carter and Lees).
   15. 1905, Stroxton, *Miss Stow.*† L. Herb.
   Native. Probably extinct.

GRAMINEAE

665. Phragmites Adans.
1. **P. communis** Trin. [**P. australis** (Car.) Trin ex Steud] *Reed, Henne*
   12. 1636, Boston, *Hexham*.
   Recorded for all Divs.
   Native. Much used in plastering and probably in thatching.

667. Molinia Schrank
1. **M. caerulea** (L.) Moench *Purple Moorgrass*
   1. 1815, Axholme, *Peck*.
   Not recorded for Divs. 9, 12, 16, 17, 18.
   Native. Occasionally in wet meadows as well as on moors.

668. Sieglingia Bernh.
1. **S. decumbens** (L.) Bernh. *Heath Grass*
   1851, *Watson*.
   Not recorded for Divs. 4, 8, 11, 12, 16, 17, 18.
   Native. Usually in poor grassland with *Potentilla erecta*.

669. Glyceria R. Br.
1. **G. fluitans** (L.) R. Br. *Flote-grass*
   1851, *Watson*.
   Recorded for all Divs.
   Native. Less common than *G. plicata*.

   G. fluitans x plicata = G. x pedicellata Townsend
   Not uncommon, but underworked.

2. **G. plicata** Fr.
   13. 1851, Bracebridge, *Watson*.
   Recorded for all Divs.
   Native. Very common.

3. **G. declinata** Breb.
   7. 1951, Holton le Moor, *E. J. Gibbons*. det. C. E. Hubbard.
   8. 1952, Legbourne, *E. J. Gibbons*. det. C. E. Hubbard.
   Recorded for Divs. 7, 8; probably overlooked elsewhere.

4. **G. maxima** (Hartm.) Holmberg *Reed-grass, Leed*
   1597, *Gerarde*.
   Recorded for all Divs.
   Native. Abundant and dominant in many watery places.

670. Festuca L.

1. **F. pratensis** Huds. *Meadow Fescue*
   12. 1820, Boston, *Thompson*.
   Recorded for all Divs.
   Native.

2. **F. arundinacea** Schreb. *Tall Fescue*
   12. 1856, Boston, *Thompson*.
   Recorded for all Divs.
   Native. Frequent on roadsides and in wet pasture.

3. **F. gigantea** (L.) Vill. *Tall Brome*
   8. 1851, Louth, *Watson*.
   Not recorded for Divs. 9, 12, 17, 18.
   Native. In woods, chiefly on basic soil.

6. **F. rubra** L. *Creeping Fescue*
   11. 1824, Skegness, *Sinclair*.
   Recorded for all Divs.
   Native.

7. **F. juncifolia** St.-Amans
   11. 1911, Skegness, *Druce (B.E.C. Report)*. 1943, *McClintock*.
   1957, Chapel-St.-Leonards, *Perring*.
   Native. Very rare or overlooked, maritime.

8. **F. ovina** L. *Sheep's Fescue*
   8. 1851, Louth, *Watson*.
   Not recorded for Div. 17.
   Native.

9. **F. tenuifolia** Sibth.
   5. 1891, Scotton Common, *Peacock*.†
   Recorded for Divs. 1, 2, 3, 4, 5, 6, 7, 13, 15.
   Native. Probably elsewhere on dry soils, locally abundant.

12. **F. glauca** Lam. var **caesia** (Sm.)
    6. 1956, Laughterton and Torksey, *J. E. Lousley*.
    Possibly elsewhere.

Festuca x Lolium = x Festulolium Aschers. & Graebn.

F. pratensis x L. perenne = x Festulolium loliaceum (Huds.) P. Fourn.
   3. 1896, Cadney, *Peacock*.†
   Recorded for Divs. 3, 5, 6, 9, 15, 16.
   Native. Presumably overlooked.

671. Lolium L.

1. **L. perenne** L. *Rye-grass*
    11. 1824, Skegness, *Sinclair*.
    Recorded for all Divs.
    Native. Cultivated and grown for seed.
2. **L. multiflorum** Lam. *Italian Rye-grass*
    3. 1893, Barton, *Firbank*.†
    Recorded for all Divs.
    Introduced and cultivated extensively.
3. **L. temulentum** L. *Darnel*
    8. 1851, Kenwick, *J. H. Thompson*.
    Recorded for Divs. 4, 5, 8.
    Alien.

672. Vulpia C. C. Gmel.

2. **V. bromoides** (L.) Gray *Squirrel Tail, Barren Fescue*
    13. 1851, Lincoln, *Watson*. 1856, Skellingthorpe, *Cole*.†
    Not recorded for Div. 12.
    Native. Very dry sand.
3. **V. myuros** (L.) C. C. Gmel. *Rat's-tail Fescue*
    5. 1879, Laughton, *Fowler*.
    Not recorded for Divs. 3, 8, 9, 17, 18.
    Native or colonist. Dry places and railways.

673. Puccinellia Parl.

1. **P. maritima** (Huds.) Parl. *Sea Meadow Grass*
    11. 1824, Skegness, *Sinclair*.
    Recorded for Divs. 2, 3, 4, 5, 9, 11, 12, 17, 18.
    Native. Frequent on coastal mud.
2. **P. distans** (L.) Parl. *Reflexed Meadow Grass*
    12. 1851, Fishtoft, *Watson*.
    Recorded for Divs. 1, 2, 3, 4, 9, 11, 12, 17, 18.
    Native. Not very common on saline mud in estuaries.

674. Catapodium Link

1. **C. rigidum** (L.) C. E. Hubbard *Fern Grass, Hard Poa*
    16. 1836, Stamford, *Dodsworth*.
    Recorded for all Divs.
    Native. Dry calcareous banks and railways.

2. **C. marinum** (L.) C. E. Hubbard *Darnel Poa*
    4. 1862, Cleethorpes, *Britten.* 1902, Grimsby Docks, *A. Smith.*
    1907, Grimsby and New Clee, *A. Bullock.*† 1959, *J. Gibbons.*
    11. 1878, Gibraltar Point, *Peacock's Rock Soil Flora.*
    Native. Rare and in small quantity.

675. Nardurus (Bluff, Nees & Schau.) Reichb.

1. **N. maritimus** (L.) Murb. *Mat-grass Fescue*
    15. 1962, Creeton, *J. H. Chandler.*
    16. 1903, Carlby, *Miss Stow*† (L.N.U. Meeting). First record for the British Isles.
    Native. Very rare.

676. Poa L.

1. **P. annua** L. *Annual Poa*
    1597, Lincolnshire Fens, *Gerarde.*
    Recorded for all Divs.
    Native. Abundant.

3. **P. bulbosa** L. *Bulbous Poa*
    4. 1958, Cleethorpes, *B. Watkinson.*
    Native. To be looked for elsewhere.

6. **P. nemoralis** L. *Wood Poa*
    9. 1856, Skidbrook, *Bogg.*
    Not recorded for Divs. 1, 17, 18.
    Native. Rather uncommon, calcareous soils.

9. **P. compressa** L. *Flattened Poa*
    13. 1851, Bracebridge, *Watson.*
    Not recorded for Divs. 1, 3, 17.
    Native. Uncommon.

10. **P. pratensis** L. *Meadow Grass*
    1597, Lincolnshire Fens, *Gerarde.*
    Recorded for all Divs.
    Native, very common.

11. **P. angustifolia** L.
    8. 1952, Beesby, *Hope-Simpson.*
    16. 1968, Bourne, *Z. Porter.*
    Native. Distribution not known. Not uncommon on dry soils.

12. **P. subcaerulea** Sm.
    - **5.** 1897, Scotter, *Peacock*.†
    - **9.** 1894, Mablethorpe, *Charman*.
    - 1910, Saltfleetby-all-Saints, *Mason*.
    - **14.** 1899, Dorrington, *Mason*.†
    - Native. Probably unnoticed elsewhere.

13. **P. trivialis** L.   *Rough Meadow Grass*
    - **12.** 1820, Boston, *Thompson*.
    - Recorded for all Divs.
    - Native. Abundant in shady places.

14. **P. palustris** L.
    - **14.** 1959, Haverholme, *Howitt*.
    - Colonist?

677. Catabrosa Beauv.

1. **C. aquatica** (L.) Beauv.   *Water Whorl-grass*
    - **13.** 1851, Lincoln, *Watson*.
    - Not recorded for Divs. 1, 6, 9, 14, 17, 18.
    - Native. Near springs on calcareous soil.

678. Dactylis L.

1. **D. glomerata** L.   *Cock's-foot*
    - **12.** 1820, Boston, *Thompson*.
    - Recorded for all Divs.
    - Native. Much cultivated.

679. Cynosurus L.

1. **C. cristatus** L.   *Crested Dog's-tail*
    - **12.** 1820, Boston, *Thompson*.
    - Recorded for all Divs.
    - Native. Common.

2. **C. echinatus** L.
    - **12.** 1856, Boston, *Thompson*.
    - Recorded for Divs. 4, 5, 9, 10, 11, 12.
    - Casual.

680. Briza L.

1. **B. media** L. — *Quaking Grass*
    16. 1836, Bourne, *Dodsworth*.
    Recorded for all Divs.
    Native. Indication of basic soil. Decreasing.
3. **B. maxima** L.
    4. 1894, Laceby, *A. Smith*.†
    Casual.

681. Melica L.

1. **M. uniflora** Retz. — *Wood Melick*
    8. 1856, Louth, *Bogg*.
    Not recorded for Divs. 4, 6, 9, 12, 14, 17, 18.
    Native. In many woods.
2. **M. nutans** L. — *Mountain Melick*
    2. 1856, Broughton, *Fowler*. Still (1960) in two spots far apart in the woods — *E. J. Gibbons*.
    Native. Calcareous soil

683. Bromus L.

1. **B. erectus** Huds. — *Upright Brome*
    6. 1797, Fillingham, *Dalton*.
    Not recorded for Divs. 1, 9, 17, 18.
    Native on dry calcareous pasture.
2. **B. ramosus** Huds. — *Hairy Brome*
    4. 1851, Grimsby, *Watson*.
    Not recorded for Div. 17.
    Native. Common in woods.
4. **B. inermis** Leyss.
    11. 1942, Skegness, *E. C. Wallace*.
    Casual.
5. **B. sterilis** L. — *Barren Brome*
    16. 1836, Bourne, *Dodsworth*.
    Recorded for all Divs.
    Native, very common, increasing.
6. **B. madritensis** L. — *Compact Brome*
    4. 1890, Grimsby, *A. Smith*.
    11. 1909, Skegness, *H. C. Brewster*.†
    Casual.

7. **B. diandrus** Roth
   4. 1897, Grimsby, *A. Smith.* Alien.
   Casual.

8. **B. rigidus** Roth
   4. 1897, Grimsby, *A. Smith.*
   Dock casual.

9. **B. tectorum** L.
   4. 1897. Grimsby, *A. Smith and E. V. Woods.*†
   Casual.

10. **B. mollis** L. *Lop-grass*
    13. 1851, Thorpe, *Watson.*
    Recorded for all Divs.
    Native.

11. **B. ferronii** Mabille
    11. 1949, Skegness, *E. J. Gibbons* (det. C. E. Hubbard).
    Native. Very local, maritime.

12. **B. thominii** Hardouin
    16. 1964, Stamford, *J. H. Chandler.*
    Recorded for Divs. 1, 2, 3, 7, 9, 11, 15, 16, 17.
    Native. Not worked yet.

13. **B. lepidus** Holmberg
    15. 1962, Castle Bytham, *J. H. Chandler.*
    16. 1957, Dowsby, *Howitt.*
    Native. Probably widespread.

14. **B. racemosus** L. *Smooth Brome*
    8. 1896, Louth, *Lees.*
    Recorded for Divs. 3, 7, 8, 12, 13, 15.
    Native.

15. **B. commutatus** Schrad. *Meadow Brome*
    13. 1851, Thorpe Station, *Watson.*
    Not recorded for Divs. 1, 5, 9, 10, 11, 12, 18.
    Native.

16. **B. interruptus** (Hack.) Druce
    16. 1911, *Druce (J. of B.,* 1911).
    ?Native.

17. **B. arvensis** L.
    12. 1856, Boston, *Thompson*.
    Recorded for Divs. 3, 7, 8, 12, 13.
    Casual.

18. **B. secalinus** L.   *Rye Brome*
    1851, *Watson*.
    Recorded for Divs. 3, 11, 17, 18.
    Introduced.

20. **B. unioloides** Kunth [**B. wildenowii** Kunth]
    4. 1955-56, Cleethorpes and Grimsby, *E. J. Gibbons and D. McClintock*.
    Alien.

684. Brachypodium Beauv.

1. **B. sylvaticum** (Huds.) Beauv.   *Slender False-brome*
   1851, *Watson*.
   Not recorded for Divs. 17, 18.
   Native. Generally common in woods and thickets.

2. **B. pinnatum** (L.) Beauv.   *Heath False-brome, Shear Grass, Tor Grass*
   8. 1851, Louth, *Watson*.
   Not recorded for Div. 12.
   Native. Dominant in some parts, in others occasional. Basic soil.

685. Agropyron Gaertn.

1. **A. caninum** (L.) Beauv.   *Fibrous Twitch, Bearded Couch-grass*
   12. 1820, Boston, *Thompson*.
   Not recorded for Divs. 4, 9, 17, 18.
   Native. Not common and usually in small quantity in woods and hedges.

3. **A. repens** (L.) Beauv.   *Couch-grass, Twitch*
   14. 1790, Spanby, *Cragg*.
   Recorded for all Divs.
   Native. Vigorous and a nuisance in light acid soils.

4. **A. pungens** (Pers.) Roem. & Schult.   *Sea Couch-grass*
   17. 1877, Fosdyke, *Fowler*.
   Recorded for Divs. 2, 3, 4, 9, 11, 12, 17, 18.
   Native. Abundant all round the coast chiefly on mud.

5. **A. junceiforme** (A. & D. Love) A. & D. Love
Sand Couch-grass
   11. 1847, Skegness, *Dr. Grantham.*
   Recorded for Divs. 4, 9, 11, 12, 17, 18.
   Native. Scarcer than the preceding on seaward side of dunes.

A. pungens x junceiforme.
   11. 1942, Skegness, *E. C. Wallace.*
   1968, Gibraltar Point, *A. J. Gray.*
   Native. Intermediate habitat between parents.

686. Elymus L.
1. **E. arenarius** L. *Lyme-grass*
   11. 1805, Sutton, *Banks.*
   Recorded for Divs. 4, 9, 11, 12, 17, 18.
   Native. All round the coast among Marram but in smaller quantity.

687. Hordeum L.
1. **H. secalinum** Schreb. *Meadow Barley*
   11. 1824, Skegness, *Sinclair.*
   Recorded for all Divs.
   Native. Common on heavy soil but not everywhere.

2. **H. murinum** L. *Wall Barley*
   4. 1851, Grimsby, *Watson.*
   Recorded for all Divs.
   Native. Occasionally hard to find.

3. **H. marinum** Huds. *Squirrel-tail Grass*
   12. Freiston, *Howitt.*
   Recorded for Divs. 3, 4, 9, 11, 12, 17, 18.
   Native. Scarce and in small quantity.

H. jubatum L. *Mare's-tail Barley*
   3. 1970, S. Ferriby, *M. P. Gooseman.*
   7. 1963, Holton le Moor; and 1964, Bleasby, *E. J. Gibbons.*
   18. 1957, Whaplode Drove, *Howitt.*
   Alien. Dumps and roadsides.

THE COUNTY FLORA

688. Hordelymus (Jessen) Harz

1. **H. europaeus** (L.) Harz   *Wood Barley*
   - **2.** 1963, Brumby, *E. J. Gibbons*.
   - **13.** 1858, Doddington, *Cole*.† 1958, *L.N.U.*
   - **15.** 1963, Twyford Forest, *Cave*.
   - **16.** 1900, Careby Wood, *Stow*. 1968, *J. Gibbons*.

   Native. Rare; might occur elsewhere.

689. Koeleria Pers.

1. **K. cristata** (L.) Pers.   *Crested Hair-grass*
   - **8.** 1856, Near Louth, *Bogg*.

   Not recorded for Divs. 1, 12, 18.

   Native. Not common; old calcareous pasture and sand dune at Skidbrook.

691. Trisetum Pers.

1. **T. flavescens** (L.) Beauv.   *Yellow Oat*
   - **11.** 1823, Skegness, *Sinclair*.

   Recorded for all Divs.

   Native. Common.

692. Avena L.

1. **A. fatua** L.   *Wild Oat, Havers*
   1851, *Watson*.
   Recorded for all Divs.
   Colonist.

2. **A. ludoviciana** Durieu   *Winter Wild Oat*
   To be looked for.

3. **A. strigosa** Schreb.   *Black Oat*
   - **12.** 1856, Boston, *Thompson*.

   Dock alien.

693. Helictotrichon Bess.

1. **H. pratense** (L.) Pilg.   *Meadow Oat*
   - **12.** 1820, Boston, *Thompson*.

   Not recorded for Divs. 1, 7, 9, 11, 18.

   Native. Uncommon in old calcareous pasture. Decreasing.

2. **H. pubescens** (Huds.) Pilg.   *Downy Oat*
   1851, *Watson*.
   Not recorded for Div. 1.

   Native. Much more generally common than *H. pratense*.

694. Arrhenatherum Beauv.

1. **A. elatius** (L.) Beauv. ex J. & C. Presl.   *False Oat-grass*
   12. 1820, Wainfleet, *Thompson*.
   Recorded for all Divs.
   Native. Abundant.

695. Holcus L.

1. **H. lanatus** L.   *Yorkshire Fog*
   1. 1815, Axholme, *Peck*.
   Recorded for all Divs.
   Native. Abundant.

2. **H. mollis** L.   *Creeping Soft-grass*
   1. 1815, Axholme, *Peck*.
   Not recorded for Div. 14.
   Native. In dry woods.

696. Deschampsia Beauv.

1. **D. cespitosa** (L.) Beauv.   *Tufted Hair-grass*
   13. 1856, Boultham, *Cole*.
   Recorded for all Divs.
   Native. In woods and wet meadows.

3. **D. flexuosa** (L.) Trin.   *Wavy Hair-grass*
   1. 1815, Axholme, *Peck*.
   Not recorded for Divs. 4, 9, 11, 12, 17, 18.
   Native. Dominant in some sandy areas.

4. **D. setacea** (Huds.) Hack.
   7. 1878, Linwood, *Lees*. in three peaty quakes in 1895.
   Native. The only record.

697. Aira L.

1. **A. praecox** L.   *Early Hair-grass*
   13. 1851, Lincoln, *Watson*.
   Not recorded for Divs. 9, 12, 18.
   Native. Dry sandy soils.

2. **A. caryophyllea** L.   *Silvery Hair-grass*
   13. 1851, Doddington, *Watson*.
   Not recorded for Divs. 4, 8, 9, 16, 17, 18.
   Native. Dry places, possibly overlooked.

698. Corynephorus Beauv.

1. **C. canescens** (L.) Beauv.   *Grey Hair-grass*
   - **11.** 1878, Gibraltar Point, *Lees*.
   - Native. Should be searched for.

699. Ammophila Host

1. **A. arenaria** (L.) Link   *Marram Grass*
   - 1780, no locality, *Sibthorp*.
   - **2.** Introduced at Risby Warren to bind sand, c. 1910.
   - **4.** 1894, Cleethorpes, *Peacock*.†
   - **9.** 1856, Mablethorpe, *Bogg*.†
   - **11.** 1824, Skegness, *Sinclair*.
   - **12.** *Check List*.
   - Native. Maritime sand dunes.

700. Calamagrostis Adans.

1. **C. epigejos** (L.) Roth   *Bushgrass*
   - **12.** 1597, *Gerarde*. 1799, East Fen, *Young*.
   - Not recorded for Divs. 1, 18.
   - Native. Common in Divs. 5, 7, 16.

2. **C. canescens** (Weber) Roth   *Purple Small-reed*
   - **12.** 1636, *Johnson*. 1799, East Fen, *Young*.
   - Not recorded for Divs. 3, 9, 14, 17.
   - Native. In smaller quantity than the preceding; in wet fen as well as in woods.

701. Agrostis L.

2. **A. canina** L.   *Brown Bent-grass*
   - **1.** 1815, Axholme, *Peck*.
   - Not recorded for Divs. 8, 9, 14, 17, 18.
   - Native. Probably overlooked.

3. **A. tenuis** Sibth.   *Common Bent-grass*
   - **1.** 1815, Axholme, *Peck*.
   - Recorded for all Divs.
   - Native. Abundant on acid soil.

4. **A. gigantea** Roth   *Tall Bent-grass*
   - **1.** 1967, Epworth, *L.N.U. Meeting*.
   - **16.** 1957, Bourne and Stamford, *J. H. Chandler*.
   - Native. Arable fields. Distribution not worked.

5. **A. stolonifera** L. *Creeping Bent, Fiorin Grass*
    1. 1815, Axholme, *Peck.*
    Recorded for all Divs.
    Native, abundant.

702. Apera Adans.

1. **A. spica-venti** (L.) Beauv. *Silky Bent*
    1. 1898, Haxey, *Fowler.*†   1956, Epworth, *J. Gibbons.*
    2. 1876, Frodingham, *Parsons.*
    4. 1902, Grimsby Dock, *A. Smith.*
    16. 1957, Deeping St. Nicholas, *Howitt.*
    Native. Likes peaty sand. Frequent in Div. 1 in arable fields.

2. **A. interrupta** (L.) Beauv. *Dense Silky Bent*
    11. 1942, Skegness, *E. C. Wallace.*
    18. 1908, Moulton, *H. Burchnall.*†
    Casual.

703. Polypogon Desf.

1. **P. monspeliensis** (L.) Desf. *Annual Beardgrass*
    3. 1920, Elsham, *Frith.*
    4. 1898, Grimsby, *Smith.*†   1963, Cleethorpes, *Howitt.*
    13. c. 1948, Lincoln, *F. T. and P. Baker.*
    18. 1931, Spalding, *Ridlington* (det. G. Foggitt).
    Casual.

707. Phleum L.

1. **P. bertolonii** DC. *Knotted Cat's-tail*
    3. 1894, Howsham, *Peacock.*†
    Not recorded for Divs. 1, 12.
    Native in old pasture.

2. **P. pratense** L. *Timothy Grass*
    12. 1820, Boston, *Thompson.*
    Recorded for all Divs.
    Native and cultivated.

5. **P. arenarium** L. *Sand Cat's-tail*
    4. 1851, Cleethorpes, *Watson.*   1958, *B. Watkinson.*
    9. 1922, Mablethorpe, *Mason.*   1958, *J. Gibbons.*
    11. 1900, Skegness, *Mason.*
    13. 1898, Boultham, *Sneath.*† (Alien).
    Native on coast. Scarce.

708. Alopecurus L.

1. **A. myosuroides** Huds.   *Black Twitch*
   12. 1820, Boston, *Thompson*.
   Recorded for all Divs.
   Native on clay soil.

2. **A. pratensis** L.   *Meadow Foxtail*
   14. 1790, Threckingham, *Cragg*.
   Recorded for all Divs.
   Native. Widespread. Decreasing.

3. **A. geniculatus** L.   *Marsh Foxtail*
   13. 1851, Boultham, *Watson*.
   Recorded for all Divs.
   Native. Edges of ponds.

4. **A. aequalis** Sobol.   *Orange Foxtail*
   1. 1898, Epworth, *Rev. A. Thornley*.† *Nat.* p. 336.
   6. 1974, Torksey, *B. Howitt*.
   15. 1904, Denton, *Stow*.† *Nat.*, p. 348.
   Native.

5. **A. bulbosus** Gouan   *Tuberous Foxtail*
   3. 1893, Barton, *Firbank*.†   1949, *L.N.U. Meeting*. (R. Good).
   Native.

709. Milium L.

1. **M. effusum** L.   *Wood Millet*
   8. 1856, Haugham, *Bogg*.†
   Not recorded for Divs. 1, 4, 9, 12, 14, 17, 18.
   Native. Not very common.

712. Anthoxanthum L.

1. **A. odoratum** L.   *Sweet Vernal-grass*
   12. 1820, Boston, *Thompson*.
   Recorded for all Divs.
   Native in old pasture.

2. **A. puelii** Lecoq & Lamotte   *Annual Vernal-grass*
   1. 1950, Epworth, *Dony and Welch*. (B.E.C. Excursion).
   3. 1896, Howsham, *Peacock*.
   4. Check List.
   13. 1897, Boultham, *Peacock*.
   Alien.

713. Phalaris L.

1. **P. arundinacea** L.   *Reed-grass*
   1851, *Watson*.
   Recorded for all Divs.
   Native. Locally common.

2. **P. canariensis** L.   *Canary Grass*
   3. 1865, Caistor, *Britten*.
   Not recorded for Div. 1.
   Casual. Cultivated near Boston.

3. **P. minor** Retz.
   4. 1953, Cleethorpes, *E. J. Gibbons*.
   13. 1903, Boultham, *E. E. Brown*.†
   Alien.

4. **P. paradoxa** L.
   4. 1897, Grimsby, *Wood*.
   9. 1955, Humberston, *E. J. Gibbons*.
   13. 1896, Boultham, *Sneath*.
   Casual.

714. Parapholis C. E. Hubbard

1. **P. strigosa** (Dumort.) C. E. Hubbard   *Sea Hard-grass*
   12. 1826, Freiston, *Howitt*.
   Recorded for Divs. 2, 3, 4, 9, 11, 12, 17, 18.
   Native. Frequent on coastal mud.

2. **P. incurva** (L.) C. E. Hubbard   *Early Sea Hard-grass*
   11. 1957, Skegness, *M. Smith*.
   Native. To be looked for. Northern limit of distribution.

715. Nardus L.

1. **N. stricta** L.   *Mat-grass*
   1. 1815, Axholme, *Peck*.
   Not recorded for Divs. 4, 8, 9, 11, 12, 16, 17, 18.
   Native. Uncommon. On acid heath.

716. Spartina Schreb.

1. **S. maritima** (Curt.) Fernald   *Cord-grass*
   12. 1826, Freiston, *Howitt*.
   17. 1951, Wyberton, *L.N.U. Meeting*.
   18. 1945, Gedney, *N. D. Simpson*.
   Native. Very rare. Northern limit of distribution.

2. **S. anglica** C. E. Hubbard *Common Rice Grass*
   17. 1937, Holbeach Marsh. B.E.C. Excursion.
   Recorded for Divs. 3, 4, 9, 11, 12, 17, 18.
   Introduced to prevent erosion, locally dominant and increasing.

   (**S. x townsendii** not recorded but to be looked for on edges of *S. anglica* colonies.)

717. Cynodon Rich.

1. **C. dactylon** (L.) Pers. *Bermuda-grass*
   12. 1836, Boston Dock, *Dodsworth*.† 1911, *S. J. Hurst*.
   Recorded for Divs. 12, 17.
   Alien.

718. Echinochloa Beauv.

1. **E. crus-galli** (L.) Beauv. *Cockspur*
   4. 1963, Cleethorpes, *Howitt*.
   7. 1945, Holton le Moor, *E. J. Gibbons*.
   9. 1946, Humberston, *E. J. Gibbons*.
   12. 1911, Boston, *Hurst*.†
   17. 1941, Swineshead, *F. Waite*.
   Alien in carrot fields and dumps.

720. Setaria Beauv.

1. **S. viridis** (L.) Beauv. *Green Bristle-grass*
   12. 1856, Boston, *Thompson*.
   Recorded for Divs. 2, 4, 5, 7, 9, 10, 12, 13.
   Casual.

2. **S. verticillata** (L.) Beauv. *Whorled Bristle-grass*
   1. 1894, Axholme, *S. Hudson*.
   Recorded for Divs. 1, 7, 9, 10.
   Casual.

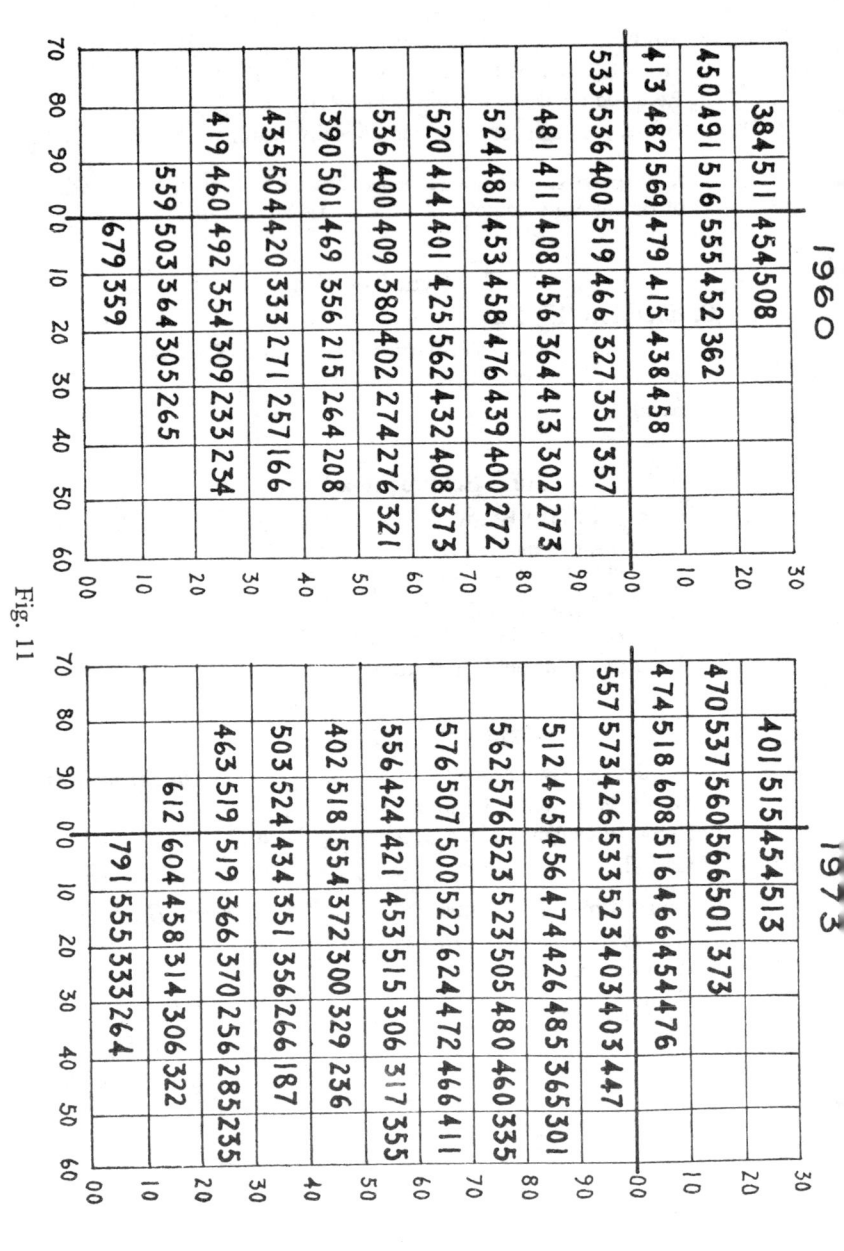

Fig. 11 Number of species recorded by 1960 and by 1973

APPENDICES

APPENDIX I
First records

There are several notable first records of plants in the British Isles found in Lincolnshire, which have since been found in other localities.

Gentiana pneumonanthe L. : 1633, Nettleton Moor, *T. Johnson (Gerarde).*

Armeria maritima ssp. *elongata* (Hoffm.) Bonnier (recognised as ssp. in 1955): 1726, Grantham, *Vincent Bacon;* 1954, Ancaster, *E. J. Gibbons.*

Viola stagnina Kit. : 1833, Boultham, *Cautley;* 1836, Dr. *J. Nicholson* (Annals. of Nat. Hist., 1839); unpubl. specimen held by E. J. Gibbons ex. Ipswich herb.

Selinum carvifolia (L.) L. : 1881, Broughton, *W. Fowler.*

Iris spuria L. : 1896, Huttoft, *F. M. Burton.*

Nardurus maritimus (L.) Murb.: 1903, Carlby, *S. C. Stow.*

Festuca juncifolia St-Amans: 1911, Skegness, *G. C. Druce.* (B.E.C. Report).

Equisetum ramosissimum Desf.: 1947, nr. Boston, *H. Airy Shaw.*

Lamiastrum galeobdolon ssp. *galeobdolon* (S. Wegmuller): 1969, Welton wood and Burwell, *M. Smith.*

Second record

Alisma gramineum Lejeune: 1955, Surfleet, *L.N.U.*

First recorded live specimen after division of genus

Aphanes microcarpa (Bois & Reut.) Rothm.: 1948, Holton-le-Moor, *E. J. Gibbons.*

APPENDIX II
Notable additions to Lees' List:

F. A. Lees, in his "Outline Flora" 1892, gives several absentees from the county, most of which have appeared since.

Vaccinium myrtillus L. : recorded 1917, 1929.

Rosa pimpinellifolia L. : recorded 1893, 1894; no recent record.

Hypericum androsaemum L. : very early records exist; also recorded 1952, but probably naturalised.

Crepis paludosa (L.) Moench. : recorded 1877; overlooked; since found in three localities.

Carex pendula Huds. : recorded 1956; since in five Divisions.

Scirpus sylvaticus L. : recorded 1964; unconfirmed record 1884.

Cardamine amara L. : not recorded in the north of the county until 1910; since in four fresh Divisions; also var. *erubescens.*

APPENDIX III
Plants extinct or thought to be extinct

Peacock, in his Rock Soil Flora Notes, 1904, lists a dozen plants extinct in the county. Some of these have been refound since (marked*).

His notes read: "In Lincolnshire though we must temporarily add E to a number of species, we have but few truly extinct plants. They are:

Silene maritima,* S. quinque-vulnera,* Lathyrus maritimus, Senecio paludosus, S. palustris, Sonchus palustris, Hypopitys,* and Lycopodium alpinum all from loss of habitat; *Crambe* and *Osmunda*,* from the purposely wrought destruction of gardeners; *Cicuta* and *Oenanthe crocata*,*: if they are absolutely gone, because they have been exterminated on account of their deadly nature".

Other plants have disappeared since as the following lists show. Unless the habitats have been completely obliterated it is dangerous to assume that they are extinct. Consequently the lists are arranged so that plants which might easily reappear are indicated.

A. Plants believed extinct
(dates given are those last recorded)

1.1.	*Lycopodium selago* L. (1815).	459.4.	*Stachys germanica* L. (1840).
1.5.	*L. alpinum* L. (1857).	465.3.	*Galeopsis segetum* Neck. (1899).
125.1.	*Agrostemma githago* L. (1950).	485.14.	*Galium parisiense* L. (1836).
137.1.	*Minuartia verna* (L.) Hiern (1913).	506.11.	*Senecio paludosus* L. (1820).
157.2.	*Halimione pedunculata* (L.) Aellen (1886).	506.16.	*S. palustris* (L.) Hook. (1810).
187.3.	*Ulex minor* Roth (1905).	506.17.	*S. integrifolius* (L.) Clairv. (1930).
207.10.	*L. japonicus Willd.* (1849).	512.5.	*Inula crithmoides* L. (1930).
225.4.	*Rosa pimpinellifolia* L. (1902)	517.1.	*Antennaria dioica* (L.) Gaertn. (1877).
243.1.	*Parnassia palustris* L. (1962).	540.5.	*Cirsium oleraceum* (L.) Scop. (1823).
247.2.	*Drosera anglica* Huds. (1893).	556.1.	*Sonchus palustris* L. (1942).
228.1.	*Cicuta virosa* L. (1849).	564.1.	*Damasonium alisma* Mill. (1923).
305.1.	*Selinum carvifolia* (L.) L. (1931).	591.1.	*Asparagus officinalis* L. (1940).
309.2.	*Peucedanum palustre* (L.) Moench (1949).	602.1.	*Colchicum autumnale* L. (1896)
319.12.	*Euphorbia portlandica* L. (1897).	616.3.	*Iris foetidissima* L. ?Introduced (1934).
358.4.	*Vaccinium oxycoccos* L. (1917).	618.1.	*Crocus nudiflorus* Sm. (1885).
364.1.	*Empetrum nigrum* L. (1893).	631.1.	*Hammarbya paludosa* (L.) Kuntze (1820).
369.1.	*Cyclamen hederifolium* Ait. (1920).	632.1.	*Liparis loeselii* (L.) Rich. (1884).
385.1.	*Gentianella campestris* (L.) Börner (1890).	640.3.	*O. sphegodes* Mill. (Doubtful locality).
402.1.	*Mertensia maritima* (L.) Gray (1884).	663.56.	*Carex diandra* Shrank (1900).
407.2.	*Cuscuta epilinum* Weihe (1860).	663.65.	*C. divulsa* Stokes (1870).
426.1.	*Limosella aquatica* L. (1856).	663.69.	*C. elongata* L. (1920).
440.4.	*Orobanche alba* Steph. ex Willd. (1897)	663.81.	*C. dioica* L. (1933).

APPENDICES

B. Plants not seen for some years; careful search might refind them
(dates given are those last recorded)

1.2.	*Lycopodium inundatum* L. (1967).	365.3.	*Limonium bellidifolium* (Gouan) Dumort. (1966).
2.1.	*Selaginella selaginoides* (L.) Link (1948).	370.7.	*Lysimachia thyrsiflora* L. (1950).
24.2.	*Thelypteris dryopteris* (L.) Slosson (1960).	372.4.	*Anagallis minima* (L.) E. H. L. Krause (1877).
48.1.	*Myosurus minimus* L. (1920).	380.1.	*Cicendia filiformis* (L.) Delarb. (1887).
83.1.	*Iberis amara* L. (1928).		
98.2.	*Barbarea stricta* Andrz. (1950).	417.1.	*Misopates orontium* (L.) Raf. (1950).
113.8.	*Viola stagnina* Kit. (1936).	434.1.	*Melampyrum cristatum* L. (1958).
133.1.	*Stellaria nemorum* L. (1920).		
167.1.	*Radiola linoides* Roth	442.4.	*Utricularia minor* L. (1961).
237.1.	*Crassula tillaea* L.-Garland (1952).	470.2.	*Teucrium scordium* L. (1953)
		548.1.	*Arnoseris minima* (L.) Schweigg & Koerte (1958).
277.2.	*Torilis arvensis* (Huds.) Link (1939).		
279.1.	*Coriandrum sativum* L. (1805).	597.1.	*Gagea lutea* (L.) Ker-Gawl. (1960).
298.1.	*Crithmum maritimum* L. (1962).	625.3.	*Epipactis purpurata* Sm. (1895).
320.14.	*Polygonum minus* Huds. (1911).	641.1.	*Himantoglossum hircinum* (L.) Spreng. (1954).
362.1.1.	*Monotropa hypopitys* L. (1946).	660.1.	*Rhynchospora alba* (L.) Vahl (1920).
362.1.2.	*M. hypophegea* Wallr. (1960).	698.1.	*Corynephorus canescens* (L.) Beauv. (1878).

C. Plants which have been recently re-recorded, some only fleetingly, which were previously believed extinct

1.4.	*Lycopodium clavatum* L.	300.5.	*Oenanthe crocata* L.
5.1.	*Osmunda regalis* L.	320.2.	*Polygonum raii* Bab.
24.2.	*Thelypteris palustris* Schott	440.2.	*Orobanche purpurea* Jacq.
26.1	*Pilularia globulifera* L.	469.2.	*Scutellaria minor* Huds.
61.1.	*Glaucium flavum* Crantz (fleeting).	576.1.	*Zostera marina* L.
121.1.	*Frankenia laevis* L.	625.1	*Epipactis palustris* (L.) Crantz
298.1.	*Crithmum maritimum* L. (fleeting).	708.4.	*Alopecurus aequalis* Sobol.

APPENDIX IV

Plants which have been added to the County List since the publication of the Atlas of the British Flora.

127.1	*Dianthus armeria* L.	319.14	*Euphorbia uralensis* Fisch. ex Link.
166.1	*Linum bienne* Mill.	343.14	*Salix nigricans* Sm.
212.8	*Potentilla norvegica* L.	461.1	*Lamiastrum galeobdolon* ssp. *galeobdolon* (S. Wegmuller)
224.1	*Acaena anserinifolia* J. R. & G.		
227.4	*Cotoneaster microphyllus* Wall ex Lindl.	503.2	*Galinsoga ciliata* (Raf.) G. Blake.
		503.1	*G. parviflora* Cav.
249.2	*Lythrum hyssopifolium* L.	560.3	*Taraxacum spectabile* Dahlst.
262.6	*Callitriche truncata* Guss.	570.3.	*Elodea nutallii* (Planch) St. John.
298.1	*Crithmum maritimum* L.		
307.2.	*Angelica archangelica* L.	655.4	*Scirpus sylvaticus* L.

APPENDIX V

Limits of distribution

The following lists are of plants in Lincolnshire at the N. and NE. or S. and SE. limits of their distribution in the British Isles. Those previously recorded (now extinct) which were limit plants have been included. (C.=Coastal).

N. and NE.

44.1.	*Pulsatilla vulgaris* Mill. (N.)	365.5.	*L. binervosum* (G. E. Sm.) C. E. Salmon (NE.)
121.1.	*Frankenia laevis* L. (N.)		
154.9.	*Chenopodium ficifolium* Sm. (N.) (NE.)	380.1.	*Cicendia filiformis* (L.) Delarb. (N.) (Ext.)
154.15.	*C. botryodes* Sm. (N.)	385.4.	*Gentianella anglica* (Pugsl.) E. F. Warb. (N.)
158.2.	*Suaeda fruticosa* Forsk. (N.) (C.)	387.1.	*Nymphoides peltata* (S. G. Gmel.) Kuntze (N.)
160.5.	*Salicornia pusilla* Woods (N.) (C.)		
165.1.	*Althaea officinalis* L. (NE.) (C.)	416.7.	*Verbascum nigrum* L. (N.)
166.1.	*Linum bienne* Mill. (NE.)	422.1.	*Kickxia spuria* (L.) Dumort. (N.)
187.3.	*Ulex minor* Roth (NE.)	422.2.	*K. elatine* (L.) Dumort. (N.)
192.3.	*Trifolium ochroleucon* Huds. (N.)	457.2.	*Prunella laciniata* (L.) L. (N.)
192.5.	*T. squamosum* L. (N.) (C.)		
192.13.	*T. subterraneum* L. (N.)	475.2.	*Campanula trachelium* L. (N.)
206.3.	*Vicia tenuissima* (Bieb) Schinz & Thell. (N.)	485.6.	*Galium pumilum* Murr. (N.)
207.2.	*Lathyrus nissolia* L. (N.)		
207.10.	*L. japonicus* Willd. (N.)	488.1.	*Viburnum lantana* L. (N.)
229.1.	*Crataegus oxyacanthoides* Thuill. (NE.)	506.11.	*Senecio paludosus* L. (N.) (Ext.)
232.7.	*Sorbus torminalis* (L.) Crantz (NE.)	514.3.	*Filago spathulata* C. Presl (N.) (Ext.)
237.1.	*Crassula tillaea* L.-Garland (N.)	556.1.	*Sonchus palustris* L. (N.) (Ext.)
262.6.	*Callitriche truncata* Guss. (NE.)	579.1.	*Ruppia spiralis* L. ex Dumort. (NE).
264.1.	*Thesium humifusum* DC (N.)	625.3.	*Epipactis purpurata* Sm. (N.) (Ext.; herb. rec. not conf.)
286.2.	*Petroselinum segetum* (L.) Koch (N.)		
287.1.	*Sison amomum* L. (N.)	644.1.	*Aceras anthropophorum* (L.) Ait.f. (NE.)
305.1.	*Selinum carvifolia* (L.) L. (N.) (Ext.)	675.1.	*Nardurus maritimus* (L.) Murb. (N.)
309.2.	*Peucedanum palustre* (L.) Moench (N.)	676.3.	*Poa bulbosa* L. (N.) (C.)
319.17.	*Euphorbia amygdaloides* L. (N.)	708.5.	*Alopecurus bulbosus* Gouan (N.) (C.)
325.13.	*Rumex pulcher* L. (NE.)	714.2.	*Parapholis incurva* (L.) C. E. Hubbard (N.) (C.)
365.3	*Limonium bellidifolium* (Gouan) Dumort. (N.)	716.1.	*Spartina maritima* (Curt.) Fernald (N.) (C.)

S. and SE.

2.1.	*Selaginella selaginoides* (L.) Link (SE.) (Ext.)	370.7.	*Lysimachia thyrsiflora* L. (SE.) (Ext.)
211.2.	*Rubus saxatilis* L. (SE.)	559.8.	*Crepis paludosa* (L.) Moench (SE.)
309.3.	*Peucedanum ostruthium* (L.) Koch (S.)		
343.1.	*Salix pentandra* L. (SE.)	607.3.	*Allium scorodoprasum* L. (SE.)
350.1.	*Andromeda polifolia* L. (SE.)		
364.1.	*Empetrum nigrum* L. (SE.) (Ext.)	657.2.	*Blysmus rufus* (Huds.) Link (SE.) (C.)

APPENDICES

APPENDIX VI

ALIENS

There are many specimens and records in the County Herbarium, Lincoln Museum, of aliens and casuals recorded between 1856 and 1935. A composite list of these is given below (List I). They include records by Peacock, Lees and Sneath, who were attracted by the Flour Mill aliens by the Fossdyke west of Lincoln 1896 and 1898, and by the Rev. H. W. Hinchliff, who found more in the West Parade brick pits, 1899. Before he became curator of the Lincoln Museum in 1907, Arthur Smith worked on the Grimsby Dock aliens, a large number of which were recorded about 1900 and verified at Kew.

In 1917 the Rev. F. S. Alston and his son Hugh discovered a chicken farm near the Tower-on-the-Moor at Woodhall Spa where they recorded a large number of plants, 1917-1919. About six of these are now established in quantity. Miss S. C. Stow recorded a few plants after an army camp had left Harrowby near Grantham, 1919, and also at her father's corn merchant store at Court Leys, Caythorpe.

Senecio squalidus (Oxford Ragwort) was recorded first by Dr. John Lowe at Anwick near Sleaford in 1856. Its subsequent history showed colonisation along the railway from Doncaster to Scunthorpe where it had become established by 1936. During World War II air raid shelters were built with gravel from Skellingthorpe and this carried *Senecio viscosus* and *Sisymbrium orientale* all around the district. Oxford Ragwort has been carried from Brigg Sugar Beet Factory along roadsides and occasionally into fields and farmyards.

Aliens from the Mediterranean have been of interest in the Grimsby, Cleethorpes and Humberstone district. A list is included of aliens recorded in 1953, 1955 and 1956 with later additions. Some of these may have been sweepings from Haith's birdseed warehouse. Skegness and Chapel St. Leonards are invaded by visitors from Nottingham who bring some aliens with them. Airfields are also producing new records, chiefly from ballast, as did the railways. *Impatiens glandulifera*, the pink Himalayan Balsam is rare; established in Boston in 1945, it has spread very little. Our rivers have not brought it from the Midlands, except at Owston Ferry on the Trent, 1968, where the salt is not too unpleasant for it. (By 1973 it had spread downstream to East Butterwick). It is also recorded at Riseholme where it is spreading upstream.

Interesting aliens are recorded from Broughton and other woods on Lord Yarborough's estate, where the shooting was let about 1900 to Mr. A. Soames. Barley sweepings from his maltings at Burton on Trent were spread as pheasant food and when subsequent felling and replanting occurred various aliens came up in the clearings and disappeared again. *Senecio squalidus*, *Lathyrus aphaca* (at Claxby), *Hemizonia pungens* (at Limber), *Asperugo procumbens* and other less known plants have been seen since 1920. Some were recorded in Broughton woods in 1935 at an L.N.U. meeting. These records were confirmed by N. Y. Sandwith and have been included in List I. Appended to that list are the more recent records excluding the Woodhall and Cleethorpes area lists which are described separately (Lists II and III) as some species are persisting. The frequently recorded established aliens are incorporated into the main County Flora.

I. List of Aliens and Casuals up to 1935

Unless otherwise stated specimens are housed in the Lincoln Museum County Herbarium arranged from the 9th London Catalogue. Nomenclature is mainly as on specimens unless updated when herbarium names are retained in brackets so as to facilitate the use of the herbarium index.

Abbreviations:

G = Grimsby Docks and Yacht Pond; *Arthur Smith* 1890-1900; verified at Kew.

Li = Lincoln; *Lees, Sneath* and *Peacock c.* 1893-7.

LF = Lincoln, Fossdyke Flour Mill; *Sneath* 1902; *F. T. Baker* 1935.

LW = Lincoln, West Parade Brick Pits, *Rev. H. W. Hinchliff* 1899.

K = Kirton, Flour Mills; *Peacock* 1895.

BO = Boston Dock, *B. Reynolds* 1912 (*J. Bot.* p. 350).

+ = No specimen.

Achillea decolorans Schrad. ex Willd. : Woolsthorpe, 1893.
A. gerberi Willd. : G.
A. lingusticat All. : G.
A. tomentosa L. : Boston 1856. +
Eremophyrum triticeum (Gaertn.) Nevski (*Agropyron prostratum* (Pall.) Beauv.) :G. +
E. bonaepartis (Spreng.) Nevski (*Agropyron squarrosum* (Roth Link): G. +
Berteroa incana (L.) DC. (*Alyssum incanum* L.) : Li; Louth 1894 (*Herb. A. Smith*); BO; G. 1933 +
Althaea hirsuta L. : Broughton 1935, *L.N.U.*
Ambrosia artemisiifolia L. : Rasen 1878, *Lees* (herb. sp. at B.M); Woodhall Spa 1917, *Alston.*
Ammi majus L. : Court Leys 1899, *Stow.*
Amsinckia lycopsoides (Lehm.) Lehm. : G; Limber 1933, *Cox;* Broughton 1903, *Wyatt;* Louth 1895, *Venables;* Gainsborough 1933, H.W.B. *Smith;* Mere 1895, *Miss Pears.*
Anacyclus radiatus Lois. : Lincoln Station 1897, *Stow;* BO +
Anagallis arvensis L. var *pallida*: Cadney 1897, *Peacock;* Irnham 1914, *Sneath.* +
Anchusa hybrida Ten. : G.
A. italica Retz : LF. (= *A. azurea* Mill.)
Anthriscus cerefolium (L) Hoffm. : Cadney 1894, *Peacock;* Wispington 1893, *Alston.*
Arabis caucasica Willd. (*A. albida* Steven ex Jacq. f.) : Boston 1884, *L. Gibbs.*
A. turrita L. : LF.
Asperugo procumbens L. : G; Li. 1893, *Goodall;* Boston 1933, *Hurst.*
Asperula arvensis L. : G; LF; Broughton 1935, *L.N.U.*
Atractylis humilis L. : Mablethorpe 1895, *B. Crow.*
Atriplex tatarica L. : Boston 1911, *Miss Trower.* +
Rhynchosinapis cheiranthos (Vill.) Dandy (*Brassica cheiranthos* (Vill.)) : Cadney 1894, *Peacock;* Grimsby 1898; Washingborough 1908.
Hirschfeldia incana (L.) Lagr. – Fuss. (*Brassica adpressa* Boiss) : Court Leys 1899, *Stow.*
Briza maxima L. : Laceby 1900, *A. Smith.*
Bromus diandrus Roth : G.
B. japonicus Thunb. : Cadney 1908, *Peacock.*
B. madritensis L. : Skegness 1909, *H. C. Brewster;* G.
B. maximus Auct. (= *B. diandrus* Roth) : G.
B. rigidus Roth : G.
B. scoparius L. : G.
B. squarrosus L. : G.
B. tectorum L. : G.
Bunias orientalis L. : K. 1893.
Calamintha graveolens (Bieb.) Benth. (= *Acinos rotundifolius* Pers.) : G.
Camelina microcarpa Andrz. ex DC. : Li 1900, *Mason.*

APPENDICES

Campanula pyramidalis L. : Woodhall 1883, *Miss Mackinder*.
Carbenia benedicta (L.) Arcang. (= *Cnicus benedictus* L.) : G.
Carrichtera annua (L.) DC. (*C. vellae* DC.) : G.
Centaurea aurea Ait. : LW; G. 1897, *E. V. Wood*. +
C. triumfetti All. (*C. axillaris* Willd.) : LF.
C. calcitrapa L. : LF; G; Low Toynton 1895.
C. iberica Trev. ex Spreng. : G.
C. intybacca Lam. (=*Cheirolophus intybaceus* (Lam.) Dostél)): L W, *Hinchliff*.
C. melitensis L. : G.
C. spinosa L. : Cleethorpes 1885, *H. Friend*. +
C. pallescens Del. : G.
C. salmantica L. (=*Mantisalca salmantica* (L.) Briq. & Cavill.) : LF.
Cephalaria syraica (L.) Roem. & Schult : G.
C. transylvanica (L.) Roem & Schult : G.
Chrysanthemum coronarium L. : Li, *G. A. Grierson*.
Cicer arietinum L. : BO. +
Collomia grandiflora Dougl. ex Lindl. : Winthorpe 1912, *Reynolds*; + Coningsby 1893, *Alston*; Trusthorpe 1890, *B. Chapman*; + Epworth 1894. +
Coriandrum sativum L. : Louth, *Lewin*; Cadney, *Smith*. +
Crepis setosa Haller f. : G; Grimsby 1899, *Parker*. +
Cynodon dactylon (L.) Pers. : Boston Dock 1836, *Dodsworth*, 1911, *S. J. Hurst*.
Cardamine heptaphylla (Vill.) *O. E. Schulz* (*Dentaria pinnata* Lam.): Ruskington 1907, *Mason* +

Echinospermum lappula L. Lehm. (=*Lappula squarrosa* (Retz.) Dumort.) : BO +
Echium italicum L. : LW. +
E. lusitanicum L. : G.
E. plantagineum L. : LF; G.
Elymus caput Medusae L. : G. 1893, *Smith*.
Eruca vesicaria (L.) Cav. (*E. sativa* Mill.) : G; Louth 1896, *J. Larder*.
Erucaria hispanica (L.) Druce (*E. myagroides* (L.) Halacsy) : Ruskington 1907, *Mason* (Ver. 1917).
Erysimum hieracifolium L. : G; Howsham 1901, *Booth*; Leverton 1894, *Mason*.
Conringia orientalis (L) Dunmort. (*Erysimum orientale* (L.) Crantz, non Mill.) : G; LF; Cadney 1897, *Peacock*; W. Allington 1904, *Wynne*.

Falcaria vulgaris Bernh. : Grimsby 1929, *S. Cox* (see County List).
Filago arvensis L. : Laughton 1934, *H. B. Willoughby Smith*.

Genista ovata Walsdt. & Kit. (=*G. tinctoria* L.) : Scopwick 1896, *Mason*.
Gilia capitata Sims : Mere 1895, *Mason*.
Glaucium phoeniceum Crantz (=*G. corniculatum* (L.) Rudolph) : G; LF; Gainsborough 1901, *Fowler*.
Gypsophila paniculata L. : G.
G. porrigens (L.) Boiss (=*G. pilosa* Huds.) : G; LW.

Hemizonia pungens (Hook. & Am.) Torr. & Grey : Limber 1932, *L.N.U.*
Herniaria hirsuta L. : G; Mablethorpe 1893, *Miss Mackinder*.
Hesperis matronalis L. : Haxey 1894, *C. C. Bell*; Howsham 1903, *Peacock*.
Hordeum hexastichon L. : BO.
Hypecoum pendulum L. : G.
Hyoscyamus reticulatus L. : Broughton 1935, *L.N.U.*
Hyssopus officinalis L. : Epworth 1901, *C. C. Bell*.

Isatis tinctoria L. : Great Coates 1900; Freiston 1899, *Peacock*.
Iberis umbellata L. : Cleethorpes 1894; Cadney 1903, *Peacock*.

Lagurus ovatus L. : Laceby 1900, *A. Smith*.
Lallemantica iberica (Bieb.) Fisch & Mey. : G.
Lappula echinata Fritsch. (=*L. squarrosa* (Retz.) Dumort.) : G; Mere 1896, *Mason*; Brigg 1917, *Claye*; BO. +

Lathyrus angulatus L. : Li.
L. annuus L. : Li; Broughton 1935, *L.N.U.*
L. aphaca L. : LF; Gainsborough 1893, *Lees;* Colsterworth 1924, *Miss Marshall.*
L. erythrinus C. Presl. (=*L. cicera* L.) : G; LW.
L. hirsutus L. : Wyberton, *S. J. Hurst.*
Lavatera trimestris L. : Broughton 1935, *L.N.U.*
Lepidium draba L. (=*Cardaria draba* (L.) Deav.) : Grantham 1935, *Stow.* +
Lepidium perfoliatum L. : G.
Linaria chalepensis (L.) Mill. : G.
Lolium remotum Schrank : Brandon 1901, *Stow.*
L. temulentum L. : Kenwick 1850, *Thompson;* Kirton 1899, *Peacock.*
L. temulentum var. *arvense* : Gainsborough 1865, *Charters.* +
Lonicera xylosteum L. : Belton 1896, *Woolward;* Colsterworth 1924, *Miss Marshall.*
Lycopsis orientalis L. : Li. +

Malcolmia africana (L.) R. Br. : G.
Malva aegyptica L. : G.
M. nicaeensis All. : Li.
M. verticillata L. : G; Scrivelsby 1917, *Alston.*
M. parviflora L. : Brigg 1898, *Claye.* +
Marrubium alysson L. : G.
Medicago denticulata Willd. (=*M. polymorpha* L.) : G; Rasen 1879, *Allen.*
M. denticulata var. *apiculata* Willd. : Howsham 1895, *W. Booth.*
M. falcata L. : G; LF.
M. orbicularis (L.) Bartal. : LF.
M. silvestris Fr. (=*M. x varia* Martyn) : G; Canwick 1886, *Sneath;* Brigg 1909, *Claye.*
Melilotus caerulea (L.) Desr. (=*Trigonella caerulea* (L.) Ser.) : LF.
Neslia paniculata (L.) Desf. : Hatton 1894. +
Nigella damascena L. : Middle Rasen 1894, *Mrs. Tryon.*
Odontospermum aquaticum (L.) Schultz Bip. (=*Astensius aquaticus* (L.) Less.) : LW. +
Ornithopus ebracteatus Brot. : G. (=*O. pinnatus* (Mill.) Druce : Grimsby 1897, *E. W. Wood*).
Papaver hybridum L. : Mere 1895, *Miss Pears;* Leadenham 1903, *Stow.*
Paronychia argentea Lam. : G +
Phacelia ciliata Benth. : G +
Phalaris minor Retz. : LF 1903, *E. E. Brown.*
P. paradoxa L. : G; LF 1903, *E. E. Brown.*
Plantago arenaria Waldst. & Kit. : G.
P. lagopus L. : BO +
Polycarpon tetraphyllum (L.) L. : G 1899, *G. Parker.*
Polypogon monspeliensis (L.) Desf. : G; Broughton 1935, *L.N.U.*
Portulaca oleracea L. : S. Kelsey 1897, *H. C. Brewster.*
Poterium polygamum Waldst. & Kit. : Louth 1896, *Lees* +
Ranunculus falcatus L. (=*Ceratocephela falcata* (L.) Pers.) : G.
Raphanus landra Moretti ex DC. : K.
Roemeria hybrida (L.) DC. : G.
Salsola tragus auct. (=*S. pestifer* A. Nels.) : Fulbeck 1901, *Stow.*
Salvia controversa auct. (=*S. verbenaca* L.) : G 1899, *G. Parker.*
S. officinalis L. : G 1897, *E. V. Wood.*
S. verticillata L. : G, *A. Smith;* BO, *Hurst.*
Scorpiurus subvillosus L. (=*S. muricatus* L.) : G.
S. vermiculatus L. : LW.
Scandix australis L. : G.
Scleria bracteata Cav. : N. Kelsey 1892.
Secale cereale L. : K.
S. orientale L. (=*Eremopyrum orientale* (L.) Jaub. & Spach) : K.

Senecio squalidus L. : Broughton 1935, *L.N.U.*
S. viscosus L. : LF 1935, *F. T. Baker.*
Setaria glauca auct. (= *S. intescens* (Weigel) Hubbard) : BO +
S. verticillata (L.) Beauv. : Belton 1894, *S. Hudson.*
S. viridis (L.) Beauv. : Crosby 1895, *Mason;* Fulbeck 1901, *Stow.*
Sideritis montana L. : G; LF; Louth 1892, *J. Larder* +; W. Ashby 1900, *Alston.*
S. romana L. : W. Ashby 1900, *Alston.* +
Silene conica L. : G; LF 1857, *G. A. Grierson.*
S. dichotoma Ehrh. : G; Epworth 1895, *S. Hudson;* Morton 1853, *Lowe;* Louth 1900, *Mason;* Broughton 1920, *P. Havelock.* +
S. italica (L.) Pers. : G; Broughton 1905, *Peacock.*
S. muscipula L. : G.
S. nutans L. : G 1891, *A. Smith;* Stickney 1915, *Stewart* and *Miss Hammond.* +
S. pendula L. : Kirkstead 1893, *Alston.*
Stipa pennata L. : Allington 1923, *Stow.*
Tolpis barbata (L.) Gaertn. : LW.
Tragopogon porrifolius L. : Barton 1893; Immingham, Woodhall, Donington.
Trifolium badium Schreb. : LF +
T. resupinatum L. : LF.
T. spumosum L. : LF.
Trigonella arabica Del. : G.
T. caerulea (L.) Ser. : G; BO 1903, *Peacock.*
T. foenum-graecum L. : G. (*C. Parker*); LF.
Triticum triunciale (L.) Rasp. (= *Aegilops triuncialis* L.) : Li.
Valerianella carinata Lois. : G.
Veronica chamaepitys Griseb. non Pers. (= *V. grisebachii* Walters) : G.
Vicia hybrida L. : LF.
V. laevigata Sm. (now = *V. lutea* L.) : LF (see County List).
V. monantha Retz. : LF.
V. narbonensis L. (*V. serratifolia* Jacq.) : LF; Broughton 1900, and 1935, *L.N.U.*
V. pannonica Crantz : LF 1935. + ; Wyberton 1935, *Hurst.*
V. pseudocracca Bertol. : LF; Caythorpe 1900, *Stow.*
V. pyrenaica Pourr. : LF +
V. villosa Roth : 1935.
Wiedemannia erythrotricha (Boiss.) Benth. : Li; K.
Ziziphora taurica Bieb. : G.

Additions Post-1935

(* indicates confirmed by D. McClintock).

Amsinckia intermedia Fisch. & Mey. : Burwell 1972, *I. Weston.* *
Carthamus tinctorius L. : Sleaford Dump 1970, *Z. Porter.*
Cicerbita macrophylla (Willd.) Wallr. : Brocklesby 1966, *R. May* (see County List).
Coronilla Emerus L. : Waddington 1971, *I. Weston.* *
Echinops viscosus DC. non Schrad. ex Reichb. (= *E. spinosissimus* Turra) : Rauceby 1963, *L & B. Howitt.*
Euphorbia uralensis auct. (= *E. esula* L.) : Kirmington 1968, *F. Lammiman.*
Geranium endressi Gay : Woodhall Spa 1960, *J. Bell.*
Geropogon glaber L. (= *Tragopogon hybridus* L.) : Nettleham 1969, *W. Heath.* *
Ionopsidium acaule (Desf.) Reichb. : Nettleham 1969, *W. Heath.* *
Matricaria decipiens (Fisch. & Mey.) C. Koch : Sleaford Dump 1970, *Z. Porter.*
Nicandra physalodes (L.) Gaertn. : Nettleham 1971, *W. Heath.*
Tragopogon porrifolius L. : Riseholme 1972, *I. Weston.*
Xanthium spinosum L. : Donington 1937, *F. L. Kirk;* Immingham 1967.

II List of plants found in 1917, 1918 and 1919 by the Rev. F. S. Alston on a chicken farm at Woodhall Spa.

Those marked * have been re-recorded since 1960 and are well established. Any natives are new to the locality and have also been brought in with the chicken food. Some were first records for the British Isles and many were first records for the County (*Trans. L.N.U.* V 1919, 52-56).

Achillea nobilis L.
Alyssum incanum L. (=*Berteroa incana* (L.) DC.) *
A. maritimum (L.) Lam. (=*Lobularia maritima* (L.) Desv.)
Amaranthus retroflexus L.
Ambrosia artemisiifolia L.
Amsinckia lycopsoides (Lehm.) Lehm. (1919)
Anchusa procera Bass ex Link. (=*A. officinalis* L.)*
Anthemis ruthenica Bieb. (1918)
Artemisia absinthium L.
Axyris amaranteoides L. (1919)

Ballota nigra var. ruderalis L. (=*B. nigra* L. ssp. *nigra*) (1918)
Brassica juncea (L.) Czern.
Bunias erucago L. (1918)

Carduus nutans var. macrocephalus (=*C. macronphalus* Desf.)
Centaurea maculosa Lam. lusus albiflora
C. orientalis L. (1918)
C. stoebe ssp. rhenana (=*C. rhenana* Bor.)
C. trichocephala Bieb. ex Willd.
Chenopodium bonus-henricus L.
C. leptophyllum auct. (=*C. pratericola* Rydb.)
C. murale L.
C. urbicum L.
Cichorium intybus L.
Coronilla varia L. *
Cynosurus echinatus L. (1919)

Delphinium consolida L.
Dipsacus fullonum var. sativus (L.) Thell (=*D. sativus* (L.) Honck.)

Echium italicum L.
E. vulgare L.
Erigeron canadense L. (=*Conyza canadensis* (L.) Crouq.)

Glaucium corniculatum (L.) Rudolph

Grindelia squarrosa (Pursh) Dunal. (1918)

Helianthus cucumerifolius (Torr. & Gray)
H. debilis Nutt.
Hyoscyamus niger L.

Inula britannica L.

Lappula echinata Fritsch (=*L. squarrosa* (Retz.) Dumort.)
Lavatera triloba L.
Leonurus cardiaca L. *
Lepidium densiflorum Schrad.
L. ruderale L. (1918)

Malva pusilla Sm.
Marrubium peregrinum L.
Medicago falcata L. *
M. sativa L.
M. x varia Martyn (1918)
Melilotus officinalis (L.) Pall. *
M. alba Medic. *
Mentha rotundifolia (? x *longifolia*)

Oenothera biennis L. *

Polygonum polystachyum Wall. ex. Meisn. (about ½ ml. away)
Potentilla intermedia L.
P. norvegica L.

Rapistrum perenne (L.) All.
Reseda lutea L.

Salvia aethiopis L.
S. sylvestris auct. (=*S. nemorosa* L.) (1918)
S. verticillata L.
Setaria viridis (L.) Beauv.
Sisymbrium altissimum L.
Sonchus asper var. laciniatus (=*S. asper* (L.) Hill) (1918)

Verbascum chaixii Vill. *
V. lychnitis L. (1918) *

APPENDICES

III List of plants from Cleethorpes, Grimsby and Humberstone

The following is a composite list of the recordings at Cleethorpes dump 1953 (E. J. Gibbons), Humberstone 1955 (E. J. Gibbons and D. McClintock), Humberstone, and Grimsby Docks on several visits 1956 (E. J. Gibbons, D. McClintock, J. E. Lousley and B. Ward). J. Mason also has a considerable collection of aliens from these localities (1964) in his herbarium, including those from the dump by the sea wall on the Humberstone/Cleethorpes boundary 1964/65.

KEY

Locations

C = Cleethorpes; G = Grimsby; H = Humberstone; H/C = Dump on Humberstone/Cleethorpes boundary.

Dates

3 = 1953; 4 = 1954; 5 = 1955; 6 = 1956; 7 = 1957; 63 = 1963; 64 = 1964; 65 = 1965; 69 = 1969.

Determinations

det. B = by J. P. M. Brenan; det G = by R. Graham; det. H = by C. E. Hubbard; det. M = by A. Melderis; det. S = by N. Y. Sandwicth; det. W = by C. West.

Others

†= plants on J. E. Lousley's detailed combined check list of all visits to Humberstone and Grimsby 1955 and 1956; (EJG) = herbarium specimens E. J. Gibbons; (JM) = herb. specimens. J. Mason; JM = records J. Mason; (Ho) = additional records at Humberstone dump by R. C. L. and B. M. Howitt 1963.

Aegilops ligustica Savign. Coss. : (det. H) † G6.
Alopecurus agrestis L.
 (=*A. myosuroides* Huds.) : †H6, G6.
Alyssum maritimum (L.) Lam.
 (=*Lobularia maritima* (L.) Desv.) : H5, H6, H69, (Ho).
Amaranthus blitoides S. Wats. : (det. B) † H5.
A. retroflexus L. : C69, (JM).
Ammi majus L. : (EJG) C3, C63, G6 † H6, (Ho), C69, (JM).
A. visnaga (L.) Lam. : (EJG) C3 † H5, G6, C69, (JM).
Anagallis arvensis ssp. *caerulea* Hartn. (=*A. foemina* Mill.) : C3, H5, H6.
A. arvensis ssp. *foemina* (Mill.) Schinz & Thell. (=*A. foemina* Mill.) : † H6, G6.
Anethum graveolens L. : (EJG) C3, C63 (det. S) † G6, H6, C64, C69, (JM).
Anthemis cotula L. : † H6.
A. mixta L. (=*Chamaemelum mixtum* (L.) All.) : H/C64, (JM).
A. ruthenica Bieb. : † H6.

Antirrhinum orontium L.
 (=*Misopates orontium* (L.) Raf.) : † G6.
A. orontium var. *calycinum* (Lam.) Lge. : (EJG) † G6.
Artemisia absinthium L. : G6.
Asperula arvensis L. : (EJG) H6 † H63, (Ho).
Atriplex sp. : † H6.
Avena fatua L. : (det. H) † G6.
A. fatua var. *glabrata* Peterm. : (det. H) † G6.
Brachypodium distachyon (L.) Beauv. (=*Trachynia distachya* (L.) Link) : H/C64, (JM).
Brassica juncea (L.) Czern. : † H6, G6.
Bromus arvensis L. : (conf. H) † G6.
Bupleurum fontanesii Guss. ex Caruel : (EJG) C3, H6, G6, † C64, (JM).
B. lancifolium Hornem. : (EJG) C3, H5, H6 † C(JM).
Calendula officinalis L. : (Ho).
Calystegia sylvestris (Willd.) Roem & Schult. (=*C. silvatica* (Kit.) Griseb.) : † H6.

Camelina sativa (L.) Crantz. : C3, H6 † (Ho), (EJG) C3.
Cannabis sativa L. : C6, H5, H6.
Carthamnus lanatus L. : † G6.
C. tinctorius L. : † H5.
Caucalis platycarpos L.: C3 † G6, H6.
C. latifolia L. : † G6.
C. nodosa : C3, H5.
Centaurea calcitrapa L. : C3, H6, (EJG) † G6.
C. cyanus L. : †H6.
C. diffusa Lam. : (EJG) G6.
C. diluta Ait. : (EJG) C3, C6, (EJG) H5, H6, H7, G6 † (Ho), C64, C69, (JM).
C. melitensis L. : (EJG) C3 † G6.
C. pallescens var. *hyalolepis* (=*C. pallescens* Del.) : H/C64, (JM)
C. salmantica L. : (=*Mantisalca salmantica* (L.) Briq. & Cavill.): (EJG) C3, H6.
C. solstitialis L. (=*C. solstitialis* subsp. *admaii* (Willd.) Nyman) : (EJG) C3, C6, H6, G6 †
C. solstitialis var. *adamii* : H/C64, (JM).
Ceratochloa unioloides (Willd.) Beauv. (=*Bromus wildenowii* (Kunth)) : † H6, G6, C5, C6, C69, (JM).
Chenopodium murale L. : (EJG) C3, † H5, G6.
C. opulifolium Schrad. ex Koch & Ziz. : H5, H6.
C. giganteum D. Don : C63.
C. amaranticolor (Coste & Reyn.) Coste & Reyne (=*C. gigantium* D. Don) : (Ho).
Chrysanthemum carinatum Schousb. f. *annulatum* : † H6.
C. coronarium L. : (EJG) C3, † H5, H6.
C. segetum L. : H6.
Cichorium endivia L. agg. : † H6, G6.
C. intybus L. : † G6, (Ho).
C. pumilum Jacq. : G6.
Convolvulus tricolor L. : (Ho).
Coriandrum sativum L. : (Ho).
Coronopus didyma (see *Senebiera didyma*) : H6.
C. squamatus (Forsk.) Aschers. : H6.
Cynosurus echinatus L. : † G6.

Datura stramonium L. : † G6.
Delphinium orientale Gay : H5 †
Digitaria sanguinalis (L.) Scop. : C69, JM.
Diplotaxis muralis (L.) DC : H6.
Dipsacus fullonum L. : H5, (Ho).

D. sativus (L.) Hoack. : H5.

Echinochloa colonum (L.) Link : (Ho).
E. crus-galli (L.) Beauv. : H5.
Erodium moschatum (L.) L'Hérit : H6.

Fagopyrum esculentum Moench : C69 (JM).
F. sagittatum Gilib. (=*F. esculentum* Moench) : † H6.

Galium tricorne Stokes pp. (=*C. tricornutum* Dandy) : † (EJG) H6, G6, (Ho).
Geropogon glaber L. (=*Tragopogon hybridus* L.) : † H5, G6.
Guizotia abyssinica (L.f.) Cass. : (EJG) H6 † G6, (Ho), C69, (JM).

Hedypnois polymorpha DC. (=*H. rhagadioloides* (L.) F. W. Schmidt) : H/C64, (JM).
Helianthus annuus L. : † H5, G6, (Ho).
Helminthia echioides (L.) Gaertn (=*Picrio echoides* L.) : † H5, H6, G6.
Hieracium maculatum Sm. : (det. W) † G6.
Hordeum murinum L. (forms) : † G6.
Hyoscyamus niger L. : † H6.

Lathyrus aphaca L. : H6, (Ho), C64, (JM).
Lavatera trimestris L. : † G6.
Lepidium chalepense L. (=*Cardaria chalepense* (L.) Hand.-Mazz) : † G6.
L. draba L. (=*Cardaria draba* (L.) Desv) : † G6.
L. sativum L. : H5, H6 † (Ho).
L. ruderale L. : (Ho).
Linaria purpurea (L.) Mill. : C69, JM.
Linum usitatissimum L. : † H5, H6, G6.
Lithospermum arvense L. : H6, (Ho).
Lobularia maritima (L.) Desv. :
Lolium multiflorum Lam. : C3, H5, H6.
L. perenne L. (forms) † G6 (det. H)
L. rigidum Gaudin : (det. H) † H6, G6.
L. rigidum Gaudi. var. ? x *temulentum* : (det. H) † H6, G6.
L. temulentum L. : † H6, G6, (det. H)

L. temulentum var. *arvense* : G6.
Lycopersicum sp. (=*L. esculentum* Mill.) : H5.
Lythrum hyssopifolia L. : H6.
L. junceum Bank & Soland. : C63, (EJG), H7, (Ho).

Malcolmia maritima (L.) R. Br. : (Ho).
Malope trifida Cav. : † G6.
Malva parviflora L. : (EJG) H6, † *M. pusilla* Sm. : (EJG) H6.
Matricaria decipiens (Fisch. & Mey.) C. Koch : (EJG) C4 † G6.
Medicago hispida Gaertn. (=*M. polymorpha* L.) : H5, H6 †
M. tribuloides Desr. (=*M. truncatula* Gaertn.) : † H6.
M. tuberculata (Retz.) Willd. (=*M. turbinata* (L.) All.)
Melilotus alba Medic. : H6.
M. arvensis Wallr. (=*M. officinalis* (L.) Pall.) : H5, H6.
M. infesta Guss. : H6 †
M. indica (L.) All. : † H6, G6, (Ho), C69, (JM).
Mentha pulegium L. : (EJG) C3 (det. G).
Microlonchus salmanticus (L.) DC. (=*Mantisalca salmantica* (L.) Briq. & Cavill.) : C3 (EJG) † H6.
Monerma cylindrica (Willd.) Coss & Durieu (see *Ceratochloa*) : (EJG) † H6.
Myagrum perfoliatum L. : † H6, G6, (Ho).

Neslia paniculata (L.) Desv. : H6 †
Nicandra physalodes (L.) Gaertn. : (Ho).

Ononis mitissima L. : (EJG) (det. S), H5 † (Ho).
Ononis sp. forms with broad leaves : † H6; narrow leaves : † H6.

Panicum laevifolium Hack. : H/C65, (JM).
P. miliaceum L. : (EJG), H5, H6, (Ho).
Papaver setigerum DC : H6.
P. somniferum L. : H6, G6, (Ho).
Phalaris canariensis L. : C3 † H5, H6, G6.
P. minor Retz. : (EJG) C3.
P. paradoxa L. : (EJG) H5, H6 † G6, C64, (JM).
Phleum subulatum (Savi) Aschers & Graebn : (det. H) † G6.

Picris echioides L. (forms) : (det. S) (EJG) H6.
P. sprengerina (L.) Poir. : (EJG) H5 (det. S) G6.
Plantago lagopus L. : † H5.
P. indica : C69 (JM).
P. psyllium (=*P. afra* L.) : (EJG) C3 † H5, H6, H/C64 (JM).
Polygonum lapathifolium L. : † H6.
P. patulum Bieb. : † H5, H6, G6.
P. pulchellum Lois. : † H5, H6, G6.
Polypogon monspeliensis (L.) Desf. : (Ho).

Ranunculus sardous Crantz. : † H6.
R. muricatus L. : (EJG) H6 †.
R. marginatus var. *trachycarpus* (Fisch. & Mey.) Arzavour : † H6.
Raphanus sativus L. (purpled) : (Ho).
Rapistrum perenne (L.) All. : H5.
R. rugosum (L.) All. (many forms) : C63, H5, H6 † G6.
Reseda alba L. : (Ho).
Ridolfia segetum (L.) Moris : † H5, H6, G6.
Rhagadiolus edulis Gaertn. (=*R. stellatus* (L.) Gaertn.) : † H6.
Rumex pulcher ssp. *divaricatus* (L.) Murb. : † G6.

Salvia reflexa Hornem. : (EJG) C3, (det. S), † H5, (EJG) G6, (Ho), C69 JM.
Scandix pecten-veneris L. : H/C64 (JM).
Scolymus hispanicus L. : (EJG) C3, † H5, H6, G6, (Ho).
Scorpiurus muricatus L. agg. : H/C64 (JM).
Secale cereale L. : † H6.
Senebiera didyma (L.) Pers. (=*Coronopus didymus* (L.) Son.) : H6.
Senecio squalidus L. : † G6.
S. viscosus L. : † G6.
Setaria italica (L.) Beauv. : (EJG) C3, (det. H) H5, H6.
S. viridis (L.) Beauv. : (EJG) H5, (Ho), C69 JM.
Sinapis alba L. : G6.
Silene dichotoma Ehrh. : † G6.
S. gallica L. : † H5.
S. muscipula L. : C63, (Ho).
S. sedoides Poir : H5.
Sideritis montana L. : † G6.
Sisymbrium altissimum L. : † H6, G6.
S. orientale L. : † H6.
Solanum sarrachoides Sendtn. : † (EJG) G6.

S. nigrum L. : (Ho).
Stachys annua (L.) L. : † G6.

Torilis nodosa (L.) Gaertn. (a large alien form) : † G6, C69 (JM).
Trachyspermum copticum (L.) Link : C69, (JM).
Trifolium alexandrinum L. : † H6.
T. echinatum Bieb. : † H5.
T. lappaceum L. : † H5, H6, C64, (JM).
T. resupinatum L. : † H6.
T. scabrum L. : † H6.
Trigonella foenum-graecum L. : H6.

Urochloa panicoides Beauv. : (det. H) H5, H6 †

Vaccaria pyramidata Medic.
(= *V. hispanica* (Mill.) Ramchert) : † H6, G6, C63, H63 (Ho).
Veronica persica Poir. : † G6.
Vicia bithynica (L.) L. : † H5 (det. M).
V. dasycarpa auct. (= *V. varia* Host) : † G6.
V. purpurascens DC. : H6.
V. villosa Roth : (Ho).

Xanthium spinosum L. : H/C65 (JM)

Zacintha verrucosa Gaertn.
(= *Crepis zacinthe* (L.) Babc.) : † H6.

APPENDIX VII

Fossil and Peat Records

The only plant rock fossil so far recorded in the Jurassic rocks of the County is a single pinnule of *Laccopteris* in Leadenham Quarry 1955, by P. Cambridge. Interglacial records of plants have been noted by Peacock from Freshney Bog in the 1890's through the work of John Cordeaux who sent material to Denmark for identification, but these hardly come into a Flora of Lincolnshire. They included *Betula nana* (Dwarf birch) and *Dryas octopetala* (Mountain avens). Interglacial peat has recently been found at Tattershall but results of analysis are not yet available.

Excavations at Dragonby in 1968 (Hayes) have shown vegetative deposits with pollen from various trees and plants among Iron Age and pre-Roman workings. Probably "moist mixed oakwood flourished in the area". The plants were shown to be:

Alnus glutinosa (L.) Gaertn. (Alder).
Corylus avellana L. (Hazel).
Betula verrucosa Ehrh. (Birch).
Fraxinus excelsior L. (Ash).
Ilex aquifolium L. (Holly).
Taraxacum officinale agg. (Dandelion).
Crepis paludosa (L.) Moench (Marsh Hawk's Beard).
Plantago coronopus L. (Buckshorn Plantain).
Urtica dioica L. (Nettle).
Scleranthus annuus L. (Annual Knawel).

Chenopodium album L. (Fat hen).
Galium sp., *Silene* sp. (Bedstraw and Campion).
Bidens tripartita L. (Tripartite Bur-Marigold).
Rumex acetosella agg. (Sheeps' Sorrel).
Angelica sylvestris L. (Angelica).
Triticum sp. (Wheat).
Calluna vulgaris L. (Ling).
Erica sp. (Heather sp.)
Pteridium aquilinum (L.) Kuhn. (Bracken)

CHAPTER 9

LINCOLNSHIRE NATURALISTS' UNION

The Lincolnshire Naturalists' Union was formed in 1893 by a small number of keen naturalists after the publication of an *Outline Flora of Lincolnshire* in 1892. The Union deserves the greatest praise for all its work over the years. Rev. E. A. Woodruffe-Peacock and his helpers produced the Critical Catalogue of Lincolnshire Plants in *The Naturalist* from 1894-1900 and collected the Lincolnshire Herbarium from past and contemporary botanists.

As the Victoria County History Vol. 1 was not being published, Peacock offered to complete the Check List of Lincolnshire Plants which is most useful in showing the number and distribution of Lincolnshire plants. This was published by the Union in 1909.

John Cordeaux in 1894 stated the aim of the Union as that of "bringing a united band of specialists to undertake a thorough and systematic investigation of the Natural History capabilities of the County." Since its formation Lincolnshire records for all branches of Natural History have been published in *The Naturalist* and in its *Transactions*. The annual *Transactions* have recorded the pattern of change over the County and form a historical and scientific basis for future publications.

As John Cordeaux, the first President, said in 1898: "There is no County in England in which the fauna and flora have been so greatly altered. Large numbers of birds, insects and plants have been altogether destroyed, or in the former case driven away by enclosure or drainage. It becomes, therefore, an imperative duty that we should use our best endeavours to preserve what is left and to take care that our earlier mammals, nesting birds and surviving plants are not utterly destroyed or unnecessarily banished." This warning is now even more apt. The challenge was really taken up in 1943 when the Lincolnshire Nature Reserves Investigation Sub-Committee was set up. Three years later it became The Nature Reserves and Wild Life Conservation Committee of the Union. In December 1948 the Articles of the Lincolnshire Naturalists' Trust Limited were signed (now the Lincolnshire Trust for Nature Conservation Limited). Under the secretaryship of Mr.

A. E. Smith, from 1948 to 1969, and now Mr. D. N. Robinson, the Trust flourished and over 40 reserves have been set up.

In a county like Lincolnshire where one sees the inroads on wild life made by efficient farming and the industrial development of Humberside it can be shown that the aim set out by the first President has indeed been maintained by its members. The *Transactions* contain a wealth of Botanical records and the Union's Field Meetings, of which over 400 have been held to 1973, provide many of the opportunities for collecting them.

Location of Field Meetings, 1893-1973

This list gives the number and location of each Field Meeting by Division (number in **bold** type) and place, records being published in the following year.

1893
1. **9.** Mablethorpe.
2. **10.** Woodhall Spa.

1894
3. **13.** Lincoln, Boultham.
4. **4.** Cleethorpes.
5. **8.** Burwell Wood.

1895
6. **2.** Broughton-Twigmoor-Manton.
7. **14.** Sleaford.
8. **7.** Linwood Warren.
9. **13.** Lincoln, Skellingthorpe-Canwick.

1896
10. **15.** Grantham-Ancaster-Syston.
11. **16.** Bourne.
12. **4.** Great Coates (Freshney Bog).

1897
13. **5.** Gainsborough, Lea-Scotter Common.
14. **10.** Holbeck-Tetford-Somersby.
15. **17.** Boston, Wyberton Marsh.
16. **7.** Linwood.

1898
17. **15.** Grantham, Colsterworth-Stoke Rochford.
18. **1.** Epworth.
19. **10.** Woodhall-Tumby Wood.
20. **13.** Hartsholme.

1899
21. **9.** Somercotes-Saltfleetby.
22. **13.** Stapleford Moor.
23. **12.** Freiston Shore.
24. **6—7.** Sudbrooke-Newball Wood.

1900
25. **7—10.** Horncastle-Baumber-Sturton.
26. **16.** Careby-Little Bytham.
27. **2.** Frodingham-Scunthorpe.
28. **9.** Mablethorpe.
29. **13.** Harmston-Coleby-Navenby.

1901
30. **6.** Torksey.
31. **10.** Revesby.
32. **16—18.** Spalding-Crowland.

1902
33. **1.** Epworth.
34. **13.** Caythorpe-Leadenham.
35. **3—7.** Caistor-Normanby-Pelham's Pillar.

1903
36. **15—16.** Careby Wood-Holywell.
37. **11.** Sutton-Huttoft.
38. **7.** Claxby Wood.
39. **15.** Lincoln.

1904
40. **3.** Barton on Humber.
41. **15.** Holywell.
42. **11.** Spilsby.
43. **16.** Rippingale.

THE LINCOLNSHIRE NATURALISTS' UNION

1905
44. 8. Louth, Acthorpe.
45. 16. Stamford.
46. 5. Scotter.
47. 7. Moortown-South Kelsey.
48. 13. Lincoln, Swanpool-Hartsholme.

1906
49. 9. Somercotes.
50. 8. Donington on Bain.
51. 16. Dunsby Wood.
52. 1. Crowle.
53. 8. Louth, Maltby.

1907
54. 6. Welton.
55. 4. Irby on Humber.
56. 18. Spalding-Holbeach.
57. 6. Newton Cliffs.
58. 15. Ropsley.

1908
59. 7. Holton le Moor.
60. 14. Sleaford.
61. 11. Alford.
62. 2. Broughton.
63. 4. Roxton.

1909
64. 13. Nocton-Potterhanworth.
65. 5. Gainsborough-Blyton.
66. 16. Bourne.
67. 12. Fishtoft.

1910
68. 4. Freshney Bog.
69. 16. Crowland.
70. 11. Welton Wood.
71. 2. Scunthorpe.
72. 16. Stamford.

1911
73. 7. Linwood.
74. 3. Barton on Humber.
75. 11. Gibraltar Point.
76. 10. Coningsby.

1912
77. 3. Barnetby-Somerby.
78. 15. Grantham-Ropsley Rise.
79. 2. Burton Stather.
80. 14. Sleaford-Wilsford.

1913
81. 11. Langton-Dalby.
82. 6. Fillingham.
83. 18. Long Sutton.
84. 2. Scunthorpe, Risby.
85. 9. Humberstone.

1914
86. 4. Croxby Pond.
87. 15. Irnham.
88. 8. Donington on Bain.
89. 6. Cherry Willingham.

1915
90. 16. Newell Wood.
91. 1. Haxey.
92. 10. Woodhall Spa.

1916
93. 2. Manton.
94. 13. Skellingthorpe.
95. 11. Willoughby.
96. 7. Holton-le-Moor.

1917
97. 2. Appleby.
98. 15. Great Ponton.

1918
99. 10. Tattershall.

1919
100. 5. Scotter.
101. 14—15. Ancaster.
102. 3. Melton Ross.

1920
103. 13. Harmston.
104. 3. Barton on Humber.
105 10. Scrivelsby.

1921
106. 9. Humberstone.
107. 8. Benniworth Haven.
108. 12. Freiston.

1922
109. 11. Well Vale.
110. 7. Claxby Wood.
111. 16. Bourne.

1923
112. 3—4. Habrough-Newsham Lake.
113. 15. Grantham Twyford Forest.
114. 5—7. South Kelsey, Brandy Wharf.
115. 7. Bardney.

1924
116. 5. Scotter.
117. 13. Doddington.
118. 17. Kirton.
119. 10. Horncastle.

1925
120.	10.	Salmonby.
121.	4.	Freshney Bog.
122.	18.	Cowbit.
123.	8.	South Elkington.

1926
124.	2.	Manton.
125.	18.	Holbeach.
126.	6.	Welton.
127.	14.	Sleaford-Rauceby.

1927
128.	5.	Thonock-Gainsborough.
129.	14.	Threckingham.
130.	8.	Burwell.
131.	4—9.	Cleethorpes-Tetney.
132.	8.	Louth.

1928
133.	9.	Saltfleet-Theddlethorpe.
134.	3.	Nettleton.
135.	7.	Linwood.
136.	10.	Horncastle-West Ashby.
137.	11.	Skegness-Gibraltar Point.

1929
138.	7.	Stainton-Rand.
139.	10.	Holbeck.
140.	8.	Binbrook, Scallows Hall.
141.	14.	Ancaster-Heydour.
142.	12.	Freiston Shore.

1930
143.	8.	Muckton.
144.	14.	Sleaford.
145.	3.	Nettleton.
146.	6.	Newton Cliffs.
147.	16.	Stamford.

1931
148.	12.	Boston-Hobhole.
149.	5.	Laughton.
150.	18.	Cowbit.
151.	2.	Appleby-Broughton.
152.	4.	Irby Dales.

1932
153.	2.	Alkborough.
154.	13.	Potterhanworth.
155.	11.	Ingoldmells.
156.	3.	Limber.
157.	10.	Tattershall.

1933
158.	3—4.	Roxton-Newsham.
159.	10.	Old Bolingbroke, Sowdale.
160.	7.	Goltho.
161.	16.	Dunsby Wood.
162.	1.	Epworth.
163.	9.	Tetney.

1934
164.	4.	Freshney Bog.
165.	15.	Belton.
166.	7.	Linwood.
167.	11.	Welton Wood.
168.	5.	Thonock.
169.	4.	Cleethorpes.

1935
170.	16.	Crowland.
171.	2.	Broughton.
172.	10.	Revesby.
173.	13.	Blankney.
174.	8.	Donington on Bain.
175.	11.	Skegness-Gibraltar Point.

1936
176.	3.	Limber.
177.	15.	Stoke Rochford.
178.	5.	Scotter.
179.	11.	Ingoldmells.
180.	13.	Hartsholme.
181.	3.	Nettleton Mine.

1937
182.	15—16.	Holywell-Grimsthorpe.
183.	3.	Limber.
184.	11.	Well Vale.
185.	5.	Norton Place and Waddingham Common.
186.	14.	Grantham, Saltersford.
187.	4.	Cleethorpes.

1938
188.	7.	Claxby Wood.
189.	14.	Haverholme Park.
190.	4.	Immingham.
191.	10.	Tattershall.
192.	2.	Scunthorpe.
193.	11.	Skegness-Gibraltar Point.

1939
194.	2.	Twigmoor.
195.	4.	Croxby Pond.
196.	17.	Kirton.
197.	14.	Sleaford-Ancaster.

1940
198.	8—10.	Haugham-Tetford.

1941
199.	7.	Linwood.
200	14.	Aswarby Thorns.

1942
201.	7.	Stainton & Rand Woods.
202	13.	Nocton Wood.

THE LINCOLNSHIRE NATURALISTS' UNION

1943
203. 8. Stainton le Vale.
204. 18. Cowbit.
205. 7. Snarford.
206. 13. Hartsholme.

1944
207. 4. Freshney Bog.
208. 13. Caythorpe.
209. 6. Fiskerton.
210. 6. Riseholme.

1945
211. 5. Gainsborough, Thonock
212. 15. Harlaxton.
213. 8. Haugham Pastures.
214. 13. Potterhanworth.
215. 6. Kettlethorpe.

1946
216. 3. Limber.
217. 11. Well Vale.
218. 15. Stoke Rochford.
219. 10. Woodhall Spa.
220. 12. Freiston Shore.
221. 6. Fiskerton.
222. 4—9. Cleethorpes-Tetney Haven.

1947
223. 6—7. Barlings.
224. 10. Tumby.
225. 8. Benniworth.
226. 14. Aswarby.
227. 7. Holton le Moor.
228. 11. Gibraltar Point.

1948
229. 3. Limber-Brocklesby.
230. 15. Syston Park.
231. 10. Holbeck-Salmonby-Tetford.
232. 2. Scunthorpe.
233. 4. Cleethorpes.
234. 13. Blankney Park.

1949
235. 11. Gibraltar Point.
236. 10. Woodhall Spa.
237. 3. Barton on Humber.
238. 13. Skellingthorpe.
239. 14—15. Ancaster.
240. 4. Cleethorpes.

1950
241. 5. Thonock.
242. 2. Scawby-Twigmoor.
243. 9. Saltfleetby-Theddlethorpe.
244. 16. Grimsthorpe Park.
245. 13. Stapleford Woods.
246. 2. Twigmoor-Manton.
247. 11. Gibraltar Point.
248. 15. Denton Park.

1951
249. 3. Limber-Brocklesby.
250. 15. Holywell Wood.
251. 7. Osgodby Moor.
252. 17. Kirton and Frampton Marshes.
253. 4—9. Cleethorpes-Humberstone.
254. 7. Linwood Warren.

1952
255. 2. Scawby-Twigmoor.
256. 11. Well Vale-Welton Wood.
257. 14. Rauceby-Copper Hill.
258. 5. Laughton Common.
259. —. (Notts.)
260. 11. Gibraltar Point.

1953
261. 7. Hainton.
262. 2. Scawby.
263. 15. Stoke Rochford
264. 8. Binbrook-Swinhope.
265. 7. Bardney.
266. 13. Skellingthorpe.
267. 4—9. Cleethorpes-Humberstone.
268. 3. Elsham.

1954
269. 3. Pelham's Pillar.
270. 15. Pickworth-Holywell.
271. 6. Fillingham.
272. 2. Alkborough.
273. 9. Saltfleet.
274. 13. Canwick-Branston.
275. 11. Gibraltar Point.

1955
276. 8. Girsby.
277. 2. Twigmoor.
278. 4. Irby Dales.
279. 15. Stubton Hall.
280. —. (Northants).
281. 17. Bicker Haven & Surfleet.
282. 13. Lincoln, Railway Ballast Pits.
283. 15. Belton Park.

1956
284. 2. Manton.
285. 7. Willingham Forest.
286. 16. Bourne.
287. 11. Swaby.
288. 10. Tattershall.
289. 9. Tetney Haven.
290. 2. Broughton.

1957
- 291. 6. Knaith Woods.
- 292. 14. Haverholme.
- 293. 15. Easton Park.
- 294. 9. Saltfleetby.
- 295. 10. Revesby.
- 296. 13. Skellingthorpe.

1958
- 297. 8. Tathwell.
- 298. 14. Aswarby.
- 299. 13. Doddington.
- 300. 3. Elsham.
- 301. 3. Barton on Humber.
- 302. 10. Woodhall Spa.

1959
- 303. —. (Northants).
- 304. 7. Baumber, Stourton Park.
- 305. 13. Martin.
- 306. 5. Scotton Common.
- 307. 9. Grainthorpe Haven.
- 308. 8. South Elkington.

1960
- 309. 10. Tetford.
- 310. 2. Twigmoor.
- 311. 6. Dunholme.
- 312. 16. Grimsthorpe.
- 313. 7. Bardney.
- 314. 16. Kirkby Underwood.

1961
- 315. 6. Riseholme.
- 316. 10. Tumby.
- 317. 11. Gibraltar Point.
- 318. 14. Haverholme.
- 319. 18. Tydd Gowt.
- 320. 13. Doddington.

1962
- 321. 3. Pillar Woods.
- 322. 15. Irnham.
- 323. 8. Muckton.
- 324. 8. Biscathorpe.
- 325. 1. Epworth.
- 326. 2. Scawby.

1963
- 327. 8. Benniworth.
- 328. 15. Castle Bytham-Creeton.
- 329. 7. Tealby.
- 330. 13. Hartsholme.
- 331. 2. Alkborough.
- 332. 10. Tumby.

1964
- 333. 2. Normanby Park.
- 334. 11. Gibraltar Point.
- 335. 15. Denton Reservoir.
- 336. —. (Geology).
- 337. 14. Ancaster-Wilsford.
- 338. 13. Stapleford Moor.
- 339. 12. Freiston Shore.
- 340. 10. Woodhall Spa.
- 341. 5. Thonock.

1965
- 342. 10. Tattershall.
- 343. 11. Welton Wood.
- 344. 15. Syston Park.
- 345. 9. Saltfleet.
- 346. 7. Wragby.
- 347. 13. Leadenham.
- 348. 15. Morkery Wood.
- 349. 11. Gibraltar Point.

1966
- 350. 13. Hartsholme.
- 351. 7. Hatton.
- 352. 7. Wickenby.
- 353. —. (Rutland).
- 354. 15. Woolsthorpe.
- 355. 3. Elsham-Wrawby Moor.
- 356. 2. Burton Stather.
- 357. 5. Scotter.
- 358. —. (Yorks.).
- 359. 4. Roxton.
- 360. 14—15. Ancaster.

1967
- 361. 9. Tetney Blow Wells.
- 362. 8. Burwell.
- 363. 2. Risby Warren.
- 364. —. (Northants).
- 365. 15. Sewstern Lane.
- 366. 10. Kirkby Moor.
- 367. 1. Epworth Turbary.
- 368. 13. Potterhanworth.
- 369. 11. Huttoft Bank.
- 370. 7. Willingham Forest.
- 371. 8. Benniworth Haven.

1968
- 372. 2. Broughton Far Wood.
- 373. 16. Baston Fen.
- 374. 14. Dembleby Thorns.
- 375. 11. Claxby Mill Hill.
- 376. 7. Scotgrove Wood.
- 377. 5. Kirton Lindsey.
- 378. 1. Haxey Turbary and Bird's Wood.
- 379. 7. Sotby.
- 380. 8. Skidbrooke-North Somercotes.
- 381. 11. Gibraltar Point.
- 382. 15. Stoke Rochford.

THE LINCOLNSHIRE NATURALISTS' UNION

1969
383. **5.** Laughton.
384. **2.** Alkborough.
385. **8.** Burwell.
386. **15.** Twyford Forest.
387. **14.** Ancaster-Wilsford.
388. **1.** Crowle Waste.
389. **3.** Barrow on Humber.
390. **7.** Goltho, Great West Wood.
391. **4.** Swallow Vale.

1970
392. **7.** Newball Wood.
393. **16.** Dobbin Wood.
394. —. (Yorks.).
395. **7.** Osgodby Moor.
396. **12.** Benington Marsh.
397. **3.** Goxhill Marsh.
398. **11.** Gibraltar Point.
399. **11.** Welton Wood.

1971
400. **11.** Welton Woods.
401. **1.** Crowle, Dirtness.
402. **7.** Sotby Green Lane.
403. **7.** Claxby Wood.
404. **17.** Frampton Marsh.
405. **7.** Panton Lodge.
406. **11.** Gibraltar Point.
407. **3.** Brocklesby, Mausoleum Woods.

1972
408. **2.** Scunthorpe, Sweeting Thorns.
409. **3.** Hendale Wood.
410. **10.** Kirkby Moor.
411. **8.** Wykeham Hall.
412. **10.** Snipe Dales.
413. **7.** Scotgrove Wood.
414. —. (Geology).
415. **17.** Surfleet-Gosberton.
416. **16.** Dunsby Woods.

1973
417. **16.** Aslackby, Temple Wood
418. **2.** Brumby Common.
419. **7.** Hardy Gang Wood.
420. **8.** Donington on Bain.
421. **5.** Laughton Common.
422. **16.** Dunsby Woods.
423. —. (Yorks.).
424. **14.** Aswarby Thorns.
425. **9.** Horse Shoe Point.
426. **7.** Scotgrove Wood.

BIBLIOGRAPHY

Abbreviation	Full Title
B.E.C. Rep.	Botanical Society and Exchange Club of the British Isles Reports.
J.B.	Journal of Botany
L.T.N.C. News	Lincolnshire Trust for Nature Conservation Newsletter
Nat.	The Naturalist
Phyt.	The Phytologist
Proc. B.S.B.I.	Proceedings of Botanical Society of the British Isles
Proc. L.S.	Proceedings of Linnean Society
Sci. Gossip	Science Gossip
Trans. L.N.U.	Transactions of the Lincolnshire Naturalists' Union
Wats.	Watsonia

ALFORD NATURALISTS' SOCIETY : Reports 1885. *Field Club* **1,** 1890, **11,** 1891.

ALLEN, D. : John Martyn's Bot. Soc. Bibliog. Anal. (re Vincent Bacon) *Proc. B.S.B.I.* **6,** Pt. 4, 310, 1967.

ALLEN, T. (pub. SAUNDERS) : History of the County of Lincoln (Leeds 1830), 62-3, list 63 plants from Turner, Dillwyn and Weir; 1833-1834 (83 plants), **I,** 62, **II,** 351; 1838, **I,** 118.

ALSTON, F. S. : Annotated copy of Haywards Pocket Book 1890-1920 in libr. Herb. Brit. Mus. (Nat. Hist.).
Some plants found near Woodhall Spa 1917-1919, *Trans. L.N.U.* **5,** 52-6, 1919. (See also Druce, G. C.).
The Ash, Pres. Add., *Trans. L.N.U.* **5,** 141-9, 1921.

ALSTON, A. H. G. : *Equisetum ramosissimum* Desp. as a British Plant, *Wats.* **1,** 149, 1949.

ANDERSON, A. H. L. : Notes from a Lincolnshire Garden, (London, 1903).

ANDERSON, M. L. A. : Bird notes from Lea, 1899. (*Calluna*) *Nat.* 1900, 157.

ANDERSON, Sir C. : A short guide to the County of Lincoln, 1847. (List of c. 100 sp. by Rev. J. K. Miller, Dr. T. P. J. Grantham and Mrs. Peel), 59-66.

ARNOLD, F. H. : *Lathyrus tuberosus, Sci. Gossip* **XXIV,** 244, 1888.

BAILEY, C. : On the Lincolnshire locality for *Selinum carvifolia, Manchester Lit. & Philos. Soc. Proc.*, **XXII,** 1883.

BAKER, F. T. : Additions and Corrections to the Comital Flora (Lincs.), *B.E.C. Rep.*, **X.,** 564, 1934.
Botanical and other Reports *Trans. L.N.U.*, **8-9,** 1933-36.
Note *Viola stagnina Trans. L.N.U.*, **9,** 109, 1936.

BAKER, W. F. : Formation of a Lincolnshire Naturalists' Union, *Nat.*, 1893, 255.

BIBLIOGRAPHY

BALL, M. E. : Lincs. Meadows and the Green-Winged Orchid, *L.T.N.C. News*, Nov. 1965, 3.
Lime Woodlands in Lincolnshire, *L.T.N.C. News.*, Mar. 1967, 4-7.
Alder Carrs of the South Wolds, *L.T.N.C. News.* Sept. 1967, 8-10.

BALL, P. W. and TUTIN, T. G. : Notes on annual species of *Salicornia* in Britain, *Proc. B.S.B.I.*, **4**, 193, 1959.

BANKS, J. : (d. 1820) MS. Notes in Hudson's Flora Anglica I. in libr. Brit. Mus. (488 e 21).

BARNES, F. A. and KING, C. A. M. : A preliminary survey at Gibraltar Point, Lincs., *Gibraltar Pt. Bird Obs. & Field Research Station Rep.*, 1951, 41-59.
Salt marsh development at Gibraltar Point, Lincs., *East Midland Geographer*, **2**, 20-31, 1961.

BAYLIS, E. : A New and Complete Body of Practical Botanic Physic, 1791.

BAYLEY, R. S. : Notitiae Ludae or Notices of Louth, 1834, 281.

BEEBY, W. H. : On the Flora of South Lincolnshire, *J.B.*, **XXII**, 17. 1884.

BENNETT, A. : in *J.B.*, **XXII**, 301, 1884 (*Zostera nana*); **XXIII**, 50, 1885 (new British and Irish Carices); **XXXVII**, 244, 326, 359, 1889 (*Selinum*); **XLVII**, 432, 1909, (*Sonchus palustris*); **XLI**, 1903 and *B.E.C.*, **II**, 1904 (*Nardurus*).
in *Nat.*, 1921, 112 (*Potamogeton berchtoldii*); 1922, 197 (*Limonium* and *Atriplex*).
in *Norfolk and Norwich Nats. Soc.*, **VI**, 459, 1899, (*Senecio paludosus* and *S. palustris* in East Anglia); **VIII**, 234, 1906 (*Limonium bellidifolium*).

BERKELEY, M. J.: *Allium carinatum*, a new British Phanerogam, *Gardener's Chron.*, 1867, 973. *J.B.*, **V**, 1867, 314.

BLACKWOOD, J. W. : Lincs. Grasslands, *L.T.N.C. News.*, Oct., 1968, 4-8.
Sea buckthorn Survey in Lincolnshire, *L.T.N.C. News.*, Apr. 1969, 12-13.
Lincolnshire Woodland Survey, *L.T.N.C. News.*, Sept. 1971, 9-12.
The Distribution of Nature Reserves in the Natural Regions of Lincolnshire. *Trans. L.N.U.*, **XVIII**, 1-6, 1972.

BLACKSTONE, J. : Specimen Botanicum quo Plantarum Angliae, 1746.

BLAIR, P. : *Pharmaco-botanologia* (including Dr. Stukeley's list of rare Boston plants), 1723, 8-28, 48; 1725, 121, 123; 1727, 144, 149, 185, 195.

BOGG, T. W. : Marked L.C., 1858 in libr. Kew Herb.

BOSTON NAT. HIST. SOC. REPORTS : 1949-1953, Flora of Boston Docks, 1953.

BOTANICAL LOCALITY RECORD CLUB : New Lincs. County Records, 1873-1883 (incl. many of F. A. Lees).

BOTANICAL SOCIETY OF THE BRITISH ISLES : Atlas of British Flora, ed. Perring, F. H. and Walters, S. M., 1962. (Records of E. J. Gibbons, etc.).
Year Book, 1951, 69: Report of Scunthorpe Meeting.

BRADFORD CITY LIBRARY : Catalogue of the Lees Botanical Collection in Ref. Lib.

BRADFORD NAT. SOC. : Rep. of Exhib. *Cakile maritima* and *Eryngium maritimum* at Skegness, *Nat.*, **VIII**, 29, 1882.

BRADLEY, C. R. S. : Some notes on *Iris spuria*. *Dorset Nat. Hist & Archaeol. Soc.*, **LXIV**, 118, 1943.

BRERETON, W. : Travels in Holland, England, 1634-5. *Cheltenham Soc. (Arch)*, **1**, 1844.

BREWSTER, H. C. : (d. 1916) History of South Kelsey. MS. Vol. XI in libr. Lincoln Cathedral.

BRITTEN, J. : Botany section in *White's Hist. Gaz. and Directy. of Lincs.*, 1872 (list of 773 sp.); 1875, 1882 (list of 864 sp.).
additional notes on the London Flora. *Bot. Chron.*, 1864, 57.
in *J.B.*, 1872 (*Gentiana pneumonanthe*, Nettleton); **XIII**, 276, 1875, (*Pyrola minor* in Lincs.); **LIII**, 310, 1915 (*Asperugo pro cumbens*).
Anemone appenina. Phyt. NS. **VI**, 511, 592, 1863.
Plants at Barnetby and Caistor. *Nat.* OS **1**, 84, 1864 (*Gentiana pneumonanthe*).

BRITTEN, J. and BOULGER, G. S. : A bibliographical index of deceased British and Irish botanists, 2nd ed. (London, 1931).

BROWN, J. : Midland phanerogamia. *Nat.*, **XVIII**, 325, 1943.

BUNTING, W. : *et al.* An outline study of Hatfield Chase (Thorne, 1969).

BURBIDGE, F. W. : British Epiphytic Orchids (on *Liparis looseli* nr. Lincoln), *Gardeners Chron.*, Feb. 1884, 144.

BURTON, F. M. : in *Nat.*, 1900, 63 (*Iris spuria*); 1899, 330 (*Galeopsis speciosa*).

BURGESS, J. T. : An account of some of the rarer plants in and around Spilsby. *Nat.* 1893, 325.

BUDDLE, A. : Methodus Nova (MS. Flora 1700-1708) Sloane MSS. 2970-2980 B.M.

CAMDEN, W. : Britannia (London), trans. E. Gibson (county list by Ray), 1695, 481; 1722, **II**, 574; 1753, **I**, 574; 1772, **I**, 434; trans. R. Gough (list of rare plants based on Ray, additions by Edward Forster and others), 1789, **II**, 282; 1806, **II**, 393 unaltered. Copy in Lindsey County Library annotated by Sir J. Banks.

CARRINGTON, J. T. : *Hippophae rhamnoides, Sci. Gossip* NS. **VIII**, 29, 1901.

CARRINGTON, B. : Notice of Plants, *Bot. Soc. Edinburgh*, 1849. *Bot. Gaz.* **I**, 323, 1849. (copy of this list in F. A. Lees MS. Outline Flora).

CARTER, C. S. : in *Nat.* 1901, 192 (*Lathrea squamaria* nr. Louth); 1904, 216, 363 (Lincs. plant notes); 1905, 216; 1910, 347 (*Mimulus langsdorffii*); 1918, 28 (*Nierembergia gracilis* and *Sisymbrium altissumum* at Theddlethorpe sandhills); 1923, 285 (*Claytonia perfoliata* and *Crepis taraxacifolia*); 1927, 245 (*Ceterach officinarum* and *Arrhenatherum tuberosum* in Lincs.)

CHANDLER, J. H. : Uffington Gravel Pits: an ecological study, Pres. Add. *Trans. L.N.U.* **16**, 203-215, 1966.

CHESTERS, C. G. C. : A preliminary account of the vegetation of the foreshore from Seacroft to Gibraltar Point, *Gibraltar Pt. Bird Obs. and Field Study Centre Rep.*, 1950, 48-57.

CLAPHAM, TUTIN and WARBURG : Flora of the British Isles (Cambridge, 1952 and 1962).

CLARKE, J. A. : Fen Sketches, 1852, 158-9, (orig. in Wisbech Advertiser).

BIBLIOGRAPHY

CLAYE, A. N. : *Vaccinium oxycoccus* and *V. myrtilis* in Lincs., *J.B.* **LV**, 257, 1917; *Selbourne Mag.*, **XXIX**, 24, 1918.
Some Ecological Features and Problems in Plant life. Pres. Add. *Trans. L.N.U.*, **5**, 169-175, 1922.

CORDEAUX, J. : Lincolnshire Natural History, Pres. Add., *Trans. L.N.U.*, 1895, 1-12; also second Pres. Add., *Trans. L.N.U.*, **1**, 132-144, 1907 (also *Lincs. Notes & Queries*, **V**, 15-26, 1896.)
Lincs. Agriculture one hundred years ago, *Nat.*, 1895, 317, (plant list, A. Young).
et al. Recent Notes from Lincs. *Nat.*, 1898, 261.

COX, S. A. : Wild life of a Lincs. Bog. *Countryside* NS., **VIII**, 459, 1930.
Monotropa hypopitys; Trans. L.N.U., **8**, 214-6, 1934.

COX, T. : Magna Britannia II, 1720, 1457 (ex J. Ray).

CRAGG, J. : MS. with refs. to Hill's Herbal, Meyrick's Herbal & Sowerby's Botany, 1790 *ex* libr. W. A. Cragg.

CRAGG, W. A. : A History of Threekingham with Stow, 1913.

CROW, B. : *Anchusa officinalis* near Louth, *Nat.*, 1900, 324.

CULPEPPER, N. : The English Physician, 1655.

DALLMAN, A. A. : *Senecio squalidus* in Yorks. and Lincs. *N.W. Nat.*, **XII**, 61, 1937.

DANDY, J. E. : List of British Vascular Plants (London 1958).
Nomenclatural changes in the list of Vascular Plants, *Wats.*, **7**, 157-178, 1969.

DAVY, J. BURTT : Review of F. A. Lee's Outline Flora in White's Directory, 1892, *J.B.*, **XXXI**, 1893, 123-4.
Asplenium adiantum nigrum, Field Club, **1**, 47, 1890 (Alford Naturalists' Soc.).
in *Nat.*, 1890, 116 (plant notes, Mablethorpe); 1891, 65 (notes of Rev. W. Fowler Limestone plants, Alford plants); 1892, 41 (Alford plants); 48 (Gibraltar Point plants); 300, (*Viburnum lantana* see also *J.B.*, **XXX**, 281, 1892).

DE ASTON SCHOOL NATURE CLUB REPORTS : 1909, 1910, 1936, 1943.

DERHAM, W. : Philosophical letters between John Ray and Dr. Lister, 1718 (Burwell notes, 18).
Select remains of the learned John Ray, 1760 (ed. G. Scott) (contains Ray's Itinerary, 1661, Spalding, Tattershall, etc., 137).

DILLENIUS, J. J. : Joannis Raii Synopsis Methodica, 1724, (3rd ed.).

DODSWORTH, J. : MS. notes and annotated copy of Smith's English Flora (c. 1838) in libr. Brit. Mus. (Nat. Hist.).

DRAYTON, M. : Polyolbion, 1662. (ed. R. Hooper, 1876), 148.

DRUCE, G. C. : Sibthorp's Lincolnshire Plants. *J.B.*, **XLVIII**, 257, 1910; **XLIX**, 66, 1911.
A Hybrid *Galeopsis J.B.*, **XLII**, 89, 1904.
Viola arenicola. Nat., 1929, 50.
Potamogeton coloratus J.B., **XLIX**, 1911, 276.
Teucrium scordium, 1876 (Cowbit) *B.E.C. Rep.*, 1911.
Plant Notes for 1917 and new County records (Woodhall Spa aliens), *B.E.C. Rep.*, **V**, 1917-1919.
The extinct and dubious plants of Britain/re *Senecio paludosus* and *S. palustris* in Lincs. Sir J. Banks records), *B.E.C. Rep.*, **V**, 733-4, 1919.
The Comital Flora, 1932.

DRURY, E. B. : A Sketch, Historical and Description of the Minster and Antiquities of Lincoln, 1831, 87 (appended plant list.)
DUGDALE, J. : A new British Traveller, 1819.
EDWARDS, E. : *Cnicus oleraceum* in Lincs. *Phyt.*, **II**, 115, 1845.
EDWARDS, L. : Survey of the Witham, 1769.
EDEES, E. S. : Brambles of Lincs., *Proc. B.S.B.I.*, **6**, Pt. 3, 209-214, 1966.
ELLER, I. : History of Belvoir Castle, 1841 (plant list from Nicholl's Hist. of Leics.)
ELLIOT, W. : *Allium carinatum* in Lincs., *Bot. Soc., Edinburgh*, **XI**, 224, 1871.
EVANS, A. H. : Flora of Fenland as compared with that of Bogs, Marshes and Mosses of Scotland, *Bot. Soc., Edinburgh*, **XXIV**, 164, 1911.
FISHER, H. : New County Records, *J.B.*, **XXXII**, 22, 53, 1894.
FOWLER, W. : MS. Notebook (transcript in possession of Miss E. J. Gibbons).
  in *Nat.* 1877-8, 129 (Lincs. coast plants); 1879, 149 (maritime); 1883, 54 (general); 1887, 349 (bog and moorland); 1888, 111 (marsh and water); 1889, 353 (sand and clay); 1890, 169 (limestone); 1891 (general); 1899, 362 (*Galeopsis speciosa*); 1900, 298 (*Schoenus nigricans*).
  New British Umbel — *Selinum*, *J.B.*, NS., **XI**, 1882.
  in *Phyt.* N.S. **II**, 303, 1857. (absence of *Digitalis purpurea*); 1858, 331, (rare plants Winterton); 416, (*Teucrium chamaedrys*); 597, (*Ophrys* and *Anacamptis*).
  Contributions to Botanical Locality Record Club Reports, 1875-1886.
GERARDE, JOHN : The Herball or General Historie of Plants, 1597.
GIBBONS, E. J. : Bot. Reports and Lincs. Records, *Trans. L.N.U.*, **9**, 1937 . . .
  Notes on the Lincs. Flora. Pres. Add. *Trans. L.N.U.*, **10**, 11-16, 1939.
  Lincs. records in B.S.B.I. Atlas of British Flora (1962).
GIBBONS, E. J. & LOUSLEY, J. E. : *Alisma gramineum* Lej. in Lincs. *Proc. B.S.B.I.*, **II**, 75, 1956. (see also Lousley, J. E., *ibid.*, in Britain, **II**, 346, 1957).
  An inland *Armeria* overlooked in Britain (part I), *Wats.* **IV**, 125, 1958; (part II — H. G. Baker, *Wats.*, **IV**, 136, 1958).
GIBSON, G. S. : Flora of Essex, 1862 (ref. to Lincs. *Lathyrus tuberosus*).
GILHAM, L. : *Cakile maritima* at Gainsborough. *Trans. L.N.U.*, **3**, 176, 1914.
GOULDING, R. W. : 'Louth Antiquarian and Naturalists' Society'. *Lincs. Notes and Queries*. **V**, 61-4. 1896 (also *Louth Advertiser*, July 6, 1895).
GRAY, A. J. : Townsend's Cord Grass. *Trans. L.N.U.*, **15**, 218-222, 1959.
GRIERSON, G. A. : Lessons from a limited area (ref. Freshney Bog) Pres. Add. *Trans. L.N.U.*, **6**, 9-18, 1923.
GUTCH, MRS. & PEACOCK, M. : Lincs. County Folklore, 1908.
HAWKINS, JOHN : List of Plants from 'Grantham Journal' compiled in *Trans. L.N.U.*, **4**, 44-5, 1916 by ed.
HAYES, A. J. : Interim Report of pollen analysis from Peat at Dragonby in 5th Prelim. Rep. 1968, May J., Univ. Nottm.
HEARNE, T.: (ed.) Dr. Plots account of an Intended Journey — The Itinerary of John Leland (re *Carum carvi*) 1710; **II**, 134, 1744; **II**, 169, 1769.

HEATHCOTE, J. M. : Reminiscences of Fen and Mere, 1876, (59, botany by T. Rooper).

HOW, WILLIAM : Phytologia Brittanica, 1650.

HOOKER, W. J. : *Viola lactea, Ann. Nat. Hist.*, **II**, 383, 1839.

HOOPER, M. D. : Lincolnshire Hedges, *L.T.N.C. News.* Mar. 1968, 8-10.
(see also symposium report, Hedges & Hedgerow Trees, Monks Wood Experimental Station, Nature Conservancy, 1968).

HOWARTH, E. : (ed.) List of plants collected chiefly in neighb. of Sheffield by J. Salt, 1889.

HUBBARD, C. E. : *Nardurus maritima. Proc. L.S.*, **CXLVIII**, 109, 1936.

HUDSON, WILLIAM : Flora Anglica (2 vol.) 1762, 1778.

HURST, S. J. : Aliens at Boston Dock. *B.E.C.*, **III**, 402, 1914.

IRVINE, ALEX : The London Flora, 1838 (incl. Lincs. plants)

JARVIS, A. E. : Phenological Observations, *Midland Nat.*, **III**, 176, 1880.
Bog Pimpernel and Greater Spearwort at Woodhall Spa, *Nat.* 1900, 155.
A Short Account of a Country Parish (Hatton), *Lincs. Notes & Queries*, **V**, 82, 1897.

JOHNSON, THOS. : The Herball very much enlarged and amended, 1633, 1636.

JOHNSON, T. : Mercuri botanici pas altera, 1641, 15-36.
Opuscula omnia botanica Thomae Johnsoni (ed. T. S. Ralph), 1847.

JUKES-BROWN, A. J. AND PEACOCK, E. A. W. : Sketch map of soils and Natural History Divisions of Lincs., *Nat.* 1895, 289-301 (also *Lincs. Notes and Queries*, **V**, 2-15, 1896).

KEW, H. WALLIS : Plant Notes and Ecology in *Nat. World*, **II**, 1885; **III**, 1886; in *Nat.* 1886; in *Sci. Gossip*, **XXII**, 207, 1886 (Botanical notes); **XXIII**, 31, 1887 (Burwell Wood list); NS **IV**, 59, 69, 1897 (Botanical notes, rare plants Scotter, etc.).
Martin Lister and Lincolnshire Natural History, Pres. Add. *Trans. L.N.U.*, **7**, 1-16, 1927.
Ray's Journey through Lincs., 1661. *Trans. L.N.U.*, **7**, 169-174, 1930, and *B.E.C. Rep.*, **IX**, 611, 1932.
Botany in Roscoe's *Guide to Mablethorpe*, 1890, 67 (MS. extract in Lincoln Mus.)
Reminiscences of Ten Summers, 1888, 93.

KEW, H. W. AND POWELL, H. E. : Thomas Johnson — Botanist and Royalist, 1932, 20 (ref. *Gentiana pneumonanthe*).

KIRK, F. L. : Outline Study of Nat. Hist. of Roughton Moor. *Trans. L.N.U.*, **13**, 175, 1954.

LAMBERT, M. R. AND WALKER, R. : Boston, Tattershall and Croyland, 1930, 10.

LARDER, J. : *Rosa spinosissima* and *R. villosa. Nat.*, 1893, 264.
Lincs. Coast, *Nat.*, 1887.

LEES, F. A. : Botany and Outline Flora of Lincs. in White's Directory, 1892.
Botany and Outline Flora of Lincs., 1893, unpublished; authors interleaved annotated copy of above with Additions to Outline Flora, etc. on loose sheets, in Bot. Sch. Cambr. with all Lees MS.
Marked L.C., **VII**, 1877; L.C., **IX**, 1895 in Bot. Sch. Cambr.
Lincs. Yorks. and Notts. Notes, Notebook V in Bot. Sch. Cambr.
MS. unfinished. (A Year's Contribution to Lincs. Botany, 1878 (Market Rasen and 7 mile radius) in Bot. Sch. Cambr.
Description of the Lees Herbarium of Library, Bradford, 1910. (Catalogue of the Lees Botanical Collection in Ref. Libr., Bradford Public Libr., 1909).
Recorder for Botanical Locality Record Club; *Reports*, 1873-1886 containing c. 150 new Lincs. records.
Selinum carvifolia, *Rep. Bot. Rec. Club*, **2**, 216, 1882.
On a new British Umbellifer, *J.B.*, **XX**, 129, 1882.
The North Lincs. *Lycopodium*, *J.B.*, **XXI**, 84, 1883. The Cambridge and Lincoln *Selinum*, *J.B.*, **XXXVII**, 326, 1899.
Trifolium filiforme, *Nat.*, 1897, 172.
Pyrola minor and *Cirsium dissectum*, *Nat.*, NS., **III**, 4, 1877-8.
Aceras at Gate Burton, *Sci. Gossip*, **VI**, 189, 1870.
The Volteface of Flora — A Rejoinder, *Nat.*, 1900, 229-36.

LEES, P. FOX : Additions to the N. Lincs. Flora, *Nat.*, 1892, 76.

LEY, A. : Some Lincolnshire *Rubi*, *J.B.*, **XLVI**, 53, 1908.

LIGHTFOOT, J. AND HILL, J. : M.S. Notes in Ray's Synopsis III, in libr. Bot. Dept. Oxford.

LINCOLNSHIRE NATURALISTS' UNION : 1893 ... Reports in *The Naturalist*, 1893-1905, *Transactions L.N.U.*, 1895 and 1905 ... Botanical Records and Reports: A. Smith, 1905-11, E. A. Woodruffe-Peacock, 1911-1920, S. C. Stow, 1922-32. F. T. Baker, 1933-36. E. J. Gibbons, 1937 ... Presidential Addresses of ... see E. A. Woodruffe-Peacock, H. Wallis Kew, F. Norris, E. J. Gibbons, J. H. Chandler, I. Weston, A. E. Smith.

LINCOLNSHIRE NOTES AND QUERIES : Reports in Nat. Hist. Sect., 1894-97. (1896, Lincs. Rye Grass). Series on Lincs. Folk Names for Plants, 1894-7.

LINCOLNSHIRE TRUST FOR NATURE CONSERVATION : Nature Reserves Investigation Sub-Cttee of L.N.U., 1943-48. Became Lincs. Naturalists' Trust, 1948, L.T.N.C. in 1964. Reports of and Newsletters. ed. A. E. Smith, 1949 ... Nature Reserves Handbook, 1972.

LISTER, MARTIN : A Journey to Paris in the Year 1698, (London, 1699).

LOUTH NATURALIST SOCIETY : 1884; Ant. and Nat. Soc., 1885, Nat. Ant. and Lit. Soc., 1910 — Reports of.

LOWE, J. : On the Flora of Sleaford and Neighbourhood, *Proc. B. S. Edin.*, 1856, 13, (copied into Lees, MS. Outline Flora).
Silene dichotoma, *Sci. Gossip*, **1**, *Anagallis coerulea*, **3**.

MALMESBURY, WILLIAM OF : Historia Novellae (1200). English trans., 1596, Sir Henry Saville. Account of Lincoln Fens and Hop Vine Cultivation as quoted by P. Thompson.

MARTYN, J. : Rare Plants on a Journey to the Peak, *Philosophical Trans.*, **XXXVI**, 22, 28, 1731; abr., **VI**, 333, 1733; abr., **IV, 2**, 333, 1734,

MARTYN, T. : Plantae Cantabrigiensis (other lists or rarities), 1763, 63.

MASON, J. E. : in *Nat.*, 1888, 102 (*Polystichum setiferum*), 284 (*Cirsium acaulon*) *Pinguicula vulgaris*, *Nat. Journ. & Guide*, 1897, 157.

MASON, W. W. : 6 Vol. MS. in Lincoln City Libr. and Mus.
1 Vol. annotated check list of records with additions by F. T. Baker, 1930-6 and E. J. Gibbons, 1936-70, in hand E.J.G. as L.N.U. Botanical Sec.
Scrophularia umbrosa, *Nat.*, 1908, 425.
Viola arenicola, *Trans. L.N.U.*, **7**, 71, 1929.

MELVILL, J. C.: Addenda to flowering plants of Woodhall Spa, *Nat.*, 1900, 323-324.

MERCATOR : Atlas, 1638 (Account of Fens as quoted by P. Thompson).

MERRETT, C. : Pinax Rerum Naturalium Britannicum, 1666-1667.
Account of Several Observables in Lincs., *Philosophical Trans.*, **XIX**, 350, 1696, abr., **III**, 533, 1705.

MIEGE, G. : Present State of Gt. Britain and Ireland, 1707, 1711, 1715, 1718, 1725, 1731 and 1738 . . . enlarged by S. Bolton (re *Myrica gale*, Axholme).

MILLER, J. K. : Walkeringham Plants, *Nat.*, 1895, 159.
Plant list in Anderson's Short Guide to County of Lincoln, 1847, 59.

MILLER, PHILIP : The Gardeners' Directory, 1748.

MILLER, S. H. : Handbook to the Fenland, 1889, 73 (botany ex Babington); 1890, 175.

MILLER, S. H. AND SKERTCHLY, S. B. J. : Fenland Past and Present (Wisbech 1878) (botany W. Marshall); Review *Nature*, **XVIII**, 514, 1878.

MOLL, H. : A new description of England and Wales, 1724, 194.

MURRAY, J. : Handbook for Lincolnshire, 1890, 1903 (botany by E. A. Woodruffe-Peacock).

NICHOLLS, J. I. : The list and antiquities of County of Leicester, 1795 (botany by Pulteney, R. and Crabbe, G.)

NICHOLSON, JOHN : in *Annals of Nat. Hist.* Series 1, **II**, 1839 (*Viola stagnina* and *Vicia hybrida*).

NEWMAN, L. F. AND WALWORTH, G. : A preliminary note on the ecology of part of the S. Lincs. coast, *J. Ecol.*, **VII**, 204-210, 1919.

NORRIS, F. : Scotton Common Nature Reserve, Pres. Add., *Trans. L.N.U.*, **16**, 85-9, 1965.

OLDFIELD, EDMUND : Topographical and Historical Account of Wainfleet (London, 1829), 180, 310 (good short plant list).

PARKINSON, JOHN : Theatrum Botanicum. Theatre of Plants, 1640.

PEACOCK, E. A. WOODRUFFE- :

MAJOR WORKS

Critical Catalogue of Lincolnshire Flowering Plants: From all known sources in *Nat.*, 1894-1897 and 1900 (Series of 14 papers).
A species Flora of Lincs., Victoria County Hist. Lincs., MS. in libr. Brit. Mus.
MS. Flora of Lincs. in Camb. Univ. Herb. Libr.
Old MSS. and letters to F. A. Lees, 1892-1921 (with all Lees corresp.) in libr. Bot. Sch. Cambr.
The Natural History Divisions of Lincolnshire, *Nat.*, 1895, 289-301.
Check List of Lincs. Plants, *Trans. L.N.U.*, **2**, 1-66, 1909. (See also *J.B.*, **XLVIII**, 1910).
Check List. Additions, *Trans. L.N.U.*, **2**, 118, 147, 1909.
Additions and Corrections to the Check List of Lincs. Plants, *Trans. L.N.U.*, **2**, 290-299, 1911.
Botany in Murray's Handbook for Lincs., 1903.

RECORDS AND ARTICLES

The Naturalist

Lincolnshire Plant Records, 1895-1897, 1900-1906, 1908.
L.N.U. Records and Reports of Field Meetings, 1893-1905.
Old Lincolnshire Plant Records (1724-1726), 1898, 177-179.
Aliens, 1897, 226; 1898, 209, 219, 227, 306; 1900, 222, 299.
A Study in Seed Dispersal in Lincs., 1894, 19-23.
Rarer Plants of the Walkeringham Neighbourhood, 1895, 159-171.
Notes on County Folios, 1896.
Cowgrass and Ryegrass, 1897, 21-25.
In memorium, John Cordeaux, 1899, 280-284.
Fenland Soils, 1902, 177-188.
The Henbane, 1904, 120-121.
Plants enlarging their area, 1905, 332.
The Lincolnshire Oxlip, 1905, 203-205.
Floral Competition and Cycles, 1906, 414-419.
Ecological Botany of Thorne Waste, 1907, 320-322.
Water carried species, 1908, 288.
The Rock Soil Method and *Ballota nigra* Linn. in Lincs., 1909, 39-44.
Ecology of Thorne Waste, 1920, 301-304, 353-356, 381-384; 1921, 21-25.
Book Review London Catalogue 9th Edit., 1895, 201-202.

Transactions L.N.U.

L.N.U. Field Notes and Records, **1-5**, 1905-1920.
L.N.U. Reports of Field Meetings, **2-5**, 1911-1920.
L.N.U. Botanical Reports, **2-5**, 1911-1920.
Natural Habitats and Nativeness (Pres. Add.), **1**, 92-100, 1906.
Broughton Woods, **1**, 168-173, 1907.
Our Dry Soil Pimpernels, **3**, 110-114, 1913.
The East Fen (incl. flora of), **3**, 228-236, 1915.
The Flora of Lincs.: Sequence Selection (Pres. Add.), **4**, 22-40, 1916.
Vaccinium myrtillus (Botanical Rep.), **4**, 69, 1917.

BIBLIOGRAPHY

PEACOCK, E. A. WOODRUFFE—*cont.*

Seed Dispersal, **5,** 14-37, 1919.
Lincs. Nat. Hist. Notes from Stonehouse's, Isle of Axholme, **5,** 60-62, 1919.

Journal of Botany

Plant Records and Notes, **XXXIV,** 1896; **XXXVI,** 1898; **XL,** 1902; **XLII,** 1904; **XLVI-L,** 1908-1912; **LV,** 1917; **LVII,** 1919.
Notes on the Flora of Lincolnshire, **XXXVI,** 55-60, 1898.
Iris spuria Linn. in Lincolnshire, **XL,** 101-102, 1902.
Lincolnshire Plant Notes, **XLII,** 50-51, 1904.
Primula elatior Jacq. in S. Lincs. **XLIV,** 242-243, 1906.
Natives and Aliens, **XLVI,** 340-346, 1908.
Two Lincolnshire Plants, **XLVI,** 359-360, 1908.
Weather and Plant Distribution, **XLVII,** 29-30, 1909.
Review of Check List, **XLVIII,** 166, 1910.
Flowers and Insects, **XLIX,** 164-167, 1911.
The Shepherd's Purse and Cultivation, **L,** 23-26, 1912.
Malva sylvestris L., **L,** 92-94, 1912.
Change of Climate and Woodland Succession, **L,** 247-253, 1912.
Poppy Hybrids, **LI,** 48-50, 1913.
Index Species in a Flora, **LII,** 124-127, 1914.
Rumex maritimus L., **LIII,** 363-364, 1915.
Sieglingia decumbeus in Lincolnshire, **LIV,** 359-360, 1916.
Juncus gerardi in Lincs., **LV,** 333-334, 1917.

Other Sources

Woodland Succession, *Quart. Journ. of Forestry,* **10,** 193-199, 253-263, 1916.
The Means of Plant Dispersal, *Selbourne Mag.,* **28,** 1917; **29,** 1918.
Eyes and Ears, *Selbourne Mag.,* **27,** 1916.
Lincolnshire Folk Names for Plants, *Lincs. Notes and Queries,* April, 1894-April 1897.
ed. Nat. Hist. Lincs. (originally Nat. Hist. Sect.) *Lincs. Notes and Queries,* 1894-1897.
The Lincolnshire Rye-grass, *Lincs. Notes and Queries,* **30,** 1896.
Notes on Aliens, *Sci. Gossip.* NS., **IV,** 59, 1897.
How to Make Notes for a Rock-Soil Flora, *Rural Studies Series Louth,* **5,** 1904. (review by W. Fowler *J.B.,* **XLII,** 313, 1904).
Frequency in Flora Analysis, *Rural Studies Series Louth,* **15,** 1912.
A Fox-Covert Study, *Journ. of Ecology,* **6,** 110, 1918.

BIOGRAPHICAL

Appreciation of by W. D. Roebuck, *Trans. L.N.U.,* **3,** 71-80, 1913.
Obituary of by R. W. Goulding, *Trans. L.N.U.,* **5,** 164-5, 1921.
Biographical and Bibliographical notes on by M. R. D. Seaward, *Lincs. History and Archaeology,* **6,** 113-124, 1971.*

(**The compilation of the above bibliography is based on this work, for the use of which I am grateful.*)

PEARSON, M. C.: in Nottingham and its Region (Brit. Assoc. Nottingham 1966). Vegetation, 110-124, The Vegetation of Gibraltar Point, 500-504.

PECK, W. : Topographical account of Isle of Axholme, 1815, 42. (Independent plant list).

PETERKEN, G. F. : Central Lincolnshire Woodlands *L.T.N.C. News.*, Sept. 1971, 12-15.

PLUKENET, LEONARD : marginal notes by, in Brit. Mus. copy of Ray's catalogue, c. 1685-8.
Almagestrum Botanicum (London, 1696).

PULTENEY, R. : An account of more rare English plants observed in Leicester, *Philosophical Trans.*, **XLIX**, 156, 1757.

PULTENEY, R. AND CRABBE, G. : Plant list in Nicholl's Leicester, 1895.

PRYME, ABRAHAM DE LA : Diary of 1870; ed. C. Jackson, *Surtees Soc.*, **LIV**, Extracts *Trans. L.N.U.*, **7**, 88, 1929.

RAY, JOHN : Catalogus Plantarum Angliae (London, 1670-1677.)
Historia Plantarum, **II**, 1688, **III**, 1704 (London).
ed. Synopsis Methodica Stirpium Britannicum (London, 1696).

RAVEN, C. E. : Thomas Lawson's notebook, *Proc. L.S.* **CLX**, 4, 1948.
Early English Naturalists from Neckham to Ray, 1947, 274, 300, 314.

RAWSON, R. W. : *Botrychium lunaria*, *Phyt.* NS. **1**, 188, 1855.

REYNOLDS, B. : Boston Dock aliens, *J.B.* **L**, 350, 1912. Lincs. Plants, *J.B.*, **XLVI**, 360, 1908; *J.B.*, **XLVIII**, 57, 1910.

RUSSELL, P. AND PRICE, O. : ed. England Displayed. (A Soc. of Gentlemen, 1769), 385.

DA SERRA, J. C. : An Account of the Submerged Forest on the Coast of Lincs., *Philosophical Trans.*, **LXXXIX**, 145, 1799.

SCIENCE GOSSIP (See Kew, H. W.) : Lincs. Plants: OS., **XXII**, 1886; OS., **XXIII**, 1887 (Burwell Wood); **XXIV**, OS. 1888, (244, *Lathyrus tuberosus*); NS. **IV**, 1897 (59, *Veronica montana*, rare plants; 69, Botanical notes).

SEAWARD, M. R. D. : Lincolnshire Natural History Recording Units — A Mathematical Approach, *Trans. L.N.U.*, **XVI**, 22-27, 1964.
The Ecology of Scunthorpe Heathlands with particular reference to Twigmoor Warren, *Journ. Scunth. Mus. Soc.* Series 2 [Natural History], No. 2, 1973.
F. A. Lees' Lincolnshire Moss Records, *Nat.* 1967, 77-80.
F. A. Lees' Botanical Collections (Part 2) *Nat.* 1968, 133-5.
Biographical and Bibliographical Notes on the Rev. E. A. Woodruffe-Peacock, 1858-1922, *Lincs. History and Archaeology*, **6**, 113-124, 1971.
Lincs. Polytrichales, *Nat.* July-Sept. 1972, 115.

SIMPSON, N. D. : A Bibliographical Index of the British Flora (Bournemouth, 1960), Very detailed Lincs. list, 182-187.

SMITH, A. : Notes and Reports in *Nat.*, 1901, 320; 1902, 303; 1903, 383; 1905, 281, 363 (Grimsby and other lists), *Trans. L.N.U.*, **7**, 70, 1929 and *Sci. Gossip*, 1897, 59.
Bot. Sec. reports in *Trans. L.N.U.*, **1-2**, 1905-1911.

SMITH, A. E. : ed. Lincs. Trust for Nature Conservation. Reports and Newsletters of, 1949 . . .
    Survey of Vegetation, *Gibraltar Pt. Bird Obs. and Field Study Centre Rep.*, 1949, 33-36.
    Nature Conservation in Lincs., Pres. Add. L.N.U., Lincoln, 1969.

SMITH, H. W. B. : *Sueda fructicosa*, *Trans. L.N.U.*, **7,** 71, 1928.

SMOLLET, T. : The Present State of all Nations, **II** (1768), 437.

SPENCER, N. : The Complete English Traveller, 1771, (486), 1773 (486).

STACE, C. A. : *Nardurus maritimus* (L.) Murb. in Britain, *Proc. B.S.B.I.*, **4,** 248, 1961.

STARK, A. : History of Lincoln, 1810.

STEERS, J. A. : The Coastline of England and Wales (Cambridge, 1946).

STEVENSON, N. S. : The Submerged Forest on Coast of Lincs., *Trans. L.N.U.*, **6,** 132-41, 1923.

STEWART, M. E. : *Viola persicifolia* Roth. (*Stagnina*) nr. Woodhall Spa, Lincs., *B.E.C. Rep.* 388, 1927.

STOW, S. C. : Botanical records and Field Meeting reports, *Trans. L.N.U.*, **5-8,** 1922-33.
    in *Nat.*, 1900, 241, 323; 1902, 115 (Woodhall plants); 1901, 229 (*Myosurus minimus*); 1902, 163 (Notes); 1902, 342 (*Juncus compressus*); 1903, 130 (*Sambucus ebulus*); 1904, 345 (*Alopecurus aequalis*).

STRAW, A. : Lincolnshire Soils, Lincs. Nat. Hist. Brochure 3, L.N.U., 1969.

STREATFEILD, G. S. : *Senecio integrifolius* and *Cineraria campestris. J.B.*, **XI,** 238, 1873.
    Lincolnshire and the Danes (London, 1884).

STRICKLAND, E. : MS. Flora Eboracensis, York City Libr., Acc. No. 146; records Appleby, 1822.

STUKELEY, W. : Itinerarium Curiosum 1724, 30; 1776, 32.

TAYLOR, J. M. AND SLEDGE, W. A. : *Potamogeton* x *cognatus*, *Nat.*, 1944, 121.

TEATHER, D. C. B. : Some factors limiting the distribution of *Halimione portulacoides* (L.) Aell var. *latifolia* (Gussone): submergence and soil type. *Trans. L.N.U.*, **16,** 90-8, 1965.

THOMPSON, PISHEY : Collections for A Topographical and Historical Account of Boston, etc. (list of 294 indig. plants in Skirbeck Hundred, T. A. Cammack, H. Gilson and T. Mathews). (London, 1820, 1856).

TILNEY-BASSETT, H. A. E. : Forestry Commission in Lincs. *L.T.N.C. News.*, Sept. 1971, 17-22.

TURNER, D. AND DILLWYN, L. W. : Botanists Guide through England and Wales, (London, 1805) (lists 74 notable Lincs. plants, 385-392).

TURNER, ROBERT : Botanologia; The British Physician, 1664.
    Botanologia; Fens (*Stratiotes* as *Sedum aquatile*), 1664, 1687, 83 and 1689 (see G. S. Boulger, *J.B.*, **XXXVIII,** 337, 1900).

WALCOTT, M. E. C. : The East Coast of England, 1861. (four lists 126, 138, 146, 148.)
    Marked copies of London Cats. (c. 1886) with N. and S. Lincs. lists (Lees collection, Bradford City Libr.).

WALLIS, T. W. : Autobiography, 1899 (Louth records, 1854).
WALTER, J. C. : Records of Woodhall Spa and Neighbourhood (Horncastle, 1899), 27-31.
WALTERS, S. M. : *Selinum carvifolia* in Britain, *Proc. B.S.B.I.*, **II,** 119, 1956.
WATSON, H. C. : The New Botanists Guide to the Localities of the Rarer Plants of Britain **1,** 1835, (Lincoln list, 271) **II,** 1837, (Godfrey Howitt's Frieston list of 1826, 651).
Compendium of the Cybele Britannia Part I, 1868, II 1869, III 1870 (supplement 1872).
MS. Notes on Lincs. and other counties; large register; annotated copy of J. Britten's List of White's Directory 1872, (all in libr. Kew Herb.)
Topographical Botany, Part I 1873, Part II 1874.
WAY, J. M. : Roadside Verges and the Conservation of Wildlife *L.T.N.C. News.* Mar. 1968, 4-8. (see also symposium report, Road Verges: their function and management, (London, 1969). (Monks Wood Experimental Station, Nature Conservancy.)
WEGMULLER, S. : A cyto-taxonounic study of *Lamiastrum galeobdolon* (L.) Ehrend and Potatschek in Britain, *Wats.* **8,** 271, 1971.
WEIR, GEORGE : Historical and Descriptive Sketches of the Town and Soke of Horncastle (London, 1820) (plant list by John Ward of Horncastle); author's MS. in Lincs. Libr.
WELLS, S. : History and Drainage of Great Level of Fens (Bedford Level) 1830, 421.
WELLS, T. C. E. AND BARLING, D. M. H. : *Pulsatilla vulgaris* Mill. *Journal of Ecology*, **59,** 275, 1971.
WESTON, I. : Observations on Grasses Growing in Lincs., Pres. Add. *Trans. L.N.U.*, **17,** 75-84, 1969.
WHEELER, W. H. : A History of the Fens of S. Lincs. (Boston, 1896), 484.
WHITE, WILLIAM : History, Gazetteer and Directory of Lincs. (Sheffield, 1842); and 1856: plant list identical to Allen and Saunders, 1830; 1872, 1882: plant list by J. Britten; 1892: Botany and Outline Flora of Lincs. by F. A. Lees.
WILMOTT, A. J. : Frieston and N. Shore of Wash, *B.E.C.*, **XI,** 432, 1938.
WILSON, P. J. : Benington Marsh 1970, *Trans. L.N.U.*, **17,** 219-221, 1971.
WITHERING, WILLIAM : An arrangement of British Plants, 1796. Annotated Copy of Withering's Botany, 1848 annotated by M. E. Dixon of Caistor, 1850-70, in libr. E. J. Gibbons.
WOLLASTON, T. V. : List of c. 30 plants observed in Lincs. (Gainsborough), *Phyt. OS.*, **I,** No. XXII, 522, 1843.
YOUNG, ARTHUR : A general view of the Agriculture of County of Lincoln (London, 1799), 232; 1813, 236, 263. (first good list of East Fen Flora when Sir J. Banks accomp. writer); reprint, Newton Abbot, 1970.
YERBURGH, RICHARD : Sketches Illustrative of Topography and History of Old and New Sleaford (Sleaford, 1825) (botany by J. Ray ex Camden).

INDEX

Latin and Common names of plants in the main text, and County Flora (page numbers from 80 onwards) except Appendices. Names in brackets are synonyms.

Aaron's Beard, 108
Aaron's Rod, 201
Abele, 184
Acaena, 149
  anserinifolia, 149
Acanthus, 210
  mollis, 210
Acer, 128
  campestre, 128
  platanoides, 128
  pseudoplatanus, 128
Aceras, 216
  anthropophorum, 4, 261
Achillea, 232
  millefolium, 232
  ptarmica, 40, 232
Acinos, 212
  arvensis, 4, 38, 212
Aconite, Winter, 89
Aconitum, 89
  (*anglicum*), 89
  napellus, 89
Acorus, 261
  calamus, 261
Adder's Tongue, 4, 40, 86
Adiantum, 82
  capillus-veneris, 82
Adonis, 92
  annua, 92
Adoxa, 223
  moschatellina, 1, 34, 223
Aegopodium, 170
  podagraria, 170
Aesculus, 129
  hippocastanum, 129
Aethusa, 172
  cynapium, 172
Agrimonia, 148
  eupatoria, 148
  odorata, 148
  (*procera*), 148
Agrimony, Common, 148
  Fragrant, 148
Agropyron, 279

  caninum, 35, 279
  junceiforme, 280
  pungens, 279
  repens, 279
Agrostemma, 112
  githago, 3, 112
Agrostis, 283
  canina, 283
  gigantea, 283
  stolonifera, 284
  tenuis, 37, 38, 283
Aira, 282
  caryophyllea, 282
  praecox, 282
Ajuga, 217
  reptans, 1, 35, 217
Alchemilla, 34, 148
  glabra, 148
  vestita, 148
  vulgaris, 53, 148
  xanthochlora, 148
Alder, 32, 33, 45, 182
  Grey, 182
Alder Buckthorn, 33, 130
Alexanders, 46, 168
Alfalfa, 131
Alisma, 244
  gramineum, 78, 244
  lanceolatum, 244
  plantago-aquatica, 244
Alison, Small, 101
  Sweet, 101
Alkanet, 197
Alliaria, 104
  petiolata, 104
Allium, 255
  carinatum, 255
  oleraceum, 255
  schoenoprasum, 255
  scorodoprasum, 255
  ursinum, 1, 35, 255
  vineale, 46, 255
All-seed, 124
  Four-leaved, 117

Alnus, 182
  glutinosa, 182
  incana, 182
Alopecurus, 285
  aequalis, 285
  bulbosus, 47, 285
  geniculatus, 285
  myosuroides, 285
  pratensis, 285
Althaea, 47, 123
  hirsuta, 124
  officinalis, 123
Alyssum, 101
  alyssoides, 101
Amaranthus, 118
  retroflexus, 118
Ambrosia, 225
  artemisiifolia, 225
Ammophila, 283
  arenaria, 37, 283
Amsinckia, 196
  lycopsioides, 196
Anacamptis, 261
  pyramidalis, 4, 5, 46, 216
Anagallis, 191
  arvensis, 191
  foemina, 191
  minima, 191
  tenella, 5, 38, 41, 191
Anaphalis, 230
  margaritacea, 230
Anchusa, 197
  officinalis, 197
Andromeda, Marsh, 186
Andromeda, 186
  polifolia, 4, 5, 186
Anemone, Blue, 89
  Wood, 1, 3, 32, 89
  Yellow Wood, 89
Anemone, 89
  apennina, 89
  nemorosa, 1, 3, 89
  ranunculoides, 89
Angelica, 173
  Wild, 173
Angelica, 173
  sylvestris, 173
  archangelica, 173
Antennaria, 230
  dioica, 230
Anthemis, 232
  arvensis, 232
  cotula, 232
  tinctoria, 232
Anthoxanthum, 285
  odoratum, 285
  puelii, 285
Anthriscus, 166
  caucalis, 166

  cerefolium, 166
  sylvestris, 166
Anthyllis, 135
  vulneraria, 4, 135
Antirrhinum, 203
  majus, 203
Apera, 284
  interrupta, 284
  spica-venti, 45, 284
Aphanes, 148
  arvensis, 148
  microcarpa, 37, 38, 39, 148
Apium, 168
  graveolens, 168
  inundatum, 169
  moorei, 169
  nodiflorum, 169
  repens, 169
Aquilegia, 93.
  vulgaris, 33, 93
Arabidopsis, 105
  thaliana, 105
Arabis, 103
  hirsuta, 103
  turrita, 103
Arcticum, 234
  lappa, 234
  minus, 234
  nemorosum, 234
Arenaria, 116
  leptoclados, 116
  serpyllifolia, 116
Aristolochia, 174
  clematitis, 174
Armeria, 189
  maritima, 189
Armoracia, 101
  rusticana, 101
Arnoseris, 237
  minima, 237
Arrhenatherum, 282
  elatius, 282
Arrow-grass, Marsh, 246
  Sea, 246
Arrowhead, 4, 245
Artemisia, 234
  absinthium, 234
  maritima, 234
  vulgaris, 234
Arum, 261
  maculatum, 261
Asarabacca, 174
Asarum, 174
  europaeum, 174
Ash, 29, 31, 32, 33, 45, 192
Asparagus, 251
  officinalis, 46, 53, 251
Aspen, 33, 55, 184

INDEX

Asperugo, 195
  procumbens, 195
Asperula, 220
  cynanchica, 47, 220
Asplenium, 50, 82
  adiantum-nigrum, 82
  ruta-muraria, 83
  trichomanes, 82
  viride, 83
Aster, 231
  novi-belgii, 231
  tripolium, 46, 231
Astragalus, 135
  danicus, 135
  glycyphyllos, 135
Athyrium, 83
  filix-feminia, 83
Atriplex, 120
  glabriuscula, 120
  hastata, 120
  laciniata, 120
  littoralis, 120
  patula, 120
Atropa, 200
  bella-donna, 200
Avena, 281
  fatua, 281
  ludoviciana, 281
  strigosa, 281
Avens, Mountain, 147
  Water, 1, 32, 34, 40, 147
  Wood, 32, 147
Azolla, 86
  filiculoides, 86

Baldellia, 244
  ranunculoides, 244
Ballota, 214
  nigra, 214
Balm of Gilead, 185
Balsam, Orange, 127
  Small, 128
Barbarea, 102
  intermedia, 102
  stricta, 102
  verna, 103
  vulgaris, 102
Barberry, 93
Barley, Mare's-tail, 280
  Meadow, 280
  Wall, 280
  Wood, 35, 281
Barren Strawberry, 146
Bartsia, Red, 208
  Yellow, 208
Base Horehound, 213
Basil Thyme, 4, 39, 40, 212
Beak-sedge, White, 266
Beardgrass, Annual, 284
Bear's Breech, 210

Bedstraw, Bog, 221
  Erect Hedge, 220
  Great Hedge, 220
  Heath, 37, 38, 220
  Lady's, 38, 220
  Marsh, 221
  Rough Corn, 221
  Slender, 221
  Wall, 221
Beech, 32, 183
Beet, 120
Bellbine, American, 199
Bellflower, Clustered, 5, 218
Bell Heather, 37, 187
Bellis, 231
  perennis, 231
Bent, Creeping, 284
  Dense Silky, 284
  Silky, 284
Bent-grass, Brown, 283
  Common, 38, 283
  Tall, 283
Berberis, 93
  vulgaris, 93
Bermuda-grass, 287
Berula, 171
  erecta, 171
Beta, 120
  vulgaris, 120
Betonica, 213
  officinalis, 213
Betony, Wood, 213
Betula, 182
  nana, 182
  pendula, 33, 182
  pubescens, 33, 182
Bidens, 225
  cernua, 225
  tripartita, 225
Bilberry, 37, 187
Bindweed, 199
  Larger, 199
  Pink, 199
  Sea, 46, 199
Birch, 2, 30, 32, 33, 36, 43, 45, 55, 182
  Downy, 182
  Dwarf, 182
Birdsfoot-trefoil, 135
  Greater, 135
  Slender, 135
Birdsfoot Vetch, 36, 136
Birthwort, 174
Bistort, 176
  Alpine, 176
  Amphibious, 176
Bitter-cress, Hairy, 102
  Large, 4, 102
  Narrow-leaved, 102
  Wavy, 102

Bittersweet, 201
Black Bindweed, 177
Black Bryony, 257
Black Dogwood, 130
Black Horehound, 214
Blackstonia, 193
 perfoliata, 4, 5, 193
Blackthorn, 155
Black Twitch, 285
Bladder-fern, Brittle, 83
Bladder-nut, 128
Bladderwort, Greater, 4, 210
 Intermediate, 210
 Lesser, 210
Blechnum, 82
 spicant, 50, 82
Blinks, 117
Bluebell, 1, 30, 32, 33, 34, 252
Blysmus, Broad, 265
 Narrow, 265
Blysmus, 265
 compressus, 51, 265
 rufus, 51, 265
Bog Asphodel, 2, 5, 37, 43, 56, 76, 250
Bogbean, 37, 194
Bog Myrtle, 2, 34, 37, 182
Bog Rosemary, 5, 186
Bog-rush, 41, 76, 266
Borage, 196
Borago, 196
 officinalis, 196
Boston Purslane, 121
Botrychium, 86
 lunaria, 86
Box, 129
Brachypodium, 279
 pinnatum, 1, 4, 37, 279
 sylvaticum, 35, 279
Bracken, 32, 33, 38, 50, 82
Bramble, 31, 140
 Stone, 34, 140
Brandy-bottle, 94
Brassica, 96
 napus, 96
 nigra, 97
 rapa, 96
Bristle-grass, Green, 287
 Whorled, 287
Briza, 277
 maxima, 277
 media, 4, 277
Brome, Barren, 277
 Compact, 277
 Hairy, 35, 277
 Meadow, 278
 Rye, 279
 Smooth, 278
 Tall, 35, 274
 Upright, 1, 277

Bromus, 277
 arvensis, 279
 commutatus, 278
 diandrus, 278
 erectus, 1, 4, 277
 ferronii, 278
 inermis, 277
 interruptus, 278
 lepidus, 278
 madritensis, 277
 mollis, 278
 racemosus, 278
 ramosus, 35, 277
 rigidus, 278
 secalinus, 279
 sterilis, 277
 tectorum, 278
 thominii, 278
 unioloides, 279
 (*wildenowii*), 279
Brooklime, 205
Brookweed, 41, 43, 192
Broom, 38, 131
Broomrape, Blue, 4, 209
 Greater, 38, 209
 Lesser, 209
 Red, 209
 Tall, 4, 5, 209
Bryonia, 174
 dioica, 174
Bryony, White, 174
Buckler-fern, Broad, 84
 Narrow, 84
Buckthorn, 129
 Sea, 47, 53, 161
Buckwheat, 177
Buddleja, 192
 davidii, 192
Bugle, 1, 32, 35, 217
Bugloss, 37, 38, 197
 Orange, 196
Bulrush, 2, 264
 Glaucous, 264
Bunias, 100
 orientalis, 100
Bupleurium, 168
 rotundifolium, 168
 tenuissimum, 46, 168
Burdock, Common, 234
 Great, 234
 Lesser, 234
Bur-Marigold, Nodding, 225
 Threecleft, 225
Burnet Saxifrage, 39, 170
 Greater, 170
Bur-parsley, Great, 167
 Small, 167
Bur-reed, 262
 Floating, 262

INDEX

Simple, 262
Small, 262
Bushgrass, 35, 283
Butcher's Broom, 251
Butomus, 245
  umbellatus, 4, 245
Butterbur, 228
  White, 228
Buttercup, Bulbous, 90
  Creeping, 47, 90
  Hairy, 43, 91
  Meadow, 90
  Small-flowered, 91
Butterwort, 5, 39, 41, 56, 76, 209
Buxus, 129
  sempervirens, 129

Cabbage, Hare's-ear, 98
Cakile, 98
  maritima, 54, 98
Calamagrostis, 283
  canescens, 35, 283
  epigejos, 35, 283
Calamint, 40, 212
Calamintha, 212
  ascendens, 212
Callitriche, 164
  intermedia, 164
  obtusangula, 164
  platycarpa, 164
  (*polymorpha*), 164
  stagnalis, 164
  truncata, 164
  (*verna*), 164
Calluna, 187
  vulgaris, 34, 36, 37, 38, 41, 187
Caltha, 88
  palustris, 1, 33, 88
  radicans, 88
Calystegia, 199
  pulchra, 199
  sepium, 199
  silvatica, 199
  soldanella, 46, 199
Camelina, 105
  microcarpa, 105
  sativa, 105
Campanula, Creeping, 218
  Giant, 1, 33, 34, 218
  Nettle-leaved, 34, 218
Campanula, 218
  glomerata, 5, 218
  latifolia, 1, 34, 218
  patula, 219
  rapunculoides, 218
  rapunculus, 219
  rotundifolia, 218
  trachelium, 34, 218
Campion, Bladder, 110
  Red, 1, 32, 33, 111

Sea, 46, 110
White, 111
Canadian Waterweed, 245
Canary Grass, 286
Candlestick, Red, 111
  White, 111
Candytuft, Wild, 99
Cannabis, 180
  sativa, 180
Capsella, 100
  bursa-pastoris, 100
  rubella, 100
Caraway, 170
  Corn, 169
Cardamine, 101
  amara, 4, 5, 57, 102
  bulbifera, 102
  flexuosa, 102
  hirsuta, 102
  impatiens, 102
  pratensis, 101
Cardaria, 99
  chalepensis, 99
  draba, 99
Carduus, 235
  acanthoides, 235
  nutans, 235
  tenuiflorus, 235
Carex, 266
  acuta, 269
  acutiformis, 268
  arenaria, 37, 38, 270
  binervis, 267
  caryophyllea, 51, 269
  curta, 51, 56, 271
  demissa, 267
  diandra, 270
  disticha, 270
  distans, 45, 267
  disticha, 270
  divisa, 270
  divulsa, 270
  echinata, 271
  elata, 269
  elongata, 271
  ericetorum, 51, 269
  extensa, 51, 267
  flacca, 269
  hirta, 269
  hostiana, 267
  laevigata, 35, 266
  lasiocarpa, 51, 269
  maritima, 270
  (*muricata*), 271
  nigra, 270
  otrubae, 270
  ovalis, 271
  pallescens, 35, 268
  panicea, 269
  paniculata, 270

pendula, 5, 35, 50, 268
pilulifera, 269
polyphylla, 271
pseudocyperus, 268
pulicaris, 51, 271
remota, 35, 271
riparia, 268
rostrata, 268
serotina, 267
spicata, 271
strigosa, 35, 50, 268
sylvatica, 35, 267
vesicaria, 268
Carlina, 234
vulgaris, 234
Carline Thistle, 234
Carnation Grass, 269
Carpinus, 183
betulus, 183
Carum, 170
carvi, 170
Castanea, 183
sativa, 183
Catabrosa, 176
aquatica, 276
Catapodium, 274
marinum, 275
rigidum, 274
Catchfly, Forked, 110
Italian, 111
Night-flowering, 111
Nottingham, 111
Small-flowered, 111
Striated, 110
Catmint, Wild, 215
Cat's Ear, 238
Smooth, 36, 39, 238
Spotted, 238
Cat's-foot, 230
Cat's-tail, Knotted, 284
Sand, 284
Caucalis, 167
latifolia, 167
platycarpos, 167
Celandine, Greater, 95
Lesser, 92
Celery, Wild, 43, 168
Centaurea, 236
axillaris, 237
calcitrapa, 237
cyanus, 3, 236
diluta, 237
iberica, 237
intybaca, 237
melitensis, 237
nemoralis, 1, 39, 236
nigra, 236
pallescens, 237
salmentica, 237
scabiosa, 1, 4, 39, 236
solstitialis, 237
spinosa, 237
Centaurium, 193
erythraea, 4, 193
littorale, 193
pulchellum, 193
Centaury, 4, 193
Slender, 193
Centranthus, 224
ruber, 224
Cerastium, 112
alpinum, 112
arvense, 39, 112
atrovirens, 113
(*diffusum*), 113
glomeratum, 113
holosteoides, 113
semidecandrum, 113
Ceratophyllum, 94
demersum, 94
submersum, 94
Ceterach, 83
officinarum, 83
Chaenorhinum, 203
minus, 203
Chaerophyllum, 166
temulentum, 166
Chaffweed, 191
Chamaemelum, 232
nobile, 232
Chamaenerion, 163
angustifolium, 163
Chamomile, 232
Corn, 232
Wild, 233
Yellow, 232
Charlock, 97
White, 97
Cheiranthus, 104
cheiri, 104
Chelidonium, 95
majus, 95
Chenopodium, 118
album, 118
bonus-henricus, 118
botryodes, 119
capitatum, 119
ficifolium, 119
glaucum, 119
hircinum, 118
hybridum, 119
murale, 119
polyspermum, 118
pratericola, 119
rubrum, 119
urbicum, 119
vulvaria, 118
Cherry, Bird, 155

INDEX

Wild, 155
Chervil, 166
  Bur, 166
  Rough, 166
Chestnut, Spanish, 183
  Sweet, 183
Chickweed, 113
  Greater, 114
  Jagged, 114
  Lesser, 114
  Upright, 114
  Water, 113
Chicory, 237
Chives, 255
Chrysanthemum, 233
  leucanthemum, 233
  maximum, 233
  parthenium, 233
  segetum, 233
  vulgare, 233
Chrysosplenium, 158
  alternifolium, 1, 159
  oppositifolium, 1, 34, 158
Cicendia, 193
  filiformis, 193
Cicerbita, 240
  macrophylla, 240
Cichorium, 237
  intybus, 137
Cicuta, 169
  virosa, 5, 169
Cinquefoil, Creeping, 147
  Hoary, 36, 38, 39, 146
  Marsh, 36, 37, 146
  Spring, 146
Circaea, 163
  lutetiana, 35, 163
Cirsium, 235
  acaulon, 236
  (*acaule*), 236
  arvense, 235
  dissectum, 40, 236
  eriophorum, 4, 235
  oleraceum, 235
  palustre, 235
  vulgare, 235
Cladium, 266
  mariscus, 5, 43, 45, 51, 266
Clary, Meadow, 212
  Wild, 213
Claytonia, Perfoliate, 117
Clematis, 46, 48, 90
Clematis, 90
  vitalba, 46, 90
Clinopodium, 212
  vulgare, 40, 212
Clover, Alsike, 134
  Crimson, 133
  Dutch, 134
  Red, 133
  Reversed, 134
  Sea, 47, 133
  Strawberry, 43, 134
  Sulphur, 133
  White, 134
  Zigzag, 40, 133
Clubmoss, Alpine, 80
  Common, 80
  Fir, 80
  Lesser, 80
  Marsh, 80
Club-rush, Sea, 43, 264
  Wood, 264
Cob-nut, 183
Cochlearia, 100
  anglica, 100
  danica, 100
  officinalis, 100
Cocklebur, Spiny, 225
Cock's-foot, 276
Coeloglossum, 258
  viride, 258
Colchicum, 252
  autumnale, 252
Cole, 96
Coltsfoot, 3, 227
Columbine, 33, 93
Comfrey, 195
  Rough, 196
  Tuberous, 196
Conium, 168
  maculatum, 168
Conopodium, 170
  majus, 170
Conringia, 98
  orientalis, 98
Convallaria, 250
  majalis, 34, 250
Convolvulus, 199
  arvensis, 199
Conyza, 231
  canadensis, 231
Coral-wort, 102
Cord-grass, 47, 286
Coriander, 168
Coriandrum, 168
  sativum, 168
Cornbine, 199
Corn Cockle, 3, 112
Cornflower, 3, 236
Cornish Moneywort, 205
Corn Marigold, 233
Corn Salad, 223
Corn Spurrey, 39, 116
Cornus, 165
  alba, 165
Coronilla, 136
  varia, 136

Coronopus, 99
　didymus, 99
　squamatus, 99
Corydalis, 96
　claviculata, 33, 96
　lutea, 96
Corylus, 183
　avellana, 183
Corynephorus, 283
　canescens, 47, 283
Cotoneaster, 155
　microphyllus, 155
Cotton-grass, 36, 37, 263
　Broad-leaved, 163
　Common, 263
Couch-grass, 279
　Bearded, 35, 279
　Sand, 280
　Sea, 279
Cowbane, 53, 169
Cowherb, 112
Cow Parsley, 166
Cowslip, 3, 5, 10, 190
Cow-wheat, 32
　Common, 34, 297
　Crested, 207
　Field, 207
Crab Apple, 43, 44, 156
Crambe, 98
　maritima, 54, 55, 98
Cranberry, 2, 43, 54, 55, 77, 187
Cranesbill, Bloody, 125
　Cut-leaved, 126
　Dove's-foot, 37, 126
　Dusky, 125
　Long-stalked, 125
　Meadow, 125
　Mountain, 125
　Round-leaved, 126
　Shining, 126
　Small-flowered, 126
Crassula, 158
　tillaea, 148
Craategus, 156
　monogyna, 59, 156
　oxyacanthoides, 33, 156
Creeping Jenny, 190
Creeping Soft-grass, 282
Crepis, 243
　biennis, 243
　capillaris, 243
　paludosa, 5, 243
　setosa, 213
　vesicaria, 243
Crested Dog's-tail, 276
Crested Hair-grass, 281
Crithmum, 171
　maritimum, 46, 171
Crocosmia, 256

　x *crocosmiflora,* 256
Crocus, Autumnal, 256
Crocus, 256
　nudiflorus, 256
Crosswort, 220
Crowberry, 2, 188
Crowfoot, Celery-leaved, 38, 91
　Corn, 3, 90
　Ivy-leaved, 91
　Water, 91, 92
Crow Garlic, 46, 255
Crown Vetch, 136
Cruciata, 220
　chersonensis, 220
　(*laevipes*), 220
Cuckoo Flower, 101
Cuckoo Pint, 261
Cudweed, 229
　Jersey, 229
　Marsh, 229
　Red-tipped, 229
　Slender, 38, 229
　Spathulate, 229
　Wood, 229
Currant, Black, 159
　Mountain, 159
　Red, 159
Cuscuta, 59, 199
　epilinum, 200
　epithymum, 200
　europaea, 199
　trifolii, 200
Cyclamen, 190
　hederifolium, 190
Cymbalaria, 204
　muralis, 204
Cynodon, 287
　dactylon, 287
Cynoglossum, 195
　officinale, 195
Cynosurus, 276
　cristatus, 276
　echinatus, 276
Cyperus, 266
　longus, 4, 266
Cystopteris, 83
　fragilis, 83

Dactylis, 276
　glomerata, 276
Dactylorchis, 260
　fuchsii, 260
　incarnata, 47, 260
　maculata, 260
　praetermissa, 47, 77, 260
Daffodil, 33, 256
Daisy, 231
Damasonium, 245
　alisma, 44, 245

INDEX

Dandelion, Common, 243
 Lesser, 244
Danewort, 221
Daphne, 161
 laureola, 4, 161
 mezereum, 161
Darnel, 274
Datura, 201
 stramonium, 201
Daucus, 173
 carota, 173
Dead-nettle, Cut-leaved, 214
 Red, 214
 Spotted, 215
 White, 214
Deer-grass, 263
Delphinium, 89
 ambiguum, 89
Deschampsia, 282
 cespitosa, 35, 282
 flexuosa, 36, 282
 setacea, 282
Descurainia, 105
 sophia, 105
Devil's-bit Scabious, 40, 41, 225
Dewberry, 140
Dianthus, 112
 armeria, 112
 deltoides, 112
Digitalis, 205
 purpurea, 205
Diplotaxis, 97
 muralis, 97
 tenuifolia, 97
Dipsacus, 224
 fullonum, 224
 pilosus, 1, 34, 224
Dock, Broad-leaved, 178
 Clustered, 179
 Curled, 178
 Fiddle, 178
 Golden, 179
 Great Water, 38, 178
 Marsh, 36, 179
 Red-veined, 178
Dockens, 178
Dodder, Clover, 200
 Common, 200
 Flax, 200
 Large, 199
Dogberry, 222
Dog Mercury, 30, 32, 33, 35, 174
Dogwood, 33, 165
Doronicum, 227
 pardalianches, 227
 plantagineum, 227
Dother, 39, 116
Douglas Fir, 86
Dropwort, 40, 140

Drosera, 160
 anglica, 2, 59, 160
 intermedia, 2, 56, 59, 160
 rotundifolia, 2, 37, 160
Dryas, 147
 octopetala, 147
Dryopteris, 83
 borreri, 37, 50, 84
 carthusiana, 50, 84
 dilatata, 50, 84
 filix-mas, 50, 83
 (*pseudo-mas*), 84
Duckweed, 262
 Gibbous, 262
 Great, 261
 Ivy, 261
Duke of Argyll's Tea-plant, 200
Dutch Rush, 80
Dyer's Greenweed, 40, 130
Dyer's Rocket, 106

Earthnut, 170
Echinochloa, 287
 crus-galli, 287
Echium, 198
 lycopsis, 198
 vulgare, 198
Eel-grass, 246
 Dwarf, 246
Elder, 222
Elecampane, 228
Eleocharis, 264
 acicularis, 264
 multicaulis, 265
 palustris, 265
 quinqueflora, 41, 76, 265
 uniglumis, 265
Elm, 30, 33, 180
 Cornish, 181
 East Anglian, 181
 Plot, 181
 Wych, 180
Elodea, 245
 canadensis, 245
 nutallii, 245
Elymus, 280
 arenarius, 47, 280
Empetrum, 54, 188
 nigrum, 188
Enchanter's Nightshade, 35, 163
Endymion, 252
 non-scriptus, 34, 252
Epilobium, 162, 163
 adenocaulon, 162 fi
 adnatum, 162
 hirsutum, 162
 montanum, 162
 obscurum, 162
 palustre, 162
 parviflorum, 162

roseum, 182
tetragonum, 162
Epipactis, 257
　helleborine, 35, 257
　palustris, 5, 47, 257
　purpurata, 257
Equisetum, 58, 80
　arvense, 81
　fluviatile, 81
　hyemale, 80
　palustre, 81
　ramosissimum, 81
　sylvaticum, 81
　telmateia, 1, 81
Eranthis, 89
　hyemalis, 89
Erica, 37, 38, 43, 187
　cinerea, 187
　tetralix, 187
Erigeron, 231
　acer, 231
Eriophorum, 263
　angustifolium, 51, 263
　latifolium, 263
　vaginatum, 45, 51, 263
Erodium, 126
　cicutarium, 127
　moschatum, 126
Erophila, 101
　praecox, 101
　verna, 101
Eryngium, 166
　maritimum, 46, 166
Erysimum, 104
　cheiranthoides, 104
Euonymus, 129
　europaeus, 129
Eupatorium, 231
　cannabinum, 41, 231
Euphorbia, 174
　amygdaloides, 34, 175
　cyparissias, 175
　exigua, 175
　helioscopia, 175
　lathyrus, 174
　peplus, 175
　platyphyllos, 175
　portlandica, 46, 175
　uralensis, 175
Euphrasia, 60, 208
　anglica, 38, 208
　nemorosa, 208
　officinalis, 208
　pseudokerneri, 208
Evening Primrose, 163
Ewe Bennet, 166
Eyebright, 208

Fagopyrum, 177

esculentum, 177
Fagus, 183
　sylvatica, 183
Falcaria, 170
　vulgaris, 170
False-brome, Heath, 279
　Slender, 279
Fat Hen, 118
Fen Sedge, 4, 45, 266
Fenugreek, Birdsfoot, 132
Fern Grass, 274
Fescue, Barren, 274
　Creeping, 273
　Mat-grass, 275
　Meadow, 273
　Rat's-tail, 274
　Sheep's, 273
　Tall, 273
Festuca, 273
　arundinacea, 273
　gigantea, 35, 273
　glauca, 173
　juncifolia, 273
　ovina, 37, 273
　pratensis, 273
　rubra, 273
　tenuifolia, 37, 273
　x *Festulolium*, 273
　loliaceum, 273
Feverfew, 233
Ficus, 181
　carica, 181
Fiddleneck, 196
Fiddles, 204
Field Madder, 219
Fig, 181
Figwort, 204
　Water, 204
　Yellow, 204
Filago, 229
　apiculata, 229
　germanica, 229
　(*lutescens*), 229
　minima, 38, 229
　(*pyramidata*), 229
　spathulata, 229
　(*vulgaris*), 229
Filipendula, 140
　ulmaria, 140
　vulgaris, 140
Fiorin Grass, 284
Fireweed, 163
Flax, Cultivated, 124
　Pale, 124
　Perennial, 5, 124
　Purging, 124
Fleabane, 229
　Blue, 39, 231
　Canadian, 231

INDEX

Fleawort, Field, 40, 227
  Marsh, 227
Flixweed, 105
Flote-grass, 272
Flowering Rush, 4, 245
Fluellen, 3, 203
Foeniculum, 172
  vulgare, 172
Fool's Parsley, 172
Fool's Watercress, 169
Forget-me-not, 1, 198
  Early, 198
  Tufted, 197
  Water, 197
  Wood, 32, 33, 34, 197
  Yellow and Blue, 198
Fox and Cubs, 242
Foxglove, 33, 205
Foxtail, Marsh, 285
  Meadow, 285
  Orange, 285
  Tuberous, 47, 285
Fragaria, 147
  vesca, 147
Frangula, 130
  alnus, 130
Frankenia, 110
  laevis, 46, 110
Fraxinus, 192
  excelsior, 192
Fringed Waterlily, 194
Fritillaria, 251
  meleagris, 251
Fritillary, 251
Frogbit, 38, 44, 245
Fumaria, 96
  bastardii, 96
  capreolata, 96
  muralis, 96
  officinalis, 96
  vaillantii, 96
Fumitory, Common, 96
  White Climbing, 32, 33, 77, 96
  Yellow, 96
Furze, 130
  Dwarf, 38, 130, 131

Gagea, 251
  lutea, 61, 251
Galanthus, 256
  nivalis, 256
(*Galeobdolon*), 214
Galeopsis, 215
  angustifolia, 215
  bifida, 215
  segetum, 215
  speciosa, 215
  tetrahit, 215
Galingale, 4, 266

Galinsoga, 225
  ciliata, 225
  parviflora, 225
Galium, 220
  aparine, 221
  mollugo, 220
  odoratum, 34, 220
  palustre, 221
  parisiense, 221
  pumilum, 221
  saxatile, 220
  tricornutum, 221
  uliginosum, 221
  verum, 220
Gallant Soldier, 225
Garden Cress, 98
Garlic, Crow, 46, 255
  Field, 255
Garlic Mustard, 104
Gean, 155
Genista, 130
  anglica, 130
  tinctoria, 40, 130
Gentian, Autumn, 4, 40, 194
  Field, 194
  Marsh, 2, 33, 37, 38, 56, 76, 194
Gentiana, 194
  pneumonanthe, 2, 5, 37, 194
Gentianella, 194
  amarella, 4, 5, 194
  anglica, 40, 194
  campestris, 194
Geranium, 125
  columbinum, 125
  dissectum, 126
  endressii, 125
  lucidum, 126
  molle, 37, 126
  phaeum, 125
  pratense, 125
  pusillum, 125
  pyrenaicum, 125
  robertianum, 126
  rotundifolium, 126
  sanguineum, 125
  versicolor, 125
Germander, Wall, 216
  Water, 54, 216
Geum, 53, 147
  rivale, 1, 34, 40, 147
  urbanum, 35, 147
Gipsy-wort, 211
Gladdon, 256
Glasswort, 121
Glaucium, 95
  flavum, 46, 95
Glaux, 192
  maritima, 45, 192

Glechoma, 216
　hederacea, 216
Glyceria, 272
　declinata, 272
　fluitans, 272
　maxima, 272
　plicata, 272
Gnaphalium, 229
　luteoalbum, 229
　sylvaticum, 229
　uliginosum, 229
Goat's Beard, 239
Golden Rain, 130
Golden-rod, 230
Golden Samphire, 46, 228
Golden Saxifrage, 32
　Alternate-leaved, 1, 159
　Opposite-leaved, 1, 34, 158
Goldilocks, 91
Gold of Pleasure, 105
Gooseberry, 160
Goosefoot, Fig-leaved, 119
　Glaucous, 119
　Many-seeded, 118
　Nettle-leaved, 119
　Red, 38, 119
　Stinking, 118
　Upright, 119
Goosegrass, 221
Gorse, 38, 130
Granny Bonnets, 93
Grass of Parnassus, 3, 41, 43, 76, 159
Great Burnet, 40, 149
Groenlandia, 249
　densa, 249
Gromwell, 34, 198
　Corn, 198
Ground Elder, 170
Ground Ivy, 216
Groundsel, 226
　Sticky, 226
　Wood, 226
Guelder Rose, 222
Gymnadenia, 258
　conopsea, 41, 258
(Gymnocarpium), 85
　(dryopteris), 85
　(robertianum), 85

Hair-grass, Early, 282
　Grey, 47, 283
　Silver, 282
　Tufted, 35, 282
　Wavy, 36, 37, 38, 282
Hairif, 221
Halimione, 120
　pedunculata, 46, 121
　portulacoides, 48, 120

Hammarbya, 258
　paludosa, 258
Hard Fern, 33, 82
Hardheads, 236
Harebell, 218
Hare's-ear, 168
　Slender, 46, 168
Hare's-tail, 263
Hart's-tongue Fern, 82
Havers, 281
Hawkbit, Autumnal, 238
　Hairy, 238
　Rough, 238
Hawk's-beard, Beaked, 243
　Bristly, 243
　Marsh, 243
　Rough, 243
　Smooth, 243
Hawkweed, 53, 240
　Common, 240
　Mouse-ear, 242
　Spotted, 240
Hawthorn, 32, 33, 156
Hazel, 30, 31, 32, 43, 183
Heath, Cross-leaved, 37, 41, 187
　Sea, 46, 110
Heath Grass, 272
Heather, 2, 43, 58, 187
　Bell, 187
Heath Rush, 37, 38, 253
Hedera, 165
　helix, 165
Hedge Mustard, 104
Hedge-parsley, Knotted, 167
　Spreading, 167
　Upright, 167
Helianthemum, 110
　chamaecistus, 4, 5, 110
Helictotrichon, 281
　pratensi, 281
　pubescens, 281
Hellebore, Green, 83
　Stinking, 83
Helleborine, Broad, 35, 257
　Marsh, 44, 47, 257
　Violet, 257
Helleborus, 88
　foetidus, 88
　viridis, 88
Hemlock, 168
Hemp, 180
Hemp Agrimony, 41, 44, 231
Hemp-nettle, Common, 215
　Downy, 215
　Large-flowered, 215
　Red, 215
Henbane, 201
Henbit, 214
Henne, 272

INDEX

Heracleum, 173
  sphondylium, 173
Herb Bennet, 35, 147
Herb Paris, 3, 32, 33, 35, 252
Herb Robert, 126
Herniaria, 117
  glabra, 4, 38, 117
  hirsuta, 117
Hieraceum, 240
  (anglorum), 241
  aurantiacum, 242
  brunneocroceum, 242
  calcaricola, 241
  diaphanoides, 241
  diaphanum, 241
  eboracense, 241
  exotericum, 240
  lachenalii, 241
  maculatum, 240
  murorum, 240
  perpropinquum, 53, 242
  pilosella, 242
  rigens, 242
  strumosum, 241
  submutabile, 241
  umbellatum, 242
  vagum, 242
  vulgatum, 240
Himantoglossum, 259
  hircinum, 259
Hippocrepis, 136
  comosa, 4, 40, 136
Hippophae, 161
  rhamnoides, 47, 48, 161
Hippuris, 163
  vulgaris, 163
Hoary Cress, 99
Hoary Pepperwort, 99
Hog's Fennel, 173
Hogweed, 173
Holcus, 282
  lanatus, 282
  mollis, 282
Holly, 33, 55, 129
Holosteum, 114
  umbellatum, 114
Honeysuckle, 222
  Fly, 222
Honkenya, 115
  peploides, 115
Hop, 180
Hordelymus, 281
  europaeus, 35, 281
Hordeum, 280
  jubatum, 280
  marinum, 280
  murinum, 280
  secalinum, 280
Hornbeam, 183

Horned-pondweed, 250
Horned-poppy, Yellow, 46, 95
Horn-wort, 94
Horse-chestnut, 129
Horseknobs, 236
Horsemint, 211
Horse-radish, 101
Horse-shoe Vetch, 4, 40, 136
Horsetail, Common, 81
  Great, 1, 81
  Marsh, 81
  Smooth, 81
  Wood, 81
Hottonia, 190
  palustris, 190
Hound's-tongue, 195
Houseleek, 158
Humlick, 166
Humulus, 180
  lupulus, 180
Hyacinth, Wild, 252
(Hydrilla nutallii), 245
Hydrocharis, 245
  morsus-ranae, 245
Hydrocotyle, 165
  vulgaris, 165
Hyoscyamus, 201
  niger, 201
Hypericum, 108
  androsaemum, 108
  calycinum, 108
  elatum, 108
  elodes, 109
  hirsutum, 33, 109
  humifusum, 109
  (inodorum), 108
  maculatum, 109
  montanum, 109
  perforatum, 108
  pulchrum, 109
  tetrapterum, 109
Hypochoeris, 238
  glabra, 36, 238
  maculata, 238
  radicata, 238

Iberis, 99
  amara, 99
Ilex, 129
  aquifolium, 129
Impatiens, 127
  capensis, 127
  glandulifera, 127
  noli-tangere, 127
  parviflora, 128
Inula, 228
  conyza, 4, 228
  crithmoides, 46, 228
  helenium, 228

336 THE FLORA OF LINCOLNSHIRE

Iris, Butterfly, 256
  Stinking, 256
  Yellow, 256
Iris, 256
  foetidissima, 256
  pseudacorus, 256
  spuria, 57, 256
Isatis, 99
  tinctoria, 99
Ivy, 165
Ivy-leaved Toadflax, 204

Jack-by-the-Hedge, 104
Jack-go-to-bed-at-noon, 239
Jacob's Ladder, 195
Japanese Knotweed, 177
Jasione, 219
  montana, 38, 219
Juglans, 181
  regia, 181
Juncus, 253
  acutiflorus, 254
  acutus, 253
  articulatus, 254
  bufonius, 253
  bulbosus, 254
  compressus, 253
  conglomeratus, 253
  effusus, 253
  gerardii, 253
  inflexus, 253
  (*kochii*), 254
  maritimus, 47, 253
  squarrosus, 38, 253
  subnodulosus, 254
  (*subuliflorus*), 253
Juniper, 97
Juniperus, 87
  communis, 87

Keck, 166
Kickxia, 203
  elatine, 3, 203
  spuria, 3, 203
Kidney-vetch, 4, 135
Kingcup, 88
Knapweed, Greater, 1, 4, 236
  Lesser, 1, 236
Knautia, 224
  arvensis, 1, 224
Knawel, 117
Knotgrass, 176
  Ray's, 46, 176
Koeleria, 281
  cristata, 4, 281

Laburnum, 130
Laburnum, 130
  anagyroides, 130

Lactuca, 239
  serriola, 239
  virosa, 239
Ladies' Fingers, 135
Lady Fern, 83
Lady's Mantle, 53, 148
Lady's Smock, 5, 101
Lady's Tresses, Autumn, 40, 53, 257
Lamb's Lettuce, 223
Lamb's Succory, 237
Lamiastrum, 214
  galeobdolon, 1, 34, 214
Lamium, 214
  album, 214
  hybridum, 214
  maculatum, 215
  purpureum, 214
Land Cress, 103
Lapsana, 237
  communis, 237
Larch, European, 31, 87
  Japanese, 87
Larix, 87
  decidua, 87
  leptolepis, 87
Larkspur, 89
Lathraea, 209
  squamaria, 209
Lathyrus, 138
  aphaca, 138
  japonicus, 46, 139
  latifolius, 139
  montanus, 139
  nissolia, 138
  palustris, 5, 44, 139
  pratensis, 138
  sylvestris, 139
  tuberosus, 139
Lavatera, 123
  arborea, 123
Leed, 272
Leek, Sand, 255
Legousia, 219
  hybrida, 3, 219
Lemna, 261
  gibba, 262
  minor, 262
  polyrhiza, 261
  trisulca, 261
Lemon Balm, 212
Leontodon, 238
  autumnalis, 238
  hispidus, 238
  taraxacoides, 238
Leonurus, 215
  cardica, 215
Leopard's Bane, 227
  Great, 227

INDEX

Lepidium, 98
  campestre, 98
  heterophyllum, 98
  ruderale, 98
  sativum, 98
Lettuce, Prickly, 239
  Purple, 240
  Wall, 239
Leucanthemum, 233
  maximum, 233
  vulgare, 233
Leucojum, 255
  aestivum, 255
  vernum, 255
Ligustrum, 192
  vulgare, 192
Lilac, 192
Lily of the Valley, 32, 33, 34, 76, 250
Lime, Small-leaved, 29, 31, 32, 33, 35, 55, 122
Limestone Fern, 85
Limonium, 189
  bellidifolium, 46, 189
  binervosum, 189
  humile, 189
  vulgare, 46, 189
Limosella, 204
  aquatica, 204
Linaria, 203
  purpurea, 203
  repens, 203
  vulgaris, 203
Lincolnshire Spinach, 118
Ling, 34, 187
Linum, 124
  anglicum, 5, 124
  bienne, 124
  catharticum, 124
  usitatissimum, 124
Liparis, 258
  loeselii, 57, 258
Liquorice, Wild, 135
Listera, 257
  ovata, 35, 257
Lithospermum, 198
  arvense, 198
  officinale, 34, 198
Littorella, 218
  uniflora, 218
Livelong, 157
Lobularia, 101
  maritima, 101
Loddon Lily, 255
Lolium, 274
  multiflorum, 274
  perenne, 274
  temulentum, 274
Long Leaf, 170

Lonicera, 222
  periclymenum, 222
  xylosteum, 222
Loosestrife, Purple, 38, 44, 161
  Tufted, 191
  Yellow, 38, 191
Lop-grass, 278
Lotus, 135
  corniculatus, 135
  tenuis, 135
  uliginosus, 135
Lousewort, 38, 207
Lucerne, 131
Luronium, 244
  natans, 244
Luzula, 254
  campestris, 254
  multiflora, 254
  pallescens, 254
  pilosa, 254
  sylvatica, 34, 254
Lychnis, 111
  flos-cuculi, 111
Lycium, 200
  (barbarum), 200
  chinense, 200
  halimifolium, 200
Lycopodium, 39, 80
  alpinum, 5, 80
  clavatum, 80
  inundatum, 80
  selago, 80
Lycopsis, 197
  arvensis, 197
Lycopus, 211
  europaeus, 211
Lyme-grass, 47, 280
Lysimachia, 190
  ciliata, 191
  nemorum, 34, 190
  nummularia, 190
  punctata, 191
  thyrsiflora, 191
  vulgaris, 191
Lythrum, 161
  hyssopifolia, 161
  salicaria, 161

Madwort, 195
Mahonia, 93
  aquifolium, 93
Maianthemum, 251
  bifolium, 251
Maidenhair Fern, 82
Male Fern, 83
  Golden Scaled, 84

Mallow, Common, 123
　Dwarf, 123
　Hispid, 124
　Musk, 123
　Tree, 123
Malus, 156
　sylvestris, 156
Malva, 123
　moschata, 123
　neglecta, 123
　nicaeensis, 123
　parviflora, 123
　pusilla, 123
　sylvestris, 123
　verticillata, 123
Mandrake, 174
Maple, Common, 128
　Norway, 128
Mare's-tail, 163
Marjoram, 4, 40, 211
Marram Grass, 37, 283
Marrubium, 216
　vulgare, 216
Marsh Fern, 50, 85
Marsh Mallow, 44, 47, 123
Marsh Marigold, 1, 32, 33, 88
Marshwort, Least, 169
Master-wort, 173
Mat-grass, 37, 38, 286
Matricaria, 233
　(*inodora*), 232
　matricarioides, 233
　recutita, 233
May Lily, 251
Mayweed, Rayless, 233
　Scentless, 232
　Stinking, 232
Meadow Grass, 275
　Reflexed, 274
　Rough, 276
　Sea, 274
Meadow Rue, Common, 93
　Lesser, 93
Meadow-sweet, 140
Meadow Saffron, 252
Medicago, 131
　arabica, 132
　falcata, 131
　(*hispida*), 132
　lupulina, 131
　minima, 132
　polymorpha, 132
　sativa, 131
Medick, Black, 131
　Sickle, 131
　Small, 132
　Spotted, 132
　Toothed, 132

Melampyrum, 207
　arvense, 207
　cristatum, 207
　pratense, 34, 207
Melampyrum, 207
Melica, 277
　nutans, 35, 277
　uniflora, 35, 277
Melick, Mountain, 35, 277
　Wood, 33, 35, 277
Melilot, Common, 132
　Small-flowered, 132
　Tall, 132
　White, 132
Melilotus, 132
　alba, 132
　altissima, 132
　indica, 132
　officinalis, 132
Melissa, 212
　officinalis, 212
Mentha, 210
　aquatica, 211
　arvensis, 210
　longifolia, 211
　pulegium, 210
　rotundifolia, 210
　(*rubra*), 211
Menyanthes, 194
　trifoliata, 58, 194
Mercurialis, 174
　annua, 174
　perennis, 35, 174
Mercury, 118
　Annual, 174
Mertensia, 198
　maritima, 46, 198
Mezereon, 161
Michaelmas Daisy, 231
Midland Hawthorn, 156
Mignonette, White, 106
　Wild, 41, 106
Milfoil, 232
Milium, 285
　effusum, 35, 285
Milk Parsley, 54, 173
Milk-Thistle, 240
　Spiny, 240
Milk Vetch, 135
　Purple, 37, 38, 39, 76, 135
Milkwort, Common, 108
　Heath, 39, 108
Millet, Wood, 33, 285
Mimulus, 204
　guttatus, 204
Mint, Apple-scented, 211
　Corn, 210
　Water, 211

INDEX

Minuartia, 115
  hybrida, 115
  verna, 115
Misopates, 202
  orontium, 202
Mistletoe, 33, 164
Moehringia, 116
  trinervia, 116
Moenchia, 114
  erecta, 114
Molinia, 272
  caerulea, 2, 36, 37, 38, 272
Monkey-flower, 204
Monkshood, 89
Monotropa, 188
  hypophegea, 188
  hypopitys, 188
Montbretia, 256
Montia, 177
  fontana, 117
  perfoliata, 117
  sibirica, 118
Moonwort, 86
Moschatel, 1, 34, 223
Motherdie, 166
Mother of Thousands, 204
Motherwort, 215
Mountain Ash, 156
Mountain Fern, 50, 84
Mouse-ear Chickweed, Alpine, 112
  Common, 113
  Dark-green, 113
  Field, 37, 38, 39, 112
  Little, 113
  Sticky, 113
Mouse-tail, 92
Mudwort, 204
Mugwort, 234
Mullein, Dark, 202
  Great, 201
  Moth, 202
  Twiggy, 202
  White, 202
Mustard, Black, 97
  White, 97
  Wild, 97
Muzzlejimp, 99
Mycelis, 239
  muralis, 239
Myosotis, 197
  arvensis, 198
  caespitosa, 197
  discolor, 37, 198
  ramosissima, 37, 198
  scorpioides, 197
  secunda, 197
  sylvatica, 1, 34, 197
Myosoton, 113
  aquaticum, 113

Myosurus, 92
  minimus, 92
Myrica, 182
  gale, 2, 34, 39, 182
Myriophyllum, 163
  alterniflorum, 163
  spicatum, 163
  verticillatum, 163
Myrrhis, 167
  odorata, 167

Narcissus, 256
  pseudonarcissus, 256
Nardurus, 275
  maritimus, 57, 60, 275
Nardus, 286
  stricta, 37, 286
Narthecium, 250
  ossifragum, 2, 5, 250
Navelwort, 158
Needle Furze, 130
Neottia, 258
  nidus-avis, 258
Nepeta, 215
  cataria, 215
Nettle, Roman, 180
  Small, 179
  Stinging, 179
Nightshade, Black, 201
  Deadly, 200
  Woody, 201
Nipplewort, 237
Nuphar, 94
  lutea, 94
Nymphaea, 94
  alba, 94
Nymphoides, 194
  peltata, 194

Oak, 2, 29, 30, 31, 32, 33, 36, 43, 45, 55
  Common, 184
  Durmast, 184
  Evergreen, 183
  Pedunculate, 184
  Sessile, 184
  Turkey, 183
Oak Fern, 85
Oat, Black, 281
  Downy, 281
  Meadow, 281
  Wild, 281
  Winter Wild, 281
  Yellow, 281
Oat-grass, False, 282
Odontites, 208
  verna, 208

Oenanthe, 171
  aquatica, 172
  crocata, 171
  fistulosa, 171
  fluviatilis, 172
  lachenalii, 40, 46, 171
Oenothera, 163
  biennis, 163
Onobrychis, 136
  viciifolia, 136
Ononis, 131
  repens, 46, 131
  spinosa, 40, 131
Onopordum, 236
  acanthium, 236
Ophioglossum, 86
  vulgatum, 4, 86
Ophrys, 259
  apifera, 4, 259
  fuciflora, 259
  insectifera, 35, 259
  sphegodes, 259
Orache, Babington's, 120
  Common, 120
  Frosted, 120
  Hastate, 120
  Shore, 120
Orchid, 1, 4
  Bee, 4, 40, 259
  Bird's-nest, 32, 258
  Bog, 258
  Burnt Tip, 260
  Dwarf, 260
  Early Marsh, 38, 260
  Early Purple, 3, 30, 32, 260
  Fen, 258
  Fly, 35, 40, 259
  Fragrant, 41, 76, 258
  Frog, 258
  Greater Butterfly, 3, 32, 33, 35, 258
  Green-winged, 3, 40, 260
  Heath Spotted, 260
  Lesser Butterfly, 259
  Lizard, 48, 55, 259
  Man, 4, 40, 261
  Marsh, 37, 44, 77, 260
  Pyramidal, 4, 5, 40, 46, 47, 261
  Spider, 40, 259
  Spotted, 40, 260
Orchis, 260
  mascula, 3, 260
  morio, 3, 40, 260
  ustulata, 260
Oregon Grape, 93
Origanum, 211
  vulgare, 4, 211
Ornithogalum, 252
  nutans, 252
  umbellatum, 252

Ornithopus, 136
  perpusillus, 36, 37, 136
Orobanche, 209
  alba, 209
  elatior, 4, 5, 209
  minor, 4, 209
  purpurea, 4, 209
  rapum-genistae, 38, 209
Orpine, 157
Osier, 185
Osmunda, 81
  regalis, 39, 81
Oxalis, 127
  acetosella, 3, 34, 127
  corniculata, 127
  europaea, 127
Ox-eye Daisy, 233
Oxlip, 190
Ox-tongue, Bristly, 238
  Hawkweed, 238
Oyster Plant, 46, 198

Pansy, Field, 107
  Wild, 107
Papaver, 94
  argemone, 95
  dubium, 95
  hybridum, 95
  lecoqii, 95
  rhoeas, 94
  somniferum, 95
Parapholis, 286
  incurva, 47, 286
  strigosa, 186
Parentucellia, 208
  viscosa, 208
Parietaria, 179
  diffusa, 179
  (judaica), 179
Paris, 252
  quadrifolia, 3, 35, 252
Parnassia, 159
  palustris, 3, 5, 58, 159
Parsley Piert, 39, 148
Parsnip, Cow, 173
  Wild, 173
Pasque Flower, 40, 90
Pastinaca, 173
  sativa, 173
Pea, Earth-nut, 139
  Everlasting, 33, 139
  Marsh, 139
  Sea, 46, 139
Pear, 156
Pearlwort, Annual, 115
  Ciliate, 115
  Knotted, 115
  Procumbent, 115
  Sea, 115

INDEX

Pearly Everlasting, 230
Pedicularis, 207
  palustris, 207
  sylvatica, 207
Pellitory-of-the-Wall, 179
Penny-cress, Field, 99
Penny-royal, 210
Pennywort, 158, 165
Pentaglottis, 197
  sempervirens, 197
Peplis, 161
  portula, 161
Pepper Saxifrage, 40, 172
Pepperwort, 98
  Narrow-leaved, 98
Periwinkle, Greater, 193
  Lesser, 193
Persicaria, 176
  Knotted, 176
  Pale, 176
Petasites, 228
  albus, 228
  fragans, 228
  hybridus, 228
Petroselinum, 169
  crispum, 169
  segetum, 169
Petty Whin, 36, 38, 130
Peucedanum, 173
  ostruthium, 173
  palustre, 173
Phalaris, 286
  arundinacea, 286
  canariensis, 286
  minor, 286
  paradoxa, 286
Pheasant's Eye, 92
Phleum, 284
  arenarium, 284
  bertolonii, 184
  pratense, 284
Phragmites, 272
  (*australis*), 272
  communis, 272
Phyllitis, 82
  scolopendrium, 82
Picea, 86
  abies, 86
Pickpurse, 39, 116
Picris, 238
  echioides, 238
  hieracioides, 238
Pignut, 170
Pigseye, 190
Pillwort, 85
Pilularia, 85
  globulifera, 85
Pimpernel, Blue, 191
  Bog, 37, 38, 41, 47, 191

Scarlet, 191
  Yellow, 33, 34, 190
Pimpinella, 170
  major, 170
  saxifraga, 39, 170
Pine, 30, 32, 33, 36, 37, 43
  Scots, 31, 87
Pineapple Weed, 233
Pinguicula, 209
  vulgaris, 5, 41, 44, 209
Pink, Deptford, 112
  Maiden, 112
Pinus, 87
  sylvestris, 87
Pirri-pirri-bur, 149
Plantago, 217
  coronopus, 217
  indica, 217
  lanceolata, 217
  major, 217
  maritima, 217
  media, 217
Plantain, Buck's-horn, 217
  Great, 217
  Hoary, 217
  Sea, 217
Platanthera, 258
  bifolia, 258
  chlorantha, 3, 35, 258
Ploughman's Spikenard, 4, 228
Poa, Annual, 275
  Bulbous, 47, 275
  Darnel, 47, 275
  Flattened, 275
  Hard, 274
  Wood, 275
Poa, 275
  angustifolia, 275
  annua, 275
  bulbosa, 47, 275
  compressa, 275
  nemoralis, 275
  palustris, 276
  pratensis, 275
  subcaerulea, 276
  trivialis, 276
Polemonium, 195
  caeruleum, 195
Policeman's Helmet, 128
Polycarpon, 117
  tetraphyllum, 117
Polygala, 108
  calcarea, 108
  serpyllifolia, 108
  vulgaris, 108
Polygonatum, 251
  multiflorum, 251

Polygonum, 176
  amphibium, 176
  (*aubertii*), 177
  aviculare, 176
  baldschuanicum, 177
  bistorta, 176
  convolvulus, 177
  cuspidatum, 177
  hydropiper, 176
  lapathifolium, 176
  minus, 177
  mite, 177
  nodosum, 176
  persicaria, 176
  polystachyum, 177
  raii, 46, 176
  sachalinense, 177
  viviparum, 176
Polypodium, 85
  vulgare, 50, 85
Polypody, 85
Polypogon, 284
  monspeliensis, 284
Polystichum, 84
  aculeatum, 84
  setiferum, 84
Pondweed, Blunt-leaved, 249
  Bog, 247
  Broad-leaved, 246
  Curled, 249
  Fen, 247
  Fennel, 249
  Flat-stalked, 248
  Grass-wrack, 249
  Hair-like, 249
  Lesser, 248
  Long-stalked, 248
  Opposite-leaved, 249
  Perfoliate, 248
  Reddish, 248
  Sharp-leaved, 249
  Shining, 247
  Small, 249
  Various-leaved, 247
Poplar, Balsam, 185
  Black, 184
  Black Italian, 185
  Grey, 184
  Lombardy, 184
  White, 184
Poppy, Babington's, 96
  Field, 94
  Long-head, 95
  Long Prickly-head, 95
  Opium, 95
  Round Prickly-head, 95
Populus, 184
  alba, 184
  canescens, 184

  gileadensis, 185
  italica, 184
  nigra, 184
  tremula, 184
Potamogeton, 246
  acutifolius, 249
  alpinus, 248
  berchtoldii, 249
  coloratus, 247
  compressus, 249
  crispus, 249
  (*densus*), 249
  friesii, 248
  gramineus, 247
  lucens, 247
  natans, 246
  obtusifolius, 249
  pectinatus, 249
  perfoliatus, 248
  polygonifolius, 38, 247
  praelongus, 248
  pusillus, 248
  trichoides, 249
Potentilla, 146
  anglica, 146
  anserina, 146
  argentea, 36, 146
  erecta, 146
  norvegica, 146
  palustris, 36, 146
  recta, 146
  reptans, 146
  sterilis, 146
  tabernaemontani, 146
Poterium, 149
  polygamum, 149
  sanguisorba, 39, 149
Primrose, 1, 3, 30, 32, 33, 34, 190
Primula, 190
  elatior, 190
  veris, 3, 40, 190
  vulgaris, 3, 34, 190
Privet, 192
Prunella, 213
  laciniata, 213
  vulgaris, 213
Prunus, 155
  avium, 155
  domestica, 155
  padus, 155
  spinosa, 155
Pseudotsuga, 86
  menziesii, 86
Pteridium, 82
  aquilinum, 82
Puccinellia, 274
  distans, 274
  maritima, 274

INDEX

Pulicaria, 229
  dysenterica, 229
Pulsatilla, 90
  vulgaris, 40, 90
Purple Moorgrass, 2, 36, 272
Pyrola, 188
  minor, 34, 57, 188
Pyrus, 156
  communis, 156

Quaking Grass, 277
Quercus, 183
  cerris, 183
  ilex, 183
  petraea, 33, 184
  robur, 33, 184

Radiola, 124
  linoides, 124
Radish, 97
  Sea, 97
  Wild, 97
Ragged Robin, 111
Ragwort, 226
  Broad-leaved, 227
  Great Fen, 56, 227
  Hoary, 226
  Oxford, 3, 226
  Water, 226
Rampion, 219
Ramsons, 1, 32, 35, 255
Ranunculus, 90
  acris, 90
  aquatilis, 92
  arvensis, 3, 90
  auricomus, 91
  baudotii, 92
  bulbosus, 90
  circinatus, 92
  ficaria, 92
  flammula, 91
  fluitans, 91
  hederaceus, 91
  lingua, 4, 44, 91
  parviflorus, 91
  repens, 90
  sardous, 45, 91
  sceleratus, 91
  trichophyllus, 92
Rape, 96
Raphanus, 97
  maritimus, 97
  raphanistrum, 97
  sativus, 97
Raspberry, 140
Red-leg, 176
Red Rattle, 41, 207
Reed, 2, 43, 54, 272

Reed-grass, 272, 286
Reedmace, Great, 262
  Lesser, 4, 263
Reseda, 106
  alba, 106
  lutea, 41, 106
  luteola, 106
Restharrow, 46, 131
  Spiny, 40, 131
Rhamnus, 129
  catharticus, 129
Rhinanthus, 207
  minor, 40, 207
  serotinus, 45, 207
Rhododendron, 33, 186
  ponticum, 186
Rhynchospora, 266
  alba, 51, 266
Ribbon Weed, 78
Ribes, 159
  alpinum, 159
  nigrum, 159
  (rubrum), 159
  sylvestre, 159
  uva-crispa, 160
Ribwort, 217
Rice Grass, Common, 46, 287
Rock-cress, Hairy, 103
  Tower, 103
Rocket, Eastern, 105
  London, 105
  Perennial Wall, 97
  Sand, 97
  Tall, 105
Rock-rose, 4, 39, 110
Rock Samphire, 46, 171
Rorippa, 103
  amphibia, 104
  islandica, 104
  microphylla, 103
  nasturtium-aquaticum, 103
  sylvestris, 103
Rosa, 60, 149
  afzeliana, 152
  agrestis, 155
  (andegavensis), 151
  arvensis, 149
  (caesia), 153
  canina, 150
  coriifolia, 152, 153
  (corymbifera), 152
  (desegliseri), 152
  (dumalis), 152, 153
  (dumetorum), 152
  (glauca), 152
  micrantha, 155
  (mollis), 153, 154
  (mollissima), 154
  (nitidula), 151

obtusifolia, 153
pimpinellifolia, 5, 150
rubiginosa, 154
(*scabriuscula*), 154
sherardii, 154
(*spinosissima*), 150
(*squarrosa*), 150
stylosa, 150
(*subcanina*), 152
tomentosa, 154
villosa, 153
(*vosaglaco*), 152
Rose, Burnet, 150
 Dog, 150, 152
 Downy, 153
 Field, 149
Rowan, 33, 156
Royal Fern, 33, 81
Rubus, 60, 140
 ampliflcatus, 142
 bellardii, 145
 belophorus, 143
 boraeanus, 144
 caesius, 140
 calvatus, 142
 cardiophyllus, 143
 carpinifolius, 142
 conjugens, 141
 dasyphyllus, 145
 drejeri, 144
 eboracensis, 141
 echinatoides, 144
 echinatus, 144
 falcatus, 144
 flexuosus, 145
 fruticosus, 140
 (*glandulosus*), 145
 gratus, 142
 hylocharis, 145
 idaeus, 140
 leyanus, 144
 lindebergii, 143
 lindleianus, 142
 macrophylloides, 143
 macrophyllus, 142
 mucronatoides, 144
 nemoralis, 142
 nessensis, 141
 newbouldii, 145
 pallidus, 145
 plicatus, 141
 polyanthemus, 143
 procerus, 143
 pyramidalis, 143
 radula, 144
 robii, 142
 rudis, 145
 rufescens, 145
 saxatilis, 34, 140

scabrosus, 141
scissus, 141
silvaticus, 142
sublustris, 141
ulmifolius, 143
vestitus, 144
Rumex, 178
 acetosa, 38, 178
 acetosella, 38, 178
 conglomeratus, 179
 crispus, 178
 hydrolapathum, 178
 maritimus, 179
 obtusifolius, 178
 palustris, 179
 pulcher, 178
 sanguineus, 178
 tenuifolius, 38, 39, 178
Ruppia, 250
 (*cirrhosa*), 250
 maritima, 250
 spiralis, 250
Rupturewort, Glabrous, 4, 38, 78, 117
 Hairy, 117
Ruscus, 251
 aculeatus, 251
Rush, Blunt-flowered, 254
 Bulbous, 254
 Compact, 253
 Hard, 253
 Heath, 37, 38, 253
 Jointed, 254
 Round-fruited, 253
 Saltmarsh, 253
 Sea, 253
 Sharp, 253
 Sharp-flowered, 254
 Soft, 253
 Toad, 253
Rye-grass, 274
 Italian, 274

Sagina, 115
 apetala, 115
 ciliata, 115
 maritima, 115
 nodosa, 115
 procumbens, 115
Sagittaria, 245
 sagittifolia, 245
Sainfoin, 136
St. John's Wort, Common, 108
 Hairy, 30, 33, 109
 Imperforate, 109
 Marsh, 109
 Mountain, 109
 Slender, 109
 Square-stemmed, 109
 Trailing, 109

INDEX

Salad Burnet, 38, 39, 149
Salicornia, 121
  dolichostachya, 48, 122
  europaea, 48, 122
  fragilis, 122
  perennis, 48, 121
  pusilla, 48, 122
  ramosissima, 48, 122
Salix, 185
  alba, 185
  aurita, 186
  caprea, 186
  cinerea, 186
  fragilis, 185
  pentandra, 185
  purpurea, 185
  repens, 37, 46, 186
  triandra, 185
  viminalis, 185
Sallow, Common, 186
  Eared, 186
  Great, 186
Salsify, 239
Salsola, 121
  kali, 121
Saltwort, 121
Salvia, 212
  horminoides, 213
  pratensis, 212
  verticillata, 212
Sambucus, 221
  ebulus, 221
  nigra, 222
Samolus, 192
  valerandi, 192
Samphire, 48
  Marsh, 121
Sand-spurrey, 116
Sandwort, Fine-leaved, 115
  Lesser Thyme-leaved, 116
  Sea, 115
  Three-nerved, 116
  Thyme-leaved, 116
  Vernal, 115
Sanguisorba, 149
  officinalis, 40, 149
Sanicle, 34, 166
Sanicula, 166
  europaea, 34, 166
Saponaria, 112
  officinalis, 112
Sarothamnus, 131
  scoparius, 131
Saw-wort, 40, 41, 237
Saxifraga, 158
  granulata, 3, 158
  tridactylites, 158
Saxifrage, Meadow, 3, 158
  Rue-leaved, 158

Scabiosa, 224
  columbaria, 224
Scabious, Field, 1, 224
  Small, 39, 224
Scandix, 167
  pecten-veneris, 3, 167
Schoenus, 266
  nigricans, 41, 51, 76, 266
Scilla, 252
  verna, 252
Scirpus, Bristle, 36, 264
  Floating, 264
Scirpus, 263
  cespitosus, 263
  fluitans, 264
  lacustris, 264
  maritimus, 264
  setaceus, 264
  sylvaticus, 264
  tabernaemontani, 264
Scleranthus, 117
  annuus, 117
Scrophularia, 204
  aquatica, 204
  (auriculata), 204
  nodosa, 204
  umbrosa, 204
  vernalis, 204
Scurvy-grass, Danish, 100
  Long-leaved, 100
Scutellaria, 216
  galericulata, 216
  minor, 216
Sea Aster, 46, 231
Seablite, Herbaceous, 121
  Shrubby, 47, 121
Sea Hard-grass, 286
  Early, 286
Sea Holly, 46, 166
Seakale, 98
Sea Lavender, 46, 48, 189
  Lax-flowered, 189
  Matted, 46, 54, 189
  Rock, 189
Sea Milkwort, 43, 45, 192
Sea Pink, 189
Sea Purslane, 46, 120
Sea Rocket, 98
Sedge, 45, 51
  Acute, 269
  Beaked, 268
  Bladder, 268
  Bottle, 268
  Breckland Spring, 269
  Brown, 270
  Common, 270
  Common Yellow, 267
  Curved, 270
  Cyperus, 268

Dioecious, 271
Distant, 45, 267
Divided, 270
Dwarf Yellow, 267
Elongated, 271
False Fox, 270
Flea, 37, 51, 271
Glaucous, 269
Graceful, 269
Great Pond, 268
Green Ribbed, 267
Grey, 270
Hairy, 269
Hammer, 269
Lesser Pond, 268
Long-bracted, 267
Oval, 271
Pale, 35, 268
Pendulous, 35, 268
Pill-headed, 269
Remote, 35, 271
Sand, 37, 38, 270
Slender, 269
Smooth, 35, 266
Spiked, 271
Star, 271
Tall Yellow, 41, 267
Tawny, 267
Thin Spiked, 35, 268
Tufted, 269
Tussock, 270
Two-stemmed, 270
Vernal, 269
Whitish, 271
Wood, 35, 267
Sedum, 157
  acre, 157
  album, 157
  anglicum, 157
  forsteranum, 157
  reflexum, 157
  sexangulare, 157
  telephium, 157
Selaginella, 80
  selaginoides, 5, 39, 50, 80
Self-heal, 213
Selinum, 172
  carvifolia, 172
Sempervivum, 158
  tectorum, 158
Senecio, 226
  aquaticus, 226
  erucifolius, 226
  fluviatis, 227
  integrifolius, 40, 227
  jacobea, 226
  paludosus, 5, 43, 57, 227
  palustris, 5, 43, 227
  squalidus, 3, 226

sylvaticus, 226
viscosus, 226
vulgaris, 126
Serratula, 237
  tinctoria, 40, 41, 137
Service Tree, Wild, 31, 33, 156
Setaria, 287
  verticillata, 287
  viridis, 287
Shaggy Soldier, 225
Shasta Daisy, 233
Shear Grass, 279
Sheep's-bit, 38, 219
Sheep's Parsley, 169
Shepherd's Cress, 100
Shepherd's Needle, 3, 167
Shepherd's Purse, 100
Shepherd's Weather-glass, 191
Sherardia, 219
  arvensis, 219
Shield-fern, 50
  Hard, 84
  Soft, 84
Shore-weed, 218
Sibthorpia, 205
  europaea, 205
Sieglingia, 272
  decumbens, 272
Silaum, 172
  silaus, 40, 172
Silene, 110
  alba, 111
  conica, 110
  dichotoma, 110
  dioica, 1, 33, 111
  gallica, 111
  italica, 111
  maritima, 46, 110
  noctiflora, 111
  nutans, 111
  vulgaris, 110
Silverweed, 47, 146
Silybum, 236
  marianum, 236
Sinapis, 97
  alba, 97
  arvensis, 97
Sison, 169
  amomum, 169
Sisymbrium, 104
  altissimum, 105
  irio, 105
  officinale, 104
  orientale, 105
Sium, 171
  latifolium, 171
Skullcap, 216
  Lesser, 216
Sloe, 155

INDEX

Small-reed, Purple, 35, 283
Smith's Cress, 98
Smyrnium, 168
  olusatrum, 46, 168
Snake's Head, 251
Snake-root, 176
Snapdragon, 203
Sneezewort, 40, 232
Snowberry, 222
Snowdrop, 256
Snowflake, Spring, 255
Soapwort, 112
Solanum, 201
  dulcamara, 101
  nigrum, 201
  sarrachoides, 201
Solidago, 230
  canadensis, 230
  gigantea, 230
  graminifolia, 230
  virgaurea, 230
Solomon's Seal, 251
Sonchus, 239
  arvensis, 240
  asper, 240
  oleraceus, 240
  palustris, 5, 43, 239
Sorbus, 156
  aria, 156
  aucuparia, 33, 156
  torminalis, 33, 156
Sorgh-grass, 266
Sorrel, 178
  Sheep's, 39, 178
Sour Sauce, 178
Sowbane, 119
Sow-Thistle, 240
  Blue, 240
  Corn, 240
  Marsh, 239
  Spiny, 240
Sparganium, 262
  angustifolium, 262
  emersum, 262
  erectum, 262
  minimum, 262
Spartina, 46, 286
  anglica, 48, 189, 287
  maritima, 47, 286
  (*townsendii*), 287
Spearwort, Great, 4, 44, 91
  Lesser, 91
Speedwell, Common, 205
  Buxbaum's, 206
  Field, 206
  Germander, 206
  Grey, 206
  Ivy, 206
  Marsh, 38, 205

Pink Water, 205
Slender Creeping, 206
Thyme-leaved, 206
Wall, 37, 206
Water, 205
Wood, 34, 206
Spergula, 116
  arvensis, 116
Spergularia, 116
  marina, 116
  media, 116
  rubra, 116
Spike-rush, Common, 265
  Few-flowered, 41, 67, 265
  Many-stemmed, 265
  Slender, 265
Spindle-tree, 129
Spiraea, Willow, 140
Spiraea, 140
  salicifolia, 140
Spiranthes, 257
  spiralis, 257
Spleenwort, Black, 82
  Green, 83
  Maidenhair, 82
Spruce, Norway, 31, 86
Spurge, Broad, 175
  Caper, 174
  Cypress, 175
  Dwarf, 175
  Petty, 175
  Portland, 46, 175
  Sun, 175
  Wood, 33, 34, 175
Spurge Laurel, 4, 161
Squill, Spring, 252
Squinancy Wort, 47, 220
Squirrel Tail, 274
Squirrel-tail Grass, 280
Stachys, 213
  arvensis, 213
  germanica, 213
  palustris, 213
  sylvatica, 214
Staphylea, 128
  pinnata, 128
Star Fruit, 245
Star of Bethlehem, 252
  Drooping, 252
  Yellow, 251
Star Thack, 45, 51, 266
Starwort, 164
Stellaria, 113
  alsine, 114
  graminea, 114
  holostea, 114
  media, 113
  neglecta, 114
  nemorum, 113

pallida, 114
palustris, 114
Stitchwort, Bog, 114
  Greater, 114
  Lesser, 39, 114
  Marsh, 114
  Wood, 113
Stone Bramble, 140
Stonecrop, 157
  English, 157
  Rock, 157
  White, 157
Stone Parsley, 169
Storksbill, Common, 127
  Musk, 126
Stratiotes, 245
  aloides, 4, 245
Strawberry, 147
Strawberry Blite, 119
Suaeda, 121
  (*bera*), 121
  fruticosa, 47, 121
  maritima, 121
Succisa, 225
  pratensis, 40, 41, 225
Sundew, 2, 5, 33, 37, 38, 160
  Great, 160
  Long-leaved, 160
Sweet Briar, 154
Sweet Cicely, 167
Sweet Flag, 261
Sweet Gale, 2, 34, 182
Sweethearts, 221
Sweet Woodruff, 32, 34, 220
Swine-cress, 99
  Lesser, 99
Sycamore, 128
Symphoricarpos, 222
  rivularis, 222
Symphytum, 195
  asperum, 196
  officinale, 195
  orientale, 196
  tuberosum, 196
Syringia, 192
  vulgaris, 192

Tamus, 257
  communis, 257
(*Tanacetum*), 233
  (*parthenium*), 233
  (*vulgare*), 233
Tansy, 233
Taraxacum, 243
  erythrospermum, 244
  laevigatum, 244
  officinale, 243
  palustre, 243
  spectabile, 243

Tare, Hairy, 136
  Slender, 137
  Smooth, 136
Tasselweed, Beaked, 250
  Spiral, 250
Taxus, 88
  baccata, 88
Teasel, 224
  Small, 1, 34, 224
Teesdalia, 100
  nudicaulis, 36, 39, 45, 58, 100
Teucrium, 216
  chamaedrys, 216
  scordium, 44, 57, 216
  scorodonia, 47, 216
Thale Cress, 105
Thalictrum, 93
  flavum, 93
  minus, 47, 93
Thelycrania, 165
  sanguinea, 165
Thelypteris, 84
  dryopteris, 85
  (*limbosperma*), 84
  oreopteris, 84
  palustris, 5, 85
  robertiana, 85
Thesium, 165
  humifusum, 57, 165
Thistle, Buck, 235
  Creeping, 235
  Marsh, 235
  Meadow, 40, 236
  Milk, 236
  Musk, 235
  St. Barnaby's, 237
  Scotch, 236
  Slender, 235
  Spear, 235
  Star, 237
  Stemless, 236
  Welted, 235
  Woolly, 4, 235
Thlaspi, 99
  arvense, 99
Thorn-apple, 201
Thrift, 189
  Elongated, 78, 189
Thyme, Large Wild, 211
  Wild, 212
Thymus, 211
  drucei, 212
  pulegioides, 211
Tilia, 122
  cordata, 33, 122
  platyphyllos, 122
  x *vulgaris*, 122
Timothy Grass, 284

INDEX

Toadflax, 203
  Bastard, 165
  Pale, 203
  Pointed-leaved, 203
  Purple, 203
  Round-leaved, 203
  Small, 203
Toad Pipes, 81
Toothwort, 209
Tor Grass, 1, 37, 279
Torilis, 167
  arvensis, 167
  japonica, 167
  nodosa, 167
Tormentil, Common, 146
  Trailing, 147
Touch-me-not, 127
Tower-cress, 103
Tower Mustard, 103
Townhall Clock, 223
Trachelium, 219
  caeruleum, 219
Tragopogon, 239
  porrifolius, 239
  pratensis, 239
Traveller's Joy, 90
Treacle Mustard, 104
Trefoil, Hare's-foot, 133
  Hop, 134
  Knotted, 133
  Lesser Yellow, 134
  Rough, 133
  Slender, 134
  Subterranean, 134
Trifolium, 132
  arvense, 133
  campestre, 134
  dubium, 134
  fragiferum, 134
  hybridum, 134
  incarnatum, 133
  medium, 40, 133
  micranthum, 134
  ochroleucon, 133
  ornithopodioides, 132
  pratense, 133
  repens, 134
  resupinatum, 134
  scabrum, 133
  squamosum, 133
  striatum, 133
  subterraneum, 134
Triglochin, 246
  maritima, 246
  palustris, 246
(*Trigonella*), 132
  (*ornithopioides*), 132
Tripleurospermum, 232
  maritimum, 232

Trisetum, 281
  flavescens, 281
Tsuga, 86
  heterophylla, 86
Turnip, 96
Turritis, 103
  glabra, 103
Tussilago, 227
  farfara, 3, 227
Tutsan, 108
Twayblade, 35, 37, 257
Twigrush, 266
Twitch, 279
  Fibrous, 35, 273
Typha, 262
  angustifolia, 4, 262
  latifolia, 262

Ulex, 130
  europaeus, 130
  gallii, 38, 130
  minor, 131
  nanus, 38, 39, 131
Ulmus, 180
  carpinifolia, 181
  coritana, 181
  glabra, 180
  plotii, 181
  procera, 180
  stricta, 181
Umbilicus, 158
  rupestris, 158
Urtica, 179
  dioica, 179
  pilulifera, 180
  urens, 179
Utricularia, 210
  intermedia, 210
  minor, 210
  vulgaris, 4, 210

Vaccaria, 112
  pyramidata, 112
Vaccinium, 187
  myrtillus, 5, 57, 187
  oxycoccos, 2, 5, 187
Valerian, 44, 223
  Marsh, 223
  Red, 224
Valeriana, 223
  dioica, 223
  officinalis, 223
Valerianella, 223
  carinata, 223
  dentata, 223
  locusta, 223
  rimosa, 223
Venus's Looking-glass, 3, 219

Verbascum, 201
　blattaria, 202
　chaixii, 202
　lychnitis, 202
　nigrum, 202
　phlomoides, 201
　thapsus, 201
　virgatum, 202
Verbena, 210
　officinalis, 210
Vernal-grass, Annual, 285
　Sweet, 285
Veronica, 205
　agrestis, 206
　anagallis-aquatica, 205
　arvensis, 37, 206
　beccabunga, 205
　catenata, 205
　chamaedrys, 206
　filiformis, 206
　hederifolia, 206
　montana, 34, 206
　officinalis, 205
　persica, 206
　polita, 206
　scutellata, 205
　serpyllifolia, 206
Vervain, 210
Vetch, Bithynian, 138
　Bush, 34 137
　Common, 137
　Narrow-leaved, 138
　Spring, 138
　Tuberous Bitter, 33, 139
　Tufted, 40, 137
　Wood, 33, 34, 137
　Yellow, 137
Vetchling, Grass, 138
　Meadow, 138
　Yellow, 138
Viburnum, 222
　lantana, 222
　opulus, 222
Vicia, 136
　angustifolia, 138
　bithynica, 138
　cracca, 40, 137
　hirsuta, 136
　hybrida, 137
　lathyroides, 58, 138
　lutea, 137
　sativa, 137
　sepium, 34, 137
　sylvatica, 34, 137
　tenuissima, 137
　tetrasperma, 136
Vinca, 193
　major, 192
　minor, 193

Viola, 106
　arvensis, 107
　canina, 107
　hirta, 33, 106
　lactea, 107
　odorata, 33, 106
　palustris, 38, 107
　(*persicifolia*), 107
　reichenbachiana, 106
　riviniana, 106
　stagnina, 44, 55, 77, 107
　tricolor, 107
Violet, 30
　Common Dog, 106
　Fen, 77, 107
　Hairy, 33, 106
　Heath Dog, 107
　Marsh, 38, 107
　Pale Heath, 107
　Sweet, 33, 106
　Wood Dog, 106
Viper's Bugloss, 198
　Early, 198
Viscum, 164
　album, 164
Vulpia, 274
　bromoides, 274
　myuros, 274

Wall Cress, Common, 105
Wallflower, 104
Wall-pepper, 157
Wall-Rue, 83
Walnut, 181
Wart-cress, 99
Watercress, 103
　One-rowed, 103
Water Dropwort, 171
　Fine-leaved, 172
　Hemlock, 171
　Parsley, 40, 46, 171
Water-lily, White 94
　Yellow, 94
Water-milfoil, Alternate-flowered, 163
　Spiked, 163
　Whorled, 163
Water Parsnip, 44, 171
　Narrow-leaved, 171
Water-pepper, 176
Water-Plantain, 244
　Floating, 244
　Lesser, 244
　Narrow-leaved, 244
　Ribbon-leaved, 78 244
Water Purslane, 161
Water Soldier, 4, 245
Water Violet, 190
Water Whorl-grass, 276

INDEX

Wayfaring Tree, 222
Weasel's Snout, 202
Weld, 106
Western Hemlock, 86
White Beam, 156
White Horehound, 216
Whitlow, Grass, 101
Wild Basil, 40, 212
Wild Carrot, 173
Willow, 2, 32, 33, 45
  Almond, 185
  Bay, 185
  Crack, 185
  Creeping, 37, 38, 46, 186
  Goat, 186
  Purple, 185
  White, 185
Willowherb, American, 162
  Broad-leaved, 162
  Great Hairy, 38, 162
  Hoary, 162
  Marsh, 162
  Pale, 162
  Rose-bay, 31, 163
  Short-fruited, 162
  Square-stemmed, 162
Willow Weed, 176
Winter Cress, 102
Wintergreen, Common, 34, 188
Winter Heliotrope, 228
Wire-thorn, 88
Woad, 99
Womandrake, 257
Woodbine, 222
Wood Millet, 33, 285
Woodruff, Sweet, 32, 34, 220
Woodrush, Field, 254
  Great, 34, 254
  Hairy, 254
  Many-headed, 254

Wood Sage, 38, 47, 216
Wood Sorrel, 3, 32, 34, 127
Wormwood, 234
  Sea, 234
Woundwort, Downy, 213
  Field, 213
  Hedge, 214
  Marsh, 213

Xanthium, 225
  spinosum, 225

Yarrow, 232
Yellow Archangel, 1, 30, 32, 34, 214
Yellow Bird's-nest, 188
Yellow-cress, Creeping, 103
  Great, 104
  Marsh, 104
Yellow Rattle, 40, 207
  Great, 207
Yellow Rocket, 102
  Early-flowering, 103
  Intermediate, 102
  Small-flowered, 102
Yellow Sorrel, Procumbent, 127
  Upright, 127
Yellow-wort, 4, 5, 53, 193
Yew, 2, 30, 45, 88
Yorkshire Fog, 282

Zannichellia, 250
  palustris, 250
Zostera, 246
  marina, 146
  noltii, 246

NOTES

NOTES

NOTES

NOTES

NOTES